W9-ANB-787

ONDON MATHEMATICAL SOCIETY LECTURE NOTE SERIES

Ianaging Editor: Professor J.W.S. Cassels, Department of Pure Mathematics and Mathematical Statistics,
niversity of Cambridge, 16 Mill Lane, Cambridge CB2 1SB, England

he titles below are available from booksellers, or, in case of difficulty, from Cambridge University Press.

London Mathematical Society Lecture Note Series. 207

Groups of Lie Type and their Geometries

Como 1993

Edited by

William M. Kantor
University of Oregon

Lino Di Martino
Università di Milano

CAMBRIDGE
UNIVERSITY PRESS

Published by the Press Syndicate of the University of Cambridge
The Pitt Building, Trumpington Street, Cambridge CB2 1RP
40 West 20th Street, New York, NY 10011-4211, USA
10 Stamford Road, Oakleigh, Melbourne 3166, Australia

First published 1995

Printed in Great Britain at the University Press, Cambridge

Library of Congress cataloguing in publication data available

British Library cataloguing in publication data available

ISBN   0 521 46790 X paperback

Contents

vi   *Contents*

Preface

The June 14–19, 1993, conference on Groups of Lie Type and their Geometries took place in Como, Italy, at the lovely 18th century Villa Olmo on the scenic shore of Lake Como. It brought together experts and interested mathematicians from numerous countries. The scientific program centered around invited expository lectures; there also were short research announcements, including talks by younger researchers.

The conference focused on both the structure theory and the geometry of groups of Lie type, with emphasis on recent results and open problems. Special attention was drawn not only to the interplay between group-theoretic methods and geometric and combinatorial aspects of groups of Lie type, but also to important and occasionally unexpected connections with other branches of mathematics.

Expanded versions of most of the talks appear in these Proceedings. This volume is intended to provide a stimulating collection of themes for a broad range of algebraists and geometers. Among those themes, represented within the conference or these Proceedings, are the following: (1) Subgroups of finite and algebraic groups, (2) Buildings and other geometries associated to groups of Lie type or Coxeter groups, (3) Generation, and (4) Applications.

We are grateful to the authors for their efforts in providing us with manuscripts in TEX. Roger Astley, Mathematics Editor of Cambridge University Press, has been very helpful and supportive throughout the prepration of this volume.

The organizing committee consisted of L. Di Martino (Milan), W. M. Kantor (Eugene), O. H. Kegel (Freiburg), and L. A. Rosati (Florence). The Center of Scientific Culture 'A.Volta' provided valuable assistance with the local organization. We thank the University of Milan, the Italian Ministry for University and Scientific Research, and the Italian National Research Council (C.N.R.), for their financial support of the conference.

GROUPS OF LIE TYPE
AND THEIR GEOMETRIES

Villa Olmo, 14 to 19 June 1993

Talks

M. Aschbacher, *Representations of groups on finite simplicial complexes*

M. Aschbacher, *Simple connectivity of p-group complexes*

A. Borovik, *Combinatorics of flag varieties*

F. Buekenhout, *On the dialectic of groups and incidence geometry*

D. Cartwright, *Groups acting simply transitively on the vertices of a building of type $\tilde{A}_n$*

C. Casolo, *Wielandt's complexes in finite groups*

A. Cohen, *Metric geometry for Coxeter groups*

H. Cuypers, *Extended generalized hexagons*

B. Ford, *On disconnected linear groups and restrictions of representations*

N. Gordeev, *Products of conjugacy classes in algebraic groups and generators of dense subgroups*

R. Guralnick, *Permutation polynomials and properties of Chevalley groups*

R. Lawther, *Double cosets and CLEFS*

M. Liebeck, *Subgroups of exceptional groups. I*

H. Van Maldeghem, *Ree octagons*

G. Malle, *Cyclotomic Hecke algebras*

T. Meixner, *Geometries that are extensions of buildings*

A. Pasini, *The direct sum problem for chamber complexes*

J. Saxl, *Some subgroups, and applications*

G. Seitz, *Subgroups of exceptional groups. II*

E. Shult, *Hyperplanes and embedded geometries of Lie type*

D. Testerman, *Large rank subgroups of simple groups of Lie type*

J. Tits, *Bad unipotent elements, buildings and the geometry of exceptional groups* two talks

N. Vavilov, *Intermediate subgroups of Chevalley groups*

T. Weigel, *A class of Frattini extensions of finite Chevalley groups*

J. Wilson, *Economical generating sets for classical groups*

List of participants

M. Abramson (Chicago, IL, USA)

S. Adami (Milano, Italy)

M. Aschbacher (Pasadena, CA, USA)

L. Bader (Roma, Italy)

C. Bartolone (Palermo, Italy)

C. Bennett (Columbus, OH, USA)

M. Bianchi (Milano, Italy)

J. van Bon (Medford, MA, USA)

C. Bonzini (Milano, Italy)

A. Borovik (Manchester, UK)

F. Buekenhout (Bruxelles, Belgium)

D. Cartwright (Sydney, Australia)

C. Casolo (Udine, Italy)

M. Cazzola (Coventry, UK)

Y. Chen (Torino, Italy)

A. Cohen (Eindhoven, The Netherlands)

B. Cooperstein (Santa Cruz, CA, USA)

M. Costantini (Padova, Italy)

R. Curtis (Birmingham, UK)

H. Cuypers (Eindhoven, The Netherlands)

E. D'Agostini (Bologna, Italy)

F. Dalla Volta (Milano, Italy)

A. D'Aniello (Napoli, Italy)

G. D'Este (Milano, Italy)

L. Di Martino (Milano, Italy)

J. Doyen (Bruxelles, Belgium)

E. Ellers (Toronto, Canada)

M. Enea (Erlangen, Germany)

M. Fiorini (Frascati, Italy)

B. Ford (Eugene, OR, USA)

A. Del Fra (Roma, Italy)

A. Frigerio (Padova, Italy)

D. Frohardt (Detroit, MI, USA)

A. Giambruno (Palermo, Italy)

A. Gillio (Milano, Italy)

A. Goodman (Eugene, OR, USA)

N. Gordeev (St. Petersburg, Russia)

T. Grundhöfer (Tübingen, Germany)

R. Guralnick (Los Angeles, CA, USA)

B. Hartley (Manchester, UK)

M. Herzog (Tel Aviv, Israel)

J. Humphreys (Liverpool, UK)

H. Ishibashi (Sakado, Japan)

P. Johnson (Manhattan, KS, USA)

M. Joswig (Tübingen, Germany)

A. Juhasz (Haifa, Israel)

W. Kantor (Eugene, OR, USA)

O. Kegel (Freiburg, Germany)

L. Kramer (Tübingen, Germany)

G. Kuhn (Milano, Italy)

H. Lausch (Würzburg, Germany)

R. Lawther (Pasadena, CA, USA)

M. van Leeuwen (Amsterdam, The Netherlands)

M. Liebeck (London, UK)

R. Liebler (Fort Collins, CO, USA)

S. Loewe (Braunschweig, Germany)

P. Longobardi (Napoli, Italy)

A. Lucchini (Padova, Italy)

G. Lunardon (Napoli, Italy)

K. Magaard (Detroit, MI, USA)

M. Mainardis (Trento, Italy)

M. Maj (Napoli, Italy)

H. van Maldeghem (Gent, Belgium)

G. Malle (Heidelberg, Germany)

A. Mann (Jerusalem, Israel)

R. Marinosci Micelli (Lecce, Italy)

V. Mazurov (Novosibirsk, Russia)

L. McCulloch (Dublin, Ireland)

T. Meixner (Giessen, Germany)

G. Micelli (Lecce, Italy)

V. Monti (Trento, Italy)

B. Muhlherr (Tübingen, Germany & Bruxelles, Belgium)

M. Muzychuk (Ramat-Gan, Israel)

F. Napolitani (Padova, Italy)

U. Ott (Braunschweig, Germany)

V. Pannone (Florence, Italy)

A. Pasini (Siena, Italy)

S. Pellegrini (Brescia, Italy)
G. M. Cattaneo Piacentini (Roma, Italy)
S. Pianta (Brescia, Italy)
P. Plaumann (Erlangen, Germany)
E. Plotkin (Ramat-Gan, Israel)
A. Prince (Edinburg, UK)
G. Roehrle (Bielefeld, Germany)
L. Rosati (Firenze, Italy)
J. Saxl (Cambridge, UK)
R. Scharlau (Bielefeld, Germany)
C. Scoppola (Roma, Italy)
G. Seitz (Eugene, OR, USA)
L. Serena (Firenze, Italy)
E. Shult (Manhattan, KS, USA)
V. De Smet (Gent, Belgium)
K. van Steen (Gent, Belgium)
M. Tamburini (Brescia, Italy)
P. Terwilliger (Madison, WI, USA)
D. Testerman (Middletown, CT, USA)
C. Tibiletti (Milano, Italy)
F. Timmesfeld (Giessen, Germany)
J. Tits (Paris, France)
A. Torre (Frascati-Roma, Italy)
A. Valenti (Palermo, Italy)
A. Valette (Neuchâtel, Switzerland)
N. Vavilov (St. Petersburg, Russia & Bielefeld, Germany)
J. Ward (Galway, Ireland)
T. Weigel (Freiburg, Germany)
J. Wilson (Birmingham, UK)
A. Woldar (Villanova, PA, USA)
A. Zalesskii (Minsk, Bielorussia)
V. Zambelli (Milano, Italy)
F. Zara (Amiens, France)
P. Zieschang (Kiel, Germany)

Representations of Groups on Finite Simplicial Complexes

MICHAEL ASCHBACHER

California Institute of Technology

Until recently there have been relatively few articles in the finite group theoretic literature on representations of finite groups on simplicial complexes. However in the last few years that situation has begun to change. This paper discusses some of the activity in the area. We begin with a fairly general discussion intended to give a feeling for the kind of activity now going on. Since space is limited, we touch on only a few examples of such activity, and to provide focus, we eventually concentrate on p-group complexes of finite groups. In the end we concentrate even further on a particular problem in the area of p-group complexes: the question of when p-group complexes of finite groups are simply connected. There we go into more detail.

This volume is devoted to groups of Lie type and their geometries. The p-group complexes of a group G should be viewed as geometries for G. The p-group complexes of the groups of Lie type will be featured prominently here. In particular we will see that if G is of Lie type and characteristic p then the p-group complexes of G are homotopy equivalent to the building of G.

The term "simplicial complex" is used here to mean an abstract simplicial complex. Thus a *simplicial complex K* consists of a set K of objects called *vertices* together with a collection of finite subsets of K called *simplices* such that each subset of a simplex is a simplex. The term simplicial complex is often used in the topological literature for a geometric realization of the abstract complex; the reader may be more familiar with this latter usage.

A simplex $s = \{x_0, \dots, x_k\}$ is of *dimension k* if it has $k + 1$ vertices. The *dimension* of K is the maximum dimension of a simplex of K. Most of the complexes we will consider are finite. Morphisms in the category are the *simplicial maps* which are maps of vertices which take simplices to simplices.

Examples (1) Let Δ be a graph. The *clique complex* $K(\Delta)$ of Δ is the simplicial complex whose vertices are the vertices of Δ and whose simplices are the finite cliques of Δ. Conversely given a simplicial complex L the *graph* $\Delta(L)$ of L is the graph on the vertices of L with x adjacent to y if

This work was partially supported by NSF DMS–9101237

$\{x, y\}$ is a simplex. Notice L is a subcomplex of $K(\Delta(L))$. L is said to be *connected* if its graph is connected.

(2) Let G be a finite group, p a prime, and $\Lambda_p(G)$ the commuting graph on the set of all subgroups of G of order p. The *commuting complex* $K_p(G)$ of G is the clique complex $K(\Lambda_p(G))$. The commuting complex is one of the p-group complexes of G. Those with some background in finite simple group theory know that the graph $\Lambda_p(G)$ has long been important in simple group theory, that it is elementary and well known that $\Lambda_p(G)$ is disconnected if and only if G has a strongly p-embedded subgroup, and that groups with a strongly p-embedded subgroup are of great importance in the proof of the Classification of the finite simple groups.

(3) Define a *geometric complex* to be a Tits geometry Γ on an index set I together with a set $\mathcal{C}$ of chambers of Γ (*ie.* flags of type I) such that each flag of rank at most 2 is contained in some member of $\mathcal{C}$. The complex can be regarded as a simplicial complex whose vertices are the objects of Γ and simplices are the nonempty flags contained in members of $\mathcal{C}$. Thus the geometric complex is a subcomplex of the clique complex of Γ. Those familiar with chamber systems will observe that the category of geometric complexes is isomorphic to the category of chamber systems X such that for each $x \in X$ and index $j \in I$, $\{x\} = \bigcap_i [x]_{i'}$ and $[x]_{j'} = \bigcap_{i \in j'} [x]_{i'}$. Chamber systems were introduced by Tits in part because the category of Tits geometries is too small. I find the simplicial complex point of view a more geometric and intuitive way to extend the category of geometries.

(4) Let G be a group, and $\mathcal{F} = (G_i : i \in I)$ a finite family of subgroups of G. Define $\mathcal{C}(G, \mathcal{F})$ to be the geometric complex whose vertex set is the union of the coset spaces G/G_i, $i \in I$, with a set s of vertices a simplex if and only if $\bigcap_{X \in s} X \neq \emptyset$. Call such a complex a *coset complex*.

In this example each vertex $G_i x$ has a type $\tau(G_i x) = i \in I$ and the maximal simplices are indeed of type I; *eg.* if s is a simplex then there is $x \in \bigcap_{X \in s} X$ and $s = S_{J,x} = \{G_j x : j \in J\}$, where $J = \tau(s)$, so $s = S_{J,x} \subseteq S_{I,x}$ of type I. Finally G is represented as a group of automorphisms of $\mathcal{C}(G, \mathcal{F})$ via right multiplication and G is transitive on simplices of type J for each $J \subseteq I$. Conversely any geometric complex K on I admitting a group of automorphisms transitive on simplices of each type is isomorphic to a coset complex.

(5) Let P be a poset. The *order complex* $\mathcal{O}(P)$ of P is the simplicial complex whose vertices are the members of P and whose simplices are the finite chains in P. We often write P for $\mathcal{O}(P)$.

(6) Let G be a finite group, p a prime, and $\mathcal{S}_p(G)$ the set of all nontrivial p-subgroups of G partially ordered by inclusion. We also write $\mathcal{S}_p(G)$ for the order complex of this poset, and call this complex the *Brown complex* of G at p. The subcomplex $\mathcal{A}_p(G)$ of all elementary abelian p-subgroups is

the *Quillen complex*. Thus we have two more p-group complexes associated to G. There are still others.

A *geometric realization* of K is simply an identification of the vertices of K with suitable points in some Euclidean space, with a simplex identified with the convex closure of its vertices. The realization is then regarded as a topological subspace of Euclidean space. We can then define two complexes to be *homotopy equivalent* if their geometric realizations are homotopy equivalent.

Invariants of the homotopy type of the complex are its homology groups $H_n(K)$ and its fundamental group $\pi_1(K)$. These can be defined combinatorially without reference to the geometric realization.

Examples (7) K is *contractible* if it has the same homotopy type as a point. In that case the reduced homology of K and its fundamental group are trivial. Recall K is *acyclic* if its reduced homology is trivial and K is *simply connected* if its fundamental group is trivial. So contractible complexes are acyclic and simply connected.

(8) Let G be a finite group and p a prime. The commuting complex, the Brown complex, and the Quillen complex at the prime p are all of the same homotopy type.

(9) Let K be a simplicial complex. The *barycentric subdivision* $sd(K)$ of K is the order complex of the poset of simplices of K ordered by inclusion. It is well known that $sd(K)$ has the same homotopy type as K.

One of the most useful tools available in this area is to replace a complex by another complex of the same homotopy type which may be better suited to analysis. We will see this tool used many times. Here are three lemmas which allow us to make such replacements:

LEMMA 1. *Let* $f : P \to Q$ *be a map of posets. Then*

(1) *(Quillen [Q]) If* $f^{-1}(Q(\geq q))$ *is contractible for all* $q \in Q$ *then* $f :$ $\mathcal{O}(P) \to \mathcal{O}(Q)$ *is a homotopy equivalence.*

(2) *[A] Assume for each* $q \in Q$, $f^{-1}(Q(\geq q))$ *is* $min\{1, h(q) - 1\}$-*connected,* $Q(> q)$ *is connected if* $h(q) = 0$, *and if* $h(q) = 1$ *then either* $Q(> q) \neq \varnothing$ *or* $f^{-1}(Q(\geq q))$ *is simply connected. Then* P *is simply connected if* Q *is simply connected.*

Here $-1, 0, 1$-connectivity means nonempty, connected, simply connected, respectively. $Q(\geq q) = \{x \in Q : x \geq q\}$ and $Q(> q)$, $Q(\leq x)$, and $Q(< x)$ are defined similarly. The *height* of $q \in Q$ is $h(q) = dim(Q(\leq q))$.

LEMMA 2. *Let C be a cover of a simplicial complex K by subcomplexes. The nerve $N(C)$ of C is the complex with vertex set C and $s \subseteq C$ a simplex if and only if $I(s) = \bigcap_{X \in C} X \neq \varnothing$. Assume $I(s)$ is empty or contractible for all s. Then $N(C)$ and K have the same homotopy type.*

LEMMA 3. *(Quillen [Q]) Let G be a group of automorphisms of the poset P and let Q be the order complex of some poset of subgroups of G. Assume*
  (1) For each $x \in P$, $Q \cap G_x$ is contractible (simply connected).
  (2) For each $X \in Q$, $Fix_P(X)$ is contractible (simply connected).
  (3) If $X \in Q$ and $x \in Fix_P(X)$ then $P(\leq x) \subseteq Fix_P(X)$.
  Then P and Q have the same homotopy type. (P is simply connected if and only if Q is simply connected.)

There are two directions one can go in the area. First, use techniques and ideas from combinatorial topology to prove things about finite groups and geometries. Second, use modern finite group theory and finite geometry to prove results in combinatorial topology. We give a quick example of each approach.

Recall K is simply connected if $\pi_1(K)$ is trivial. Equivalently K has no proper connected coverings. Coverings of K correspond to topological coverings of its geometric realization. Combinatorially a covering of K is a surjective local isomorphism $f : \tilde{K} \to K$ of simplicial complexes.

In simple group theory we often wish to characterize a group G as the unique group satisfying suitable hypotheses. In particular in the Classification of the finite simple groups we need to characterize each simple group via suitable hypotheses on local subgroups, preferably centralizers of involutions. In [AS1], Yoav Segev and I came up with an approach to this problem with a topological flavor. In brief, given some hypotheses $\mathcal{H}$ on the centralizer of an involution in some finite simple group G, we produce a family $\mathcal{F}$ of subgroups of G and form the coset complex $K = \mathcal{C}(G, \mathcal{F})$. For $J \subset I$ let $G_J = \bigcap_{j \in J} G_j$, be the stablizer of the simplex $S_{J,1}$ of type J. The inclusion maps $G_J \to G_K$ for $K \subset J$ define an *amalgam* $\mathcal{A}$ of groups and there is a largest group $\tilde{G}$ realizing this family and a surjective local isomorphism $\alpha : \tilde{G} \to G$. We show $\mathcal{A}$ is determined up to isomorphism; hence if $\bar{G}$ is any group satisfying hypotheses $\mathcal{H}$, there is a local isomorphism $\bar{\alpha} : G \to \bar{G}$.

Now $\tilde{G}$ acts on its coset complex $\tilde{K}$ and there is a covering $f : \tilde{K} \to K$ of complexes. Our approach is to show K is simply connected; hence f and therefore also α is an isomorphism. But then $\bar{\alpha} : G \to \bar{G}$ is a group homomorphism and as G is simple, $\bar{\alpha}$ is an isomorphism, so G is determined up to isomorphism.

Segev and I usually use a lot of knowledge of the complex K and some simple minded techniques to show K is simply connected. It would be good to have more powerful techniques.

So far the applications to finite group theory involve only low dimensional properties of complexes; *ie.* connectivity and simple connectivity are
properties of the 1-skeleton and 2-skeleton. However Alperin's Conjecture
can be stated in terms of the p-group complexes; this may turn out to be
an application which uses the full strength of the theory. Also Webb has
results which describe the p-part of the cohomology of a finite group G in
terms of the cohomology of stabilizers of simplices in any p-group complex
of G, perhaps making possible an inductive approach to the cohomology of
finite groups. See Webb's survey article [**W**] for a discussion of these topics.

Let us next turn to an example of how finite simple group theory can be
used to answer a question from combinatorial topology. Let K be a finite
acyclic simplicial complex and $G \leq Aut(K)$. Replacing K by its barycentric
subdivision, we can assume with little loss of generality that the action of G
on K is *admissible*; that is if $g \in G$ fixes a simplex then it fixes each vertex
of the simplex. The Lefschetz Fixed Point Theorem says that g has a fixed
point on K; but what about G? Constructions of Robert Oliver [**O**] show
that for most finite groups G there exist finite acyclic complexes K such
that G has no fixed points on K. But if the dimension of K is small one
can hope to prove something.

CONJECTURE. *(Warren Dicks) If G is a finite group acting admissibly on a
2-dimensional contractible complex then G has a fixed point.*

The Conjecture fails if we weaken "contractible" to acyclic; the Poincaré
dodecahedron disk is a 2-dimensional acyclic complex admitting the fixed
point free action of A_5. Its fundamental group is $SL_2(5)$.

Tensoring the simplicial chain complex with any ring R we get homology
$H_*(K, R)$ with coeficients in R. K is R-acyclic if $\tilde{H}_*(K, R) = 0$. If π is a
set of primes define K to be π-acyclic if K is F- acyclic for each field F of
characteristic p and each $p \in \pi$.

THEOREM. *(Aschbacher-Segev [**AS3**]) If G is finite group and K is $\pi(G)$-
acyclic finite complex with G admissible on K then either G has a fixed
point on K or G has a composition factor which is a rank 1 group of Lie
type or J_1.*

We also construct a family of 2-dimensional complexes admitting the
action of $G = PGL_2(q)$, which appear to be $\pi(G)$-acyclic when q is even.
We call these complexes *polygon complexes*. Each is of the form $\mathcal{C}(G, \mathcal{F})$,
where $\mathcal{F} = (G_1, G_2, G_3)$ with G_1 a Borel subgroup of G and G_2 and G_3
normalizers of suitable representatives of the 2 classes of maximal tori. More
concretely, one can think of the coset space G/G_1 as points on the projective
line $X = GF(q) \cup \{\infty\}$, G/G_2 as edges $\{x, y\} \in X \times X$, and G/G_3 as

$q + 1$-gons on X. The Poincaré dodecahedron disk is the polygon complex for $q = 4$. Using Mathematica we calculated the homology groups for $q \leq 16$ and Richard Wilson made a pseudo calculation for $q = 32$ for one of the complexes. The complexes were p-acyclic for all primes distinct from 17,1087,239,$\{67, 659, 1033\}$, for $q = 8$, 16,16,32, respectively. We find these primes mysterious.

This leaves open the Dicks Conjecture. One can also ask if A_5 is the unique simple group acting without fixed points on a 2-dimensional $\mathbf{Z}$-acyclic complex.

We now concentrate on p-subgroup complexes of finite groups. The Brown and Quillen complexes of a finite group G were introduced by Brown [**B**] and Quillen [**Q**], presumably to study group cohomology. As indicated earlier, the graph of the commuting complex has long been important in simple group theory. To get some feel for these complexes let us consider an example:

Example (10) Let G be a finite simple group of Lie type in characteristic p. Then the building of G is a finite simplicial complex; indeed the building is the coset complex of the family of maximal parabolics over a fixed Borel subgroup. It is easy to show that the building has the same homotopy type as $\mathcal{S}_p(G)$. By the Solomon-Tits Theorem, the building is *spherical*; that is it is simply connected if the dimension is at least 2, and reduced homology vanishes except in the top dimension. The top dimensional homology is the Steinberg module for G.

More generally we would like to know the homology and fundamental group of $\mathcal{S}_p(G)$, for G simple and p a prime divisor of G, and we want results which reduce problems about p-group complexes of general finite groups to problems about simple groups. In a moment, we illustrate with a discussion of simply connectivity of p-group complexes.

The problem on p-group complexes with the most history and visibility is the Quillen Conjecture:

QUILLEN CONJECTURE. *Let G be a finite group and p a prime. Then $\mathcal{S}_p(G)$ is contractible if and only if $O_p(G) \neq 1$.*

One direction is easy: if $O_p(G) \neq 1$ then $\mathcal{S}_p(G)$ is contractible. The opposite direction was proved by Quillen [**Q**] for solvable groups. It is known to be true for simple groups. In [**ASm**], Steve Smith and I show that if $p > 5$ and G has no components which are unitary groups then the Conjecture holds. If one could show the top dimensional homology of $\mathcal{A}_p(U_n(q))$ is nontrivial for p dividing $q + 1$ then the Conjecture would (essentially) be settled for $p > 5$.

In the remainder of the talk we will concentrate on the following question:

QUESTION 1. *Let G be a finite group and p a prime. When is $K_p(G)$ simply connected?*

If $O_p(G) \neq 1$ we just saw that $K_p(G)$ is contractible and hence simply connected, so the interesting case occurs when $O_p(G) = 1$. We will see that in that event a necessary condition for $K_p(G)$ to be simply connected is that $m_p(G) > 2$. (Recall $m_p(G)$ is the *p-rank* of G; *ie.* the largest n such that G has a subgroup isomorphic to n copies of the group of order p.) The answer to Question 1 seems to be that if $m_p(G) > 2$ then $K_p(G)$ is almost always simply connected. As is usual in finite group theory, a strategy for proving this conjecture involves two steps:

(1) Reduction: Reduce the simple connectivity of $K_p(G)$ for the general finite group G to the simply connectivity of $K_p(G_0)$ for certain minimal groups G_0.

(2) Analysis of Minimal Groups: Determine when $K_p(G_0)$ is simply connected for the minimal groups G_0.

The reduction step has been accomplished in [A]. Minimal groups include the simple groups. Analysis of the minimal groups is begun in [A], [S], and [D]. I regard this project as a test case for a more ambitious analysis of the behavior of the p-group complexes of finite groups; *eg.* obtaining a qualitative description of the homology of these complexes including a reduction theory, and a precise description of the homology of an appropriate set of minimal groups.

Our first set of minimal groups is the set Min_0 of groups G such that $G = HA$ with A an elementary abelian p-group, H a normal subgroup of G of order prime to p, $O_p(G) = 1$, and G is minimal subject to these constraints. The case where G is solvable was handled by Quillen in [Q]:

THEOREM. *(Quillen) If G is solvable in Min_0 then $\mathcal{A}_p(G)$ is spherical. That is $\tilde{H}_k(\mathcal{A}_p(G)) = 0$ for $k < dim(\mathcal{A}_p(G)) = m_p(G) - 1$ and $\mathcal{A}_p(G)$ is simply connected if $m_p(G) > 2$.*

One can ask if $G_0 \in Min_0$ is nonsolvable is Quillin's result still true? We conjecture at least:

CONJECTURE. *Let G be a finite group such that $G = AF^*(G)$, where A is an elementary abelian p-subgroup of rank at least 3 and $F^*(G)$ is the direct product of the A-conjugates of a simple component L of G of order prime to p. Then $K_p(G)$ is simply connected.*

The following result from [A] supplies strong evidence for the conjecture, in essence reducing it to the case where L is a group of Lie type and Lie rank 1 or a sporadic group.

THEOREM 3. *Assume G and L satisfy the hypotheses of the Conjecture and that the Conjecture holds in proper sections of G. Then*

(1) If L is of Lie type and Lie rank at least 2 then $K_p(G)$ is simply connected.

(2) If $L \cong L_2(q)$ with q even then $K_p(G)$ is simply connected.

(3) If L is an alternating group then $K_p(G)$ is simply connected.

(4) If L is a Mathieu group then $K_p(G)$ is simply connected.

The minimal groups not in Min_0 are the finite simple groups of p-rank at least 3. The next two theorems from [A] supply the reduction step which reduces the question of simple connectivity of $K_p(G)$ to the case G minimal. Given a graph Λ and a vertex x of Λ, write $\Lambda(x)$ for the set of vertices distinct from x and adjacent to x in Λ.

THEOREM 1. *Assume the Conjecture and let G be a finite group, p a prime divisor of the order of G, and $\Lambda = \Lambda_p(G)$. Assume $\Lambda(x)$ is connected for all $x \in \Lambda$ and let $\bar{G} = G/O_{p'}(G)$. Then exactly one of the following holds:*

(1) $K_p(G)$ is simply connected.

(2) $\bar{G} = \bar{G}_1 \times \bar{G}_2$ and $\bar{G}_i$ has a strongly p-embedded subgroup for $i = 1$ and 2.

(3) $\bar{G} = \bar{X}(\bar{G}_1 \times \bar{G}_2)$, for some $X \in \Lambda$, $p = 3, 5$, $\bar{G}_1 \cong L_2(8)$, $Sz(32)$, respectively, $\bar{G}_2$ is a nonabelian simple group with a strongly p-embedded subgroup, and X induces outer automorphisms on $\bar{G}_i$ for $i = 1$ and 2.

(4) $\bar{G}$ is almost simple and $K_p(\bar{G})$ and $K_p(F^*(\bar{G}))$ are not simply connected.

THEOREM 2. *Let G be a finite group, p a prime divisor of the order of G, and assume $O_p(G) = 1$, $\Lambda = \Lambda_p(G)$ is connected, and $H_1(K_p(G)) = 0$. Then $m_p(G) > 2$ and $\Lambda(x)$ is connected for each $x \in \Lambda_p(G)$.*

Theorems 1 and 2 say that, modulo the Conjecture and a short list of exceptions, $K_p(G)$ is simply connected if and only if $m_p(G) > 2$ and $\Lambda(x)$ is connected for each $x \in \Lambda_p(G)$.

The following observations expand upon these points:

Remarks (1) If $O_p(G) \neq 1$ then G is contractible and hence simply connected. Thus the restriction that $O_p(G) = 1$ in Theorem 2 causes no loss of generality.

(2) It is well known that $\Lambda_p(G)$ is disconnected if and only if G has a strongly p-embedded subgroup. (*cf.* 44.6 in [**FGT**]) Moreover we know all groups with strongly p-embedded subgroups. (*cf.* [**A**]) Thus the restriction in Theorem 2 that Λ be connected results in no loss of generality, and the groups in Cases (2) and (3) of Theorem 1 are completely described.

(3) Recall a simplicial complex is simply connected if and only if its fundamental group is trivial, while the first homology group of the complex is the abelianization π_1/π_1' of its fundamental group. Thus the hypothesis in Theorem 2 that $H_1(K_p(G)) = 0$ is weaker than simple connectivity. So Theorem 2 says that the hypothesis in Theorem 1 that $\Lambda(x)$ be connected for each $x \in \Lambda$ is necessary for simple connectivity, and that if $O_p(G) = 1$ and $K_p(G)$ is simply connected then $m_p(G) > 2$.

(4) The condition that $\Lambda(x)$ be connected has various equivalent formulations; it is roughly equivalent to $C_G(x) = \Gamma_{2,P}(C_G(x))$ for $P \in Syl_p(C_G(x))$, for those readers who are simple group theorists. All finite groups G with $m_p(G) \geq 3$ such that $\Lambda(x)$ is disconnected for some $x \in \Lambda$ are determined in [**A**]. Thus Theorems 1 and 2 constitute a fairly complete reduction to the simple case, modulo the Conjecture.

(5) Recall from the Classification of the finite simple groups that each nonabelian simple group L is an alternating group, a group of Lie type, or one of the 26 sporadic groups. Thus Theorem 3 reduces a verification of the Conjecture to the case where L is of Lie type and Lie rank 1 (*ie.* $L \cong L_2(q)$, $U_3(q)$, $Sz(q)$, or $^2G_2(q)$) or L is one of the 21 sporadic groups which are not Mathieu groups. We will see one possible approach to handling these groups in a while.

In short Theorems 1 through 3 reduce the problem of determining when $K_p(G)$ is simply connected to (a) the problem of verifying the Conjecture for rank 1 groups of Lie type and the sporadic groups, and (b) determining when $K_p(G)$ is simply connected for G a finite simple group.

We now try to give enough of an idea of the proof of Theorem 1 to see how the minimal groups arise. Recall part 2 of Lemma 1. Let H be a normal subgroup of order prime to p, $\bar{G} = G/H$, $P = \mathcal{A}_p(G)$, and $Q = \mathcal{A}_p(\bar{G})$. Let $f : P \to Q$ be the map $f(A) = \bar{A}$. Then $f^{-1}(Q(\leq \bar{A})) = \mathcal{A}_p(AH)$, so to achieve the hypotheses of Lemma 1, part 2, we need to know $\mathcal{A}_p(AH)$ is simply connected if $h(\bar{A})) > 1$; as $h(\bar{A}) = m_p(A) - 1$, this is equivalent to $m_p(A) > 2$. This can be proved by an easy induction argument if it holds for $AH \in Min_0$.

We also need to know that $Q(> \bar{A})$ is connected if $h(\bar{A}) = 0$; *ie.* if A is of order p then $m_p(C_G(A)) > 1$, which follows if $m_p(G) > 1$. Finally if $h(\bar{A}) = 1$ we need $Q(> \bar{A}) \neq \varnothing$; *ie.* if $m_p(A) = 2$ then $m_p(C_G(A)) > 2$, which holds if $m_p(G) > 2$ and $\Lambda(x)$ is connected for each x of order p in A.

So Lemma 1 allows us to conclude:

LEMMA 4. *Assume the conjecture with $m_p(G) > 2$, let H be a normal p'-subgroup of G, $\Lambda(x)$ connected for all x of order p in G, and $\mathcal{A}_p(G/H)$ simply connected. Then $\mathcal{A}_p(G)$ is simply connected.*

The converse of Lemma 4 is fairly easy to prove, so we can reduce to the case $O_{p'}(G) = O_p(G) = 1$. To reduce to the case G simple requires:

LEMMA 5. *Let $H \trianglelefteq G$ such that $K_p(C_H(x))$ is connected for all $x \in \Lambda(G) - \Lambda(H)$ and $K_p(H)$ is simply connected. Then $K_p(G)$ is simply connected.*

Lemma 5 is proved using results of Yoav Segev and the author on simple connectivity of simplicial complexes found in [**AS2**].

Let us next consider the Conjecture. The following lemma is a result of Segev in [**S**] extending a result in [**A**] used to prove Theorem 3:

LEMMA 6. *Assume the hypotheses of the conjecture and let $B = N_A(L)$. Assume $\mathcal{F}$ is a family of B-invariant proper subgroups of L such that*
  (1) $N_L(X) \cap C_L(B) \leq X$ for each $X \in \mathcal{F}$.
  (2) $\mathcal{C}(G, \mathcal{F})$ is simply connected.
  (3) $A_p(\langle X, A \rangle)$ is simply connected for each $X \in \mathcal{F}$.
  (4) $link_{\mathcal{C}(G,\mathcal{F})}(X)$ is connected for all $X \in \mathcal{X}$.
  (5) The truncation of $\mathcal{C}(G, \mathcal{F})$ at any 2-subset of $\mathcal{F}$ is connected.
  (6) If $B \neq 1$ then $C_L(B) = \langle C_X(B) : X \in \mathcal{F} \rangle$.
Then G satisfies the Conjecture.

Hypothesis (3) holds automatically in a minimal counter example to the conjecture. Hypotheses (4) and (5) hold if the coset complex $\mathcal{C}(G, \mathcal{F})$ is residually connected. If A is regular on the components of G then (6) is vacuously satisfied; if not L is of Lie type and B is of order p inducing field automorphisms on L. The critical condition is the hypothesis that the coset complex be simply connected. That is why small groups like groups of Lie type of Lie rank 1 cause problems; their subgroup structure is not rich enough to have a coset complex for which simple connectivity is easy to verify. However Segev has shown in [**S**] that:

THEOREM 4. *(Segev) Assume the hypotheses of the Conjecture with A regular on the components of G and $L \cong L_2(q)$, $Sz(q)$, or $U_3(r)$, $r \equiv 0, 1$ mod 3. Then the Conjecture holds.*

Finally let us discuss what is known about the simple connectivity of $K_p(G)$ for G a finite simple group with $m_p(G) > 2$.

If G is of Lie type and characteristic p then by Example (10), $K_p(G)$ has the homotopy type of the building $\mathcal{B}$ of G. By the Solomon-Tits Theorem [**ST**], $\mathcal{B}$ is spherical, so $K_p(G)$ is simply connected if and only if G has Lie rank at least 3.

Let $G = GL_n(q)$ with $q \equiv 1 \mod p$. Using the Solomon-Tits Theorem and Lemma 3, Quillen shows in [**Q**] that $A_p(G)(> Z)$ is spherical for Z of order p in $Z(G)$. That is a large subcomplex of $A_p(PGL_n(q))$ is spherical. If p does not divide n then $A_p(G) \cong A_p(L_n(q))$.

QUESTION 2. *Let $n \equiv 0 \mod p$ and $q \equiv 1 \mod p$. What does $A_p(L_n(q))$ look like? How close is its homology to that of $A_p(G)(> Z)$?*

K. M. Das, a graduate student at Caltech is analyzing the simple connectivity of $A_p(G)$ when G is a finite classical group and p a prime divisor of the order of G. (Recall that by the discussion in Example 10, we may as well assume p is not the natural characteristic of the group.) For example he has used Quillen's result to prove that $A_p(L_n(q))$ is simply connected when $m_p(L_n(q)) > 2$ and p does not divide q or $q - 1$. He has shown that $A_p(Sp_n(q))$ is spherical when $q \equiv 1 \mod p$, and is close to showing $Sp_n(q))$ is simply connected when p does not divide q or $q - 1$. He is in the process of analyzing $U_n(q)$ and has analogous results for that family of classical groups.

Notice the following result is a special case of Lemma 3:

LEMMA 7. *Let G be a group of automorphisms of a poset P. Assume*
(1) For each $x \in P$, $A_p(G_x)$ is simply connected, and
(2) For each $A \in A_p(G)$, $Fix_P(A)$ is simply connected and $P(\leq x) \subseteq Fix_P(A)$ for each $x \in Fix_P(A)$.
Then P is simply connected if and only if $A_p(G)$ is simply connected.

In the case of $G = GL_n(q)$, Das uses Lemma 7 with P the the truncation of the barycentric subdivision of the projective geometry of the natural module for G at the subspaces of dimension divisible by $d_p(q)$. Here $d_p(q)$ is the least positive integer d such that $q \equiv 1 \mod p$. The truncation of a building is spherical so if $dim(P) > 2$ then P is simply connected and Lemma 7 says $A_p(G)$ is simply connected. If $m_p(G) > 2$ and $dim(P) = 2$ then $n = 3d_p(q)$ and some extra work and ingenuity is necessary. Similar arguments are useful for the other classical groups, although again more ideas are required. Along the way Das builds up a library of interesting complexes and homotopy equivalences among these complexes which should be useful in future work in the field.

References

FGT. M. Aschbacher, "Finite Group Theory," Cambridge University Press, Cambridge, 1986.

A. M. Aschbacher, *Simple connectivity of p-group complexes*, Israel J. Math. **82** (1993), 1–43.

AS1. M Aschbacher and Y. Segev, *Extending morphisms of groups and graphs*, Ann. Math. **135** (1992), 297–323.

AS2. M Aschbacher and Y. Segev, *Locally connected simplical maps*, Israel J. Math. **77** (1992), 285–303.

AS3. M Aschbacher and Y. Segev, *A fixed point theorem for groups acting on finite 2-dimensional acyclic simplical complexes*, Proc. London Math. Soc. **67** (1993), 329–354.

ASm. M Aschbacher and S. Smith, *On Quillen's conjecture for the p-groups complex*, Ann. Math. **137** (1993), 473–529.

B. K. Brown, *Euler characteristic of groups: the p-fractional part*, Invent. Math. **29** (1975), 1–5.

D. K. Das, *Homotopy and homology of p-group complexes*, Thesis in preparation.

O. R. Oliver, *Fixed-point sets of group actions on finite acyclic complexes*, Comment. Math. Helv. **50** (1975), 155–177.

Q. D. Quillen, *Homotopy properties of the poset of nontrivial p-subgroups*, Adv. in Math. **28** (1978), 101–128.

S. Y. Segev, *Simply connected coset complexes for rank 1 groups of Lie type*, preprint.

ST. L. Solomon, *The Steinberg character of a finite group with BN-pair*, "Theory of Finite Groups," Benjamin, New York, 1969, pp. 213–221.

W. P. Webb, *Subgroup complexes*, Proc. Sym. Pure Math. **47** (1987), 349–365.

Pasadena, California 91125

Coxeter Groups and Matroids

Alexandre V. Borovik

K. Sian Roberts

Department of Mathematics, UMIST
PO Box 88, Manchester M60 1QD
United Kingdom

Introduction

WP-matroids are combinatorial objects introduced by I. M. Gelfand and V. V. Serganova in [**GS1**] as a generalization of the classical notion of matroid. The definition of a WP-matroid is given in terms of the Bruhat ordering on a Coxeter group.

> Let $P < W$ be a standard parabolic subgroup of a Coxeter group W. A map
>
> $$\mu : W \longrightarrow W/P$$
>
> is called a WP-matroid if
>
> $$w^{-1}\mu(u) \leq w^{-1}\mu(w)$$
>
> for all $u, w \in W$, where $\leq$ is the Bruhat ordering of the left coset space W/P.

Our approach to WP-matroids is based on a systematic use of Coxeter complexes for Coxeter groups. It will be convenient for us to identify a Coxeter group W with its Coxeter complex. This enables us to state the main result of the paper using the language of the combinatorial geometry of Coxeter complexes.

Theorem *Let W be a Coxeter group, P a finite standard parabolic subgroup of W, and $\mu : W \longrightarrow W/P$ a WP-matroid. Then*

(1) *The fibers $\mu^{-1}[m]$, $m \in W/P$, of the map μ are convex subsets of W.*

(2) *If two fibers $\mu^{-1}[m]$ and $\mu^{-1}[n]$ of μ are adjacent then their images m, n are symmetric with respect to some wall σ of the Coxeter complex W. Moreover, all common panels of $\mu^{-1}[m]$ and $\mu^{-1}[n]$ lie on σ.*

(3) *If W is infinite, then the nonempty fibers of the map μ are also infinite.*

This result provides a combinatorial version of properties of convex polyhedra associated with matroids [**GGMS**] and WP-matroids for finite Weyl groups [**GS2**]. Notice that the class of Coxeter groups covered by the theorem includes, in particular, Coxeter groups W of spherical, affine and hyperbolic types. In all these cases all proper parabolic subgroups of W are finite.

A convenient reference for the system of definitions related to Coxeter groups and complexes and Tits systems and freely used throughout the paper is M. Ronan's book [**Ron**]. But we made every effort to make the paper relatively self-contained. We have also devoted two sections of the paper to discussion of the background and motivation for our results.

1   Matroids

Matroids.   According to one of the many equivalent definitions, a pair $M = (E, \mathcal{B})$, where E is a finite set and $\mathcal{B} \subseteq 2^E$ is a set of subsets in E, is called a *matroid* if it satisfies the **Exchange Principle**:

For all $A, B \in \mathcal{B}$ and $x \in A \setminus B$ there exists $y \in B \setminus A$, such that $A \setminus \{x\} \cup \{y\}$ lies in $\mathcal{B}$.

The set $\mathcal{B}$ is called the *base set* and its elements the *bases* of the matroid M. We shall say that the base $A \setminus \{x\} \cup \{y\}$ in the statement of the Exchange Principle is obtained from the base A by the transposition (x, y). It can be easily shown that all the bases of a matroid have the same cardinality, which is called its *rank*.

If E is a *vector configuration*, i.e. a finite set of vectors in a finite-dimensional vector space V, then a well-known fact from linear algebra asserts that the set of all maximal linearly independent subsets of E is the base set of a matroid on E. Historically, matroids were introduced as means of axiomatization, in purely combinatorial terms, of the notion of linear dependence in vector spaces [**Whi**]. Easy examples show that, despite the origins of the notion, not every matroid is *representable*, i.e. isomorphic to the matroid of a vector configuration.

Matroids and Convex Polyhedra.   We shall usually identify the underlying set E of the matroid $M = (E, \mathcal{B})$ with the set $\{1, 2, \ldots, n\}$. Let now

$\mathbb{R}^n$ be the n-dimensional real vector space with the canonical base $e_1, \ldots, e_n$. We set for every $B \in \mathcal{B}$

$$\delta_B = \sum_{i \in B} e_i.$$

Let $\Delta = \Delta(M)$ be the convex hull of δ_B, $B \in \mathcal{B}$. This is a (solid) polytope.

The following beautiful theorem is due to I. M. Gelfand, M. Goresky, R. MacPherson and V. V. Serganova [GGMS].

Fact 1.1 ([GGMS], Theorem 4.1) *The points δ_A, $A \in \mathcal{B}$, form the vertex set of Δ. Two vertices δ_A, δ_B are adjacent (i.e. connected by an edge) if and only if the bases A and B of the matroid can be obtained from each other by a transposition.*

Notice that if we consider the convex polyhedron Δ^* dual to Δ (for a definition see, for example, [Grü]), then two faces δ_A^*, δ_B^* are adjacent if and only if the bases A and B can be obtained from each other by a transposition. One of the main results of the present paper, Theorem 6.1, is a combinatorial version of this remarkable property of matroids.

Maximality Principle for matroids. In what follows we shall work with a definition of matroid in terms of orderings. This approach was developed by D. Gale [Gal] as a solution of the problem of optimal assignment in applied combinatorics and later but independently used by I. M. Gelfand and V. V. Serganova in their work on stratifications of flag varieties [GS2].

Let $\mathcal{P}_k = \mathcal{P}_k(E)$ be the set of all k-element subsets in a finite set $E = \{1, 2, \ldots, n\}$. We introduce a partial ordering $\leq$ on $\mathcal{P}_k$ as follows. Let $A, B \in \mathcal{P}_k$, where

$$A = (i_1, \ldots, i_k), \; i_1 < i_2 < \ldots < i_k$$

and

$$B = (j_1, \ldots, j_k), \; j_1 < j_2 < \ldots < j_k,$$

then we set

$$A \leq B \iff i_1 \leq j_1, \ldots, i_k \leq j_k.$$

Let $W = Sym_n$ be the group of all permutations of the elements of E. Then we can associate an ordering of $\mathcal{P}_k$ with each $w \in W$ by putting

$$A \leq^w B \iff w^{-1}A \leq w^{-1}B.$$

Clearly $\leq^1$ is just $\leq$.

Fact 1.2 (Gale [Gal], see also Gelfand - Serganova [GS1]) *Let $\mathcal{B} \subseteq \mathcal{P}_k$. The set $\mathcal{B}$ is the base set of some matroid if and only if $\mathcal{B}$ satisfies the **Maximality Principle**: for every $w \in Sym_n$ the set $\mathcal{B}$ contains an element $A \in \mathcal{B}$ maximal in $\mathcal{B}$ with respect to $\leq^w$:*

$$B \leq^w A \text{ for all } B \in \mathcal{B}.$$

(We call A the w-maximal element in B).

Matroids as maps. Now let $\mathcal{B}$ be the base set of a matroid M on $E = \{1, \ldots, n\}$ of rank k. We can define a map

$$\mu : Sym_n \longrightarrow \mathcal{P}_k$$

assigning to each $w \in Sym_n$ an element $A \in \mathcal{B}$ maximal in $\mathcal{B}$ with respect to $\leq^w$. Obviously this map satisfies the inequality

$$\mu(u) \leq^w \mu(w) \tag{1}$$

for all $u, w \in Sym_n$. Since any k-element set $B \in \mathcal{P}_k$ can be made maximal in $\mathcal{P}_k$ after some reordering of the symbols $1, 2, \ldots, n$, we have $\mu[Sym_n] = \mathcal{B}$. Vice versa, the image of every map μ from Sym_n to $\mathcal{P}_k$, satisfying the above inequality, is the base set of some matroid.

2   Grassmann Varieties

In this section we follow [**GS2**] and use matroids for the description of combinatorial properties of Grassmann varieties. These applications to the geometry of Grassmannians are the main source of motivation for the theory of WP-matroids.

Grassmann varieties and matroids. Let $\mathbb{G}_{n,k}$ be the *Grassmann variety* of k-dimensional vector subspaces in $V = K^n$ with a standard basis $e_1, \ldots, e_n$. Let V^* be the dual space of V with the dual basis $e_1^*, \ldots e_n^*$.

If now $X \in \mathbb{G}_{n,k}$ is a k-dimensional subspace in $V = K^n$ then the subsets J of $E = \{1, 2, \ldots, n\}$ such that $(e_j^*|_X \mid j \in J)$ are a basis of X, form a base set of a matroid. We denote this matroid by M_X and call it the *matroid associated with X*.

Plücker coordinates and the moment map. Now we restrict our attention to the case $K = \mathbb{C}$. Let $e_1, \ldots, e_n$ be the standard basis in $V = \mathbb{C}^n$. We write vectors from $\mathbb{C}^n$ in the column form. Take a basis $x_1, \ldots, x_k$ in the subspace $X < V$ and form the coordinate matrix

$$\begin{pmatrix} x_{11} & x_{12} & \ldots & x_{1k} \\ \vdots & \vdots & & \vdots \\ x_{n1} & x_{n2} & \ldots & x_{nk} \end{pmatrix}.$$

Notice that the rows of the matrix (x_{ij}) form a vector configuration in the dual space X^*. Since a change of a base in V inflicts a linear transformation of the columns of the matrix (x_{ij}) and since linear transformations of columns do not change the pattern of linear dependence of the rows, the matroid associated with the vector configuration of the rows coincides with the matroid M_X built in the previous paragraph in a coordinate-free way.

Denote by $p^{j_1 \cdots j_k} = p^J$, $J \in \mathcal{P}_k$, the value of the minor formed by the rows $j_1 < j_2 < \ldots < j_k$ of the matrix (x_{ij}). Then the vector $(p^J)_{J \in \mathcal{P}_k}$ is uniquely determined up to multiplication by a scalar, and the quantities p^J are called the *Plücker coordinates* of X.

Notice that the base set $\mathcal{B}$ of the matroid M_X can be determined in terms of the Plücker coordinates of X:

$$\mathcal{B} = \{J \in \mathcal{P}_k,\ p^J \neq 0\}.$$

Following [GGMS] we define the *moment map*

$$\lambda : \mathbb{G}_{n,k} \longrightarrow \mathbb{R}^n$$

by

$$\lambda(X) = \frac{\sum_{J \in \mathcal{P}_k} |p^J(X)|^2 \cdot \delta_J}{\sum_{J \in \mathcal{P}_k} |p^J(X)|^2},$$

where $\delta_J = \sum_{j \in J} e_j$ and $e_1, \ldots, e_n$ is the canonical basis for $\mathbb{R}^n$.

The convexity property of the moment map. Now we introduce one of the main results of [GGMS]. Let $G = SL_n(\mathbb{C})$, H the group of diagonal matrices and B the group of upper triangular matrices in G. We introduce also the groups $N = N_G(H)$ and $W = N/H$. It is well-known that N consists of the monomial matrices and $W \simeq Sym_n$. Then (B, N) is a Tits system in G (for definitions see [Ti1] or [Bou]). The subgroups G, H and B act naturally on $V = \mathbb{C}^n$. The stabilizer P of the k-dimensional subspace $\mathbb{C}e_1 \oplus \ldots \oplus \mathbb{C}e_k$ is a standard parabolic subgroup in G. The Grassmannian $\mathbb{G}_{n,k}$ can be identified with the factor space G/P. If $X \in \mathbb{G}_{n,k}$, we denote by $H \cdot X$ the H-orbit of X and by $\overline{H \cdot X}$ its closure.

Fact 2.1 ([GGMS]) *The image under the moment map of the closure of the orbit $H \cdot X$ is the convex polyhedron associated with the matroid M_X:*

$$\lambda(\overline{H \cdot X}) = \Delta(M_X).$$

Moreover in every dimension $m \leq n - 1$ the map λ induces a one-to-one correspondence between the m-dimensional orbits of H in $\overline{H \cdot X}$ and the open m-dimensional faces of $\Delta(M_X)$.

Stratification of $\mathbb{G}_{n,k}$. Set $W_P = (N \cap P)/H$, then W_P is the stabilizer in $W \simeq Sym_n$ of the set $\{1, \ldots, k\}$. We can identify $\mathcal{P}_k$ with the factor set $W^P = W/W_P$. For any $w \in W$ we have the well-known *Bruhat decomposition* of the Grassmannian $G/P = \mathbb{G}_{n,k}$:

$$G/P = \bigsqcup_{\alpha \in W^P} wBw^{-1}\alpha P/P \qquad (2)$$

(since the subgroups B, P, N contain H, the products of cosets and subgroups in the formula are well defined). The sets $wBw^{-1}\alpha P/P$ are called *Schubert cells*. If $X \in G/P$, then, since the decomposition (2) is disjoint, for every $w \in W$ there is a unique $\alpha \in W^P$ such that

$$X \in wBw^{-1}\alpha P/P.$$

We shall denote this α by $\mu_X(w)$. Taking the intersection over all $w \in W$, we get

$$X \in \bigcap_{w \in W} wBw^1\mu_X(w)P/P. \qquad (3)$$

The expression in the right is called the *thin Schubert cell* of G/P containing X. Obviously the thin Schubert cells form a partition of G/P.

Fact 2.2 ([GGMS]) *Two points $X, Y \in \mathbb{G}_{n,k}$ belong to the same thin Schubert cell if and only if the corresponding matroids M_X and M_Y coincide.*

As a corollary we have an interpretation of thin Schubert cells in terms of the moment map: two points $X, Y \in \mathbb{G}_{n,k}$ belong to the same Schubert cell if and only if

$$\lambda(\overline{H \cdot X}) = \lambda(\overline{H \cdot Y}).$$

But we also have a matroid naturally associated with a Schubert cell:

Fact 2.3 (Borovik-Gelfand [BG1]) *The map $\mu_X : W \longrightarrow W^P = \mathcal{P}_k$ associated with a Schubert cell in (3) satisfies*

$$\mu_X(u) \leq^w \mu_X(w)$$

for all $u, w \in W$. Moreover, the map μ_X coincides with the map from Sym_n to $\mathcal{P}_k$ associated with the matroid M_X.

Matroids on Grassmannians: an approach via the Maximality Principle. Let now $X \in \mathbb{G}_{n,k}$ be a k-dimensional vector subspace in $\mathbb{C}^n$ and (x_{ij}) the coefficient matrix of a base in X.

Construction of the w-maximal base in the matroid M_X can be described in a very elementary way. A permutation $w \in Sym_n$ permutes the rows of the matrix (x_{ij}). Those rows of the original matrix which correspond to the last k linearly independent rows of the permuted matrix constitute the w-maximal basis in the matroid M_X.

3 Coxeter groups and WP-matroids

Definitions and notations in this section are mostly standard and may be found in [**Bou**] and [**Ron**].

3.1 Coxeter groups and Bruhat ordering

Bruhat Ordering. Let now W be a Coxeter group and S its set of distinguished involutive generators. Recall that the *Bruhat ordering* $\leq$ on W is defined as follows: let $u, v \in W$. Then $u \leq v$ if and only if there is a reduced expression $s_{i_1} \cdots s_{i_k}$ (where $s_{i_j} \in S$) for v such that $u = s_{i_{j_1}} \cdots s_{i_{j_m}}$ where $1 \leq j_1 < j_2 < ... < j_m \leq k$.

We now state without proof a very important result about the Bruhat ordering.

Fact 3.1 (Deodhar [Deo1]) *If we fix a reduced expression* $u = s_{i_1} \cdots s_{i_k}$ *for an element* $u \in W$ *then*

$$\{v \in W, v \leq u\} = \left\{ \begin{array}{c} s_{i_{j_1}} \cdots s_{i_{j_m}} \ \text{for all substrings} \\ i_{j_1} ... i_{j_m}, 1 \leq j_1 < j_2 < ... < j_m \leq k \end{array} \right\}.$$

Fact 3.1 shows that the relation $v \leq u$ does not depend on the choice of a reduced expression $u = s_{i_1} \cdots s_{i_k}$ in its definition. Moreover now it can be easily shown that $\leq$ is an ordering so the name "Bruhat ordering" is justified.

The Bruhat ordering on W^P. Let P be a standard parabolic subgroup in W, i.e. $P = \langle s_j, j \in J \subset I \rangle$ is generated by some set of distinguished generators for W. The following fact makes it clear how to define the Bruhat ordering on the left coset set $W^P = W/P$. (Compare with Corollary 5.12).

Fact 3.2 (see [Deo2], Lemma 2.1) (1) *Any coset* $\alpha \in W^P$ *contains a unique element* w_α *minimal in* α *with respect to the Bruhat ordering.*

(2) *Let* $\alpha, \beta \in W^P$ *be two cosets and* $a \in \alpha$, $b \in \beta$ *any representatives of* α *and* β, *correspondingly. If* $a \leq b$, *then* $w_\alpha \leq w_\beta$.

Following [Deo2], we introduce a partial ordering $\leq$ on W^P by putting $\alpha \leq \beta$ if $w_\alpha \leq w_\beta$. In view of Fact 3.2 this is equivalent to the condition that $a \leq b$ for some representatives $a \in \alpha$, $b \in \beta$.

WP-matroids.   We define a WP-*matroid* as a map

$$\mu : W \longrightarrow W^P$$

satisfying the inequality

$$w^{-1}\mu(u) \le w^{-1}\mu(w)$$

for all $u, w \in W$.

If $P = 1$, we prefer to use the term W-*matroid*. We will return to discussion of W-matroids and WP-matroids in Section 6 after developing a geometric approach to the Bruhat ordering.

3.2   Matroids as a partial case of WP-matroids

Let E be the set $\{1, 2, ..., n\}$ and as before let Sym_n represent the symmetric group on E. It is a well-known fact that $W = Sym_n$ is a Coxeter group with the canonical transpositions $s_1 = (12), s_2 = (23), ..., s_{n-1} = (n-1, n)$ as distinguished generators.

Now let $P = \langle (12), (23), ..., (k-1, k), (k+1, k+2), ..., (n-1, n) \rangle$. Then P is the stabilizer of $A = \{1, 2, ..., k\} \in \mathcal{P}_k$.

Now $W = Sym_n$ acts on $\mathcal{P}_k$ transitively and $P = Stab_A$. The 1-1 map

$$\begin{aligned} f : \quad W^P \quad &\longrightarrow \quad \mathcal{P}_k, \\ f : \quad xP \quad &\mapsto \quad xA \end{aligned}$$

commutes with the action of W on W^P and $\mathcal{P}_k$. Moreover, W^P and $\mathcal{P}_k$ are isomorphic as ordered sets (recall that we introduced the ordering $\le$ on $\mathcal{P}_k$ in Section 1). This easy fact is well-known and belongs to folklore. We had difficulties in finding an appropriate reference for a proof of it until we found one in a preprint by A. Cohen and R. Cushman [CoS].

Fact 3.3 (See [CoS], Proposition 4.2) *The map* $f : W^P \longrightarrow \mathcal{P}_k$ *preserves the orderings* $\le$ *on* W^P *and* $\mathcal{P}_k$.

For this reason f preserves the orderings $\le^w$ on W^P (where $\alpha \le^w \beta$ is understood as $w^{-1}\alpha \le w^{-1}\beta$) and $\mathcal{P}_k$, i.e.,

$$\varphi \le^w \psi \iff f(\varphi) \le^w f(\psi).$$

Thus a matroid of rank k on the set $E = \{1, 2, ..., n\}$ can be interpreted as a map $\mu : Sym_n \longrightarrow \mathcal{P}_k$ satisfying the inequality $\mu(u) \le^w \mu(w)$ for all $u, w \in W$. It can also be interpreted as a map $\mu' : W \to W^P$ satisfying the same inequality: $\mu'(u) \le^w \mu'(w)$, where $\mu' = f^{-1} \circ \mu$.

4   Tits Systems and Thin Schubert Cells

Having a definition of WP-matroid, we return to the discussion of applications of WP-matroids and can now state Fact 2.3 in its full generality.

Consider a group G with a Tits system (B, N). Let $H = B \cap N$, $W = N/H$ and $S = \{s_i, i \in I\}$ the distinguished set of generators in the Weyl group W. Recall that W is a Coxeter group with S the set of its distinguished generators.

If $J \subset I$, let $P = P_J = \langle B, s_i, i \in J \rangle$ be a standard parabolic subgroup in G and $W_P = \langle s_i, i \in J \rangle$ the corresponding parabolic subgroup in W. The factor space G/P is called a *flag space* for the group G.

Denote $W^P = W/W_P$. Using the axioms of a Tits system, one can easily prove that for any $w \in W$ there is a decomposition

$$G = \bigsqcup_{\alpha \in W^P} wBw^{-1}\alpha P$$

into a disjoint union of double cosets with respect to subgroups $wBw^{-1}\alpha P$. Taking the natural projection of G onto G/P, we consider this partition as a decomposition of G/P:

$$G/P = \bigsqcup_{\alpha \in W^P} wBw^{-1}\alpha P/P.$$

For any $g \in G$ and $w \in W$, we denote by $\mu_g(w)$ a unique element in W^P, such that

$$g \in wBw^{-1}\mu_g(w)P.$$

Taking the intersection over all $w \in W$, we have

$$gP \in K = \bigcap_{w \in W} wBw^{-1}\mu_g(w)P/P.$$

The set K is called a *thin Schubert cell* on the flag space G/P. Obviously thin Schubert cells form a partition of G/P generated by all partitions of G/P into Schubert cells. Moreover, it is easy to see that the partition into thin Schubert cells is invariant under the action of the Weyl group W.

The function $\mu_g : W \longrightarrow W^P$ does not depend on choice of a coset $gP \in K$ and thus can be denoted by μ_K.

Fact 4.1 (Borovik-Gelfand [BG1]) *In this notation the function μ_K is a WP-matroid.*

5   Coxeter Complexes

5.1   Galleries and walls

We refer to the work by J. Tits [**Ti2**] or to M. Ronan's book [**Ron**] for definitions of chamber systems, galleries, residues, panels.

The Coxeter Complex of a Coxeter Group.   Let W be a Coxeter group with a system of distinguished generators $S = \{s_i, i \in I\}$. Take the elements of W as chambers and for each $i \in I$ define *i-adjacency* by $w \sim ws_i$. This gives a chamber system over I which is called the *Coxeter complex* of the Coxeter group W; since the elements s_i generate W it is connected.

Throughout this paper W will denote both the Coxeter group and its Coxeter complex.

Geodesic galleries and distance.   Suppose $x, y \in W$ then there is a gallery of type $i_1 \ldots i_k$ from x to y if and only if y can be written as $x \tilde{s}_{i_1} \cdots \tilde{s}_{i_k}$ or $x^{-1}y = \tilde{s}_{i_1} \cdots \tilde{s}_{i_k}$ where

$$\tilde{s}_{i_j} = 1 \text{ or } s_{i_j}$$

and the gallery is

$$(x, x\tilde{s}_{i_1}, x\tilde{s}_{i_1}\tilde{s}_{i_2}, \ldots, x\tilde{s}_{i_1} \cdots \tilde{s}_{i_k} = y).$$

A gallery $(x = x_0, x_1, \ldots, x_k = y)$ is said to have *length* k and the *distance* $d(x, y)$ between x and y is the least such k. (Recall that galleries are allowed to stammer; this happens if for some j, $\tilde{s}_{i_j} = 1$). A gallery from x to y is called *geodesic* if its length is $d(x, y)$.

Fact 5.1 (see [Ron], Lemma 2.4) *If $x, y, y' \in W$ and y' is adjacent to and distinct from y then*

$$d(x, y') = d(x, y) \pm 1.$$

Reflections and Walls.   Let W be a Coxeter complex. A *reflection* s by definition is a conjugate of some s_i. Its *wall* $\sigma = \sigma_s$ consists of all panels of W fixed by s (acting on the left). A panel *lies on* σ_s if and only if s interchanges its two chambers. So if π is any *i*-panel and x is one of the two *i*-adjacent chambers for which π is a common panel then xs_i is the *i*-adjacent chamber to x and $s = xs_ix^{-1}$ is the unique reflection interchanging x and xs_i. Thus *each panel lies on a unique wall* and there is a bijective correspondence between the set of walls and the set of reflections. We say that the wall σ_s is the *wall of symmetry* of two chambers x and y if the corresponding reflection s interchanges x and y.

A gallery $(c_0, ..., c_k)$ *crosses* the wall σ_s whenever the reflection s interchanges c_{i-1} with c_i (or equivalently whenever the panel $\{c_{i-1}, c_i\}$ lies on the wall σ_s) for some $i, 1 \leq i \leq k$.

Fact 5.2 (see [Ron], Lemma 2.5) *In a Coxeter complex W,*

(1) *A geodesic gallery cannot cross a given wall twice.*

(2) *Given chambers x and y, the number of times that a gallery from x to y crosses a given wall is either even for each gallery or odd for each gallery.*

Half-complexes. Given a wall σ it follows from Fact 5.2(2) that the relation

"Chambers x and y can be connected by a gallery which intersects σ an even number of times"

is an equivalence relation with precisely two classes of equivalence which are called *half-complexes* determined by σ. They form complementary subsets of W and are said to be opposite one another. If one is denoted α the other is denoted $-\alpha$. If s is a reflection let $\pm\alpha_s$ denote the two half-complexes determined by s. We shall say that a set $X \subseteq W$ *lies on one side of the wall* σ if X lies entirely in one of two complementary half-complexes $\pm\alpha$ determined by σ.

Convex sets. A set X of chambers is *convex* if any geodesic gallery between two chambers of X lies entirely in X. Notice that obviously the intersection of convex sets is convex.

Fact 5.3 (see [Ron], Proposition 2.6) *In any Coxeter complex W,*

(1) *Half-complexes are convex.*

(2) *If α is a half-complex and x, y adjacent chambers with $x \in \alpha$ and $y \in -\alpha$ then $\alpha = \{c : d(x,c) < d(y,c)\}$.*

(3) *There are natural bijective correspondences between the set of reflections, the set of walls and the set of pairs of opposite half-complexes.*

Before continuing more terms need to be defined. Suppose X is a set of chambers in W and $Y = W \backslash X$. A panel π belongs to the *boundary* of X if there are chambers $x \in X$, $y \in Y$ for which π is a common panel. (It follows that π is also on the boundary of Y). The boundary of X is denoted ∂X. A chamber c lies on the boundary ∂X if a panel of c lies on ∂X. A *supporting half-complex* for a set of chambers X is a half-complex which contains X and is formed by the wall on which lies some panel π of ∂X the boundary of X. The panel π is called a *supporting panel*. A *supporting wall of X* is the bounding wall of a supporting half-complex for X and we say that the supporting wall *supports* X. So if σ is a supporting wall for X then X lies entirely on one side of σ and some panel of ∂X lies on σ.

Lemma 5.4 *If α is a half-complex determined by the wall σ then $\sigma = \partial \alpha$.*

Proof: σ *lies entirely on $\partial \alpha$:* Let π be a panel of σ and x, y its two chambers. The gallery (x, y) intersects σ once so x and y belong to opposite half-complexes formed by σ. This shows that $\sigma \subseteq \partial \alpha$.

$\partial \alpha$ *lies entirely on σ:* Let π be a panel on $\partial \alpha$. One of the chambers of π, say x, belongs to α, another, say y, belongs to $W \setminus \alpha = -\alpha$. Then (x, y) is a gallery which should intersect the wall σ an odd number of times. Obviously this number is 1 and thus $\pi = \{x, y\}$ lies on σ.

Hence $\sigma = \partial \alpha$.   □

Foldings. Let α be any half-complex and s the corresponding reflection. Fact 5.3(2) shows that s interchanges α and $-\alpha$. Thus we have a map $\rho_\alpha : W \to \alpha$ defined by:

$$\rho_\alpha(x) = \begin{cases} x & \text{if } x \in \alpha \\ s(x) & \text{if } x \notin \alpha. \end{cases}$$

If $x \in \alpha$ is adjacent to $y \in -\alpha$ then $\rho_\alpha(y) = \rho_\alpha(x) = x$ so ρ preserves i-adjacency for each i. We call ρ_α the *folding* of W onto α. Notice that since any gallery Γ from a chamber $c \in \alpha$ to a chamber $d \in -\alpha$ crosses the wall $\partial \alpha$ its image $\rho_\alpha(\Gamma)$ contains at least one repeated chamber and hence there is a shorter gallery from c to $\rho_\alpha(d)$ and therefore

$$d(c, \rho_\alpha(d)) < d(c, d).$$

Theorem 5.5 *If W is a Coxeter complex then for any set $X \subseteq W$ the following four statements are equivalent:*

(1) *X is the intersection of all the half-complexes containing X.*

(2) *X is convex.*

(3) *Every panel of ∂X supports X.*

(4) *X is the intersection of its supporting half-complexes.*

Proof: (1) *implies* (2): Follows from Fact 5.3(1).

(2) *implies* (3): Assume that $X \subseteq W$ is convex. Let $\pi \in \partial X$ be a panel on the boundary of X with two chambers $x \in X$ and $y \in W \backslash X$ and let σ be the wall containing π. We claim that X lies on one side of σ. Assume the contrary. Let $z \in X$ where z and x lie on opposite sides of σ and consider a geodesic gallery Γ of length k connecting x and z, $\Gamma = (x = x_0, x_1, \ldots, x_k = z)$. Since Γ is a geodesic gallery it crosses σ once only in the panel $\{x_s, x_{s+1}\}$ say. Since X is convex every chamber in Γ lies in X. Now consider ρ the folding in σ which maps x onto y. As x and z lie on opposite sides of σ, ρ leaves z invariant.

Then

$$\rho(\Gamma) = (\rho(x_0) = y, \rho(x_1), \ldots, \rho(x_s) = \rho(x_{s+1}), \rho(x_{s+1}), \ldots, \rho(z) = z)$$

is a gallery of length k from y to z and hence

$$(y, \rho(x_1), \ldots, \rho(x_s), \rho(x_{s+2}), \rho(x_{s+3}), \ldots, \rho(z) = z)$$

is a gallery of length $k - 1$ from y to z. But then

$$(x, y, \rho(x_1), \ldots, \rho(x_s), \rho(x_{s+2}), \rho(x_{s+3}), \ldots, \rho(z) = z)$$

is a gallery of length k from x to z and is therefore geodesic. As X is convex, $y \in X$. But by definition $y \notin X$. This is a contradiction. Therefore X lies entirely on one side of σ and π supports X. Since π is an arbitrary panel of ∂X, (3) holds.

(3) *implies* (4): Assume that a set $X \subseteq W$ satisfies (3). Let X' be the intersection of all half-complexes supporting X. We have to prove that $X = X'$. Assume the contrary. Then as $X \subset X'$, let $x \in X$ and $y \in X' \backslash X$. By Fact 5.3(1) X' is convex so it contains a geodesic gallery connecting x and y. So we can assume without loss of generality that the chambers $x \in X$ and $y \in X' \backslash X$ are adjacent and have a panel π in common. But then $\pi \in \partial X$ and by our assumption π supports X so X lies on one side of π. This means that the set X and the chamber y lie in opposite half-complexes determined by the panel π and by definition of X', $y \notin X'$. This is a contradiction. Therefore $X = X'$ and (4) holds.

(4) *implies* (1): Trivial.

Hence the four statements are equivalent.   □

Finite Coxeter Complexes. If W is a finite Coxeter complex let $diam(W)$ denote the maximum distance between two chambers. Two chambers are *opposite* if the distance between them is $diam(W)$.

Fact 5.6 (see [Ron], Theorem 2.15) *Let W be a finite Coxeter complex. Then the following statements are true:*

(1) *Every chamber has a unique opposite.*

(2) *If x and y are opposite chambers then every chamber in W lies on a geodesic gallery from x to y.*

5.2 The Bruhat Ordering

Foldings of galleries. Now we introduce the notion of a folding of a gallery which will play a crucial role in the proofs of the main results of the present work. Let $\Gamma = (x_0, ..., x_k)$ be a gallery of type $i_1...i_k$. This implies that for all $1 \le j \le k$,

$$x_j = \begin{cases} x_{j-1} & \text{if } \Gamma \text{ stammers at } \{x_{j-1}, x_j\} \\ x_{j-1}s_{i_j} & \text{otherwise.} \end{cases}$$

Let us denote

$$\tilde{s}_{i_j} = \begin{cases} 1 & \text{if } \Gamma \text{ stammers at } \{x_{j-1}, x_j\} \\ s_{i_j} & \text{otherwise} \end{cases}$$

and assume that Γ does not stammer at the panel $\{x_{j-1}, x_j\}$ for a fixed j. We say that the gallery

$$\Gamma' = (x_0, ..., x_{j-1}, x_{j-1}, x_{j-1}\tilde{s}_{i_{j+1}}, ..., x_{j-1}\tilde{s}_{i_{j+1}}...\tilde{s}_{i_k})$$

is obtained from Γ by folding in the panel $\pi = \{x_{j-1}, x_j\}$. Notice that Γ' necessarily has the same type as Γ. Notice also that folding the gallery $\Gamma = (x_0, ..., x_k)$ subsequently in the panels $\{x_{k-1}, x_k\}$, $\{x_{k-2}, x_{k-1}\}$,...,$\{x_0, x_1\}$ we can always fold Γ onto any of its subgalleries $\Gamma' = (x_0, ..., x_l)$, $l \le k$.

The number of $\tilde{s}_{i_j}$'s which are not equal to 1 is called the *span* of Γ and denoted $span(\Gamma)$. Notice that if Γ is geodesic then its span is equal to its length. Also notice that folding a gallery decreases its span but leaves its length the same.

We will frequently and without reference use the following useful remark on foldings. In the previous notation if $s = x_{j-1}s_{i_j}x_{j-1}^{-1}$, then s is the reflection in the wall σ containing the panel $\pi = \{x_{j-1}, x_j\} = \{x_{j-1}, x_{j-1}s_{i_j}\}$. Notice that

$$\Gamma' = (x_0, x_1, \ldots, x_{j-1}, sx_j, \ldots, sx_k);$$

indeed for $j \le l \le k$

$$sx_l = x_{j-1}s_{i_j}x_{j-1}^{-1} \cdot x_{j-1}\tilde{s}_{i_j} \cdots \tilde{s}_{i_l} = x_{j-1}\tilde{s}_{i_{j+1}} \cdots \tilde{s}_{i_l}$$

as $\tilde{s}_{i_j} = s_{i_j}$. This simple observation has the following useful geometric interpretation which will also be frequently used:

The end chambers x_k, sx_k of the galleries Γ and Γ' respectively are symmetric in the wall containing the panel $\pi = \{x_{j-1}, x_j\}$.

Theorem 5.7 *Let u, v, $w \in W$. Then the following four conditions are equivalent:*

(1) *There is a geodesic gallery $\Gamma = (x_0, \ldots, x_k)$ where $w = x_0$, $u = x_k$ and a gallery (not necessarily geodesic) $\Gamma' = (y_0, \ldots, y_k)$, $w = y_0$, $v = y_k$ of the same type $i_1 \ldots i_k$ connecting w with the chambers u and v respectively.*

(2) $w^{-1}v \le w^{-1}u$.

(3) *There is a geodesic gallery Γ and a gallery Γ' (not necessarily geodesic) from w to u, v respectively such that either $\Gamma = \Gamma'$ or Γ' is produced from Γ by one or more foldings.*

(4) *For **any** geodesic gallery Γ from w to u there is a gallery Γ' from w to v of the same type as Γ.*

Proof: (1) *implies* (2): This is Lemma 1 in [**BG1**].

(2) *implies* (3): Assume that $w^{-1}u \ge w^{-1}v$ then a reduced expression $w^{-1}u = s_{i_1} \cdots s_{i_k}$ exists such that $w^{-1}v$ can be obtained from $s_{i_1} \cdots s_{i_k}$ by deleting some number of s_{i_j}'s. We know that $i_1 \ldots i_k$ is the type of a geodesic gallery from w to u, Γ say. Then

$$\Gamma = (w, ws_{i_1}, ws_{i_1}s_{i_2}, \ldots, ws_{i_1} \cdots s_{i_k} = u).$$

If s_{i_j} is deleted from the reduced expression for $w^{-1}u$ the gallery Γ_1 is obtained where

$$\Gamma_1 = (w, ws_{i_1}, \ldots, ws_{i_1} \cdots s_{i_{j-1}}, ws_{i_1} \cdots s_{i_{j-1}}, ws_{i_1} \cdots s_{i_{j-1}}s_{i_{j+1}}, \ldots, ws_{i_1} \cdots s_{i_k})$$

which is also of type $i_1 \ldots i_k$ and is obtained from Γ by folding in the panel

$$(ws_{i_1} \cdots s_{i_{j-1}}, ws_{i_1} \cdots s_{i_{j-1}}s_{i_j}).$$

Repetition of this procedure produces a gallery Γ' of type $i_1 \ldots i_k$ which connects w and $w \cdot w^{-1}v = v$. Therefore Γ' is obtained by folding Γ a finite number of times. Hence (3) holds.

(3) *implies* (1): This is obvious since by definition folding of a gallery does not change its type.

Therefore the three conditions (1), (2) and (3) are equivalent.

(2) *implies* (4): We can assume without loss of generality that $w = 1$. Then this is immediate from Theorem 3.1 and the definition of the folding of a gallery.

(4) *implies* (1): This is trivial.

Therefore the four conditions (1), (2), (3) and (4) are equivalent.   □

The relation

$$u \leq^w v \Longleftrightarrow w^{-1}u \leq w^{-1}v$$

will be called the *Bruhat ordering with center* w. Clearly $\leq^1$ is just the Bruhat ordering. We will use mostly the geometric form of the definition (parts (1), (3) and (4) of Theorem 5.7).

The following simple lemma will be used later:

Lemma 5.8 *If* $\Gamma = (x_0, ..., x_k)$ *is a geodesic gallery from* $w = x_0$ *to* $u = x_k$ *and* $v = x_l$, $0 \leq l \leq k$, *then* $v \leq^w u$.

Proof: Let $\Gamma' = (x_0, ..., x_l)$ be the subgallery from $w = x_0$ to $v = x_l$ obtained from Γ by subsequent foldings in the panels

$$\{x_{k-1}, x_k\}, \{x_{k-2}, x_{k-1}\}, ..., \{x_l, x_{l+1}\}.$$

Then by Theorem 5.7,

$$w^{-1}v \leq w^{-1}u$$

and hence $v \leq^w u$.   □

5.3   Bruhat Ordering on W/P

Let $W = \langle s_i, i \in I \rangle$ be a Coxeter group and $J \subseteq I$. Denote $P = \langle s_j, j \in J \rangle$. Then J-residues of W are precisely the left cosets of P in W because by definition J-residues are classes of the equivalence relation:

$$w \sim u \Longleftrightarrow w = us_{j_1} \cdots s_{j_k} \text{ for some } j_1, ..., j_k \in J,$$

or, what is the same, $w = uv$ for some $v \in P$. So we identify the set of all J-residues with the left factor set $W^P = W/P$. Now we introduce the Bruhat ordering on W^P for which we need the following theorem (which is a geometrical form of Fact 3.2):

Fact 5.9 (see [Ron], Theorem 2.9) *Given any* $w \in W$ *and any* J-residue φ:

(1) *There is a unique chamber of* φ *nearest* w *(we call it* $proj_\varphi w$*).*

(2) *For any chamber* $x \in \varphi$, *there is a geodesic gallery from* w *to* x *via* $proj_\varphi w$.

Now if φ, ψ are two J-residues and $w \in W$ we set

$$\varphi \leq^w \psi \Longleftrightarrow proj_\varphi w \leq^w proj_\psi w.$$

Corollary 5.10 *For all* $u \in \varphi$, $proj_\varphi w \leq^w u$.

Proof: Immediate consequence of Fact 5.9(2) and Lemma 5.8.   □

Lemma 5.11 *Let φ, ψ be two J-residues and $x \in \varphi$, $y \in \psi$ be chambers. Then if $x \leq^w y$ it follows that $proj_\varphi w \leq^w proj_\psi w$.*

Proof: Let Γ be a geodesic gallery from w to y which contains $proj_\psi w$. By Fact 5.9(2) such a gallery exists. Since $x \leq^w y$, by Theorem 5.7 there is a gallery Γ' from x to w of the same type as Γ obtained by foldings of Γ. Denote $x_0 = proj_\varphi w$ and $y_0 = proj_\psi w$. Now let y'_0, y' be the images in Γ' of y_0, y respectively. Then $y' = x$. As Γ and Γ' are of the same type it follows that the subgallery of Γ' from y'_0 to y' has the same type as the subgallery of Γ from y_0 to y. Since y_0 and y belong to ψ the subgallery of Γ from y_0 to y is of J-type and so the subgallery of Γ' from y'_0 to y' is also of J-type. This implies that $y' = x$ and y'_0 belong to the same J-residue which is φ as $x \in \varphi$ by hypothesis.

By Theorem 5.7, $y'_0 \leq^w y_0$. Also by Corollary 5.10, x_0 is the w-minimal element of φ so as $y'_0 \in \varphi$, it follows that $x_0 \leq^w y'_0$.

Therefore $x_0 \leq^w y_0$ and hence $proj_\varphi w \leq^w proj_\psi w$.   □

Corollary 5.12 $\varphi \leq^w \psi \Longleftrightarrow x \leq^w y$ *for some $x \in \varphi$, $y \in \psi$.*

Fact 5.13 (see [Ron], Theorem 2.16) *Let x, w be two chambers of the Coxeter complex W, and assume that for any chamber $v \in W$ which is adjacent to w,*

$$d(x, w) > d(x, v).$$

Then W is finite and x is opposite w in W.

Lemma 5.14 *Assume that the J-residue $\varphi \subset W$ is finite. Let w be any chamber in W and u the chamber in φ opposite $proj_\varphi w$. Then $x \leq^w u$ for all $x \in \varphi$.*

Proof: We shall prove that x lies on a geodesic gallery Γ from w to u, then by Lemma 5.8, $x \leq^w u$.

Let Γ' be a geodesic gallery from w to x via $proj_\varphi w$. Such a gallery exists by Fact 5.9(2). Also let Γ'' be a geodesic gallery from $proj_\varphi w$ to u via x which exists by Fact 5.6(2). As any sub-gallery of a geodesic gallery is also geodesic, if we connect Γ' to the sub-gallery of Γ'' from x to u we get a geodesic gallery from w to u via x.   □

The following lemma reduces in most interesting cases the study of WP-matroids to the partial case of W-matroids.

Lemma 5.15 *Assume that the standard parabolic subgroup P is finite. If $\varphi \in W^P$ is a left P-coset in W, denote by $max_w\varphi$ the w-maximal element of φ (it exists by Lemma 5.14). In this notation a map*

$$\mu : W \to W^P$$

is a WP-matroid if and only if the map

$$\mu' : \quad W \quad \to \qquad W,$$
$$\mu' : \quad w \quad \mapsto \quad max_w\mu(w)$$

is a W-matroid.

Proof: By Corollary 5.12,

$$\mu(u) \leq^w \mu(w) \Leftrightarrow max_u\mu(u) \leq^w max_w\mu(w) \Leftrightarrow \mu'(u) \leq^w \mu'(w).$$

Hence μ is a WP-matroid if and only if μ' is a W-matroid.   □

6 W-matroids

In this section we restrict our attention to W-matroids. It is justified by Lemma 5.15.

We shall on several occasions use the following **Metric Property** of WP-matroids which is immediate from their definition: if $\mu : W \to W$ is a W-matroid then

$$d(w, \mu(w)) \geq d(w, \mu(u))$$

for all $u, w \in W$ and in the case of equality $\mu(u) = \mu(w)$.

The following result can be considered as a generalization of Fact 1.1.

Theorem 6.1 (A. Borovik and S. Roberts) *Let W be a Coxeter complex and $\mu : W \longrightarrow W$ a W-matroid. Assume that m, n are chambers in $M = \mu(W)$ such that the full preimages $\mu^{-1}[n]$, $\mu^{-1}[m]$ of m and n have a common panel π. Then m and n are symmetric in the wall σ containing the panel π. Moreover, m and $\mu^{-1}[m]$ lie on opposite sides of σ.*

Proof: Let u and v be two chambers in W such that $\mu(u) = m$ and $\mu(v) = n$ with u and v adjacent cells whose common panel π lies on the wall σ say.

(1) *m and n lie on different sides of σ.* Suppose for contradiction that m and n lie on the same side of σ as v. Since u and v are adjacent we have by Fact 5.1 that $d(u, n) = d(v, n) \pm 1$. But since v and n lie on the same side of σ it follows that u and n lie on opposite sides of σ so $d(u, n) = d(v, n) + 1$.

Now from the metric property of W-matroids, $d(u, \mu(u)) > d(u, \mu(v))$, so $d(u, m) > d(u, n)$ and therefore $d(u, m) > d(v, n) + 1$. Similarly

$$d(u, m) = d(v, m) + 1,$$

so $d(v, m) + 1 > d(v, n) + 1$, which contradicts to the metric property $d(v, n) > d(v, m)$. Therefore m and n lie on different sides of σ.

(2) σ *separates u from $\mu(u) = m$ and separates v from $\mu(v) = n$.* For contradiction suppose that u lies on the same side of σ as $\mu(u)$ and v lies on the same side of σ as $\mu(v)$. As before using Fact 5.1 and the metric property of W-matroids

$$d(u, m) > d(u, n) = d(v, n) + 1$$

and

$$d(v, n) > d(v, m) = d(u, m) + 1$$

which is again a contradiction. Therefore σ separates u from m and separates v from n.

(3) m *and n are symmetric in σ.* As $\mu(v) = n \leq^u \mu(u) = m$ a gallery Γ' from u to n may be obtained by foldings of a geodesic gallery Γ from u to m. As before using Fact 5.1 and the metric property of W-matroids, we have

$$d(u, m) > d(u, n) = d(v, n) - 1$$

and

$$d(v, n) > d(v, m) = d(u, m) - 1.$$

This implies $d(u, m) = d(v, n)$ and hence

$$d(u, n) = d(v, m) = d(u, m) - 1 = d(v, n) - 1.$$

Now

$$span(\Gamma') < span(\Gamma) = d(u, m)$$

and as Γ' connects u to n,

$$span(\Gamma') \geq d(u, n) = d(u, m) - 1 = span(\Gamma) - 1.$$

So $span(\Gamma') = span(\Gamma) - 1$ and Γ' is a geodesic gallery. This means that Γ' is obtained from Γ by just one folding in the panel π. But then m and n are symmetric in the wall σ which contains π. This proves the theorem. $\square$

The following lemma will be used in the proof of Theorem 6.3.

Lemma 6.2 *Let m, n be chambers of a Coxeter complex W. Assume that m and n are symmetric in a wall σ. Let $\alpha, -\alpha$ be two half-complexes bounded by σ which contain the chambers m, n respectively. Then*

$$\alpha = \{w \in W, m \leq^w n\}$$

and

$$-\alpha = \{w \in W, n \leq^w m\}.$$

Proof: It suffices to prove that if $w \in \alpha$ then $m \leq^w n$. Indeed let Γ be a geodesic gallery connecting w to n where $w \in \alpha$. Then Γ meets the wall σ in some panel π say. Now if Γ' is the gallery obtained from Γ by folding in the panel π then Γ' connects w to m as m and n are symmetric in σ. This proves that $m \leq^w n$. $\square$

Theorem 6.3 (A. Borovik and S. Roberts) *In a W-matroid $\mu : W \to W$ the full preimage $\mu^{-1}[m]$ of any $m \in \mu[W]$ is convex.*

Proof: Let $M = \mu[W]$. Fix $m \in M$ and define

$$N = \{n \in M, \ \mu^{-1}[m] \ \text{and} \ \mu^{-1}[n] \ \text{have a common panel}\}.$$

If $n \in N$, then by Theorem 6.1 the chambers m and n are separated by the wall of symmetry of m and n, σ_n say. Now it follows from the definition of a W-matroid that $n \leq^w m$ for any $w \in \mu^{-1}[m]$ and all $n \in N \subseteq M$. Let α_n, $n \in N$ be the half-complex bounded by σ_n which does not contain m. Then by Lemma 6.2,

$$w \in \bigcap_{n \in N} \alpha_n$$

and as w is an arbitrary chamber in $\mu^{-1}[m]$,

$$\mu^{-1}[m] \subseteq \bigcap_{n \in N} \alpha_n.$$

Now if π is any panel on $\partial(\mu^{-1}[m])$ it follows from the proof of Theorem 6.1 that π belongs to one of σ_n. But then, as for all $n \in N$, α_n is bounded by σ_n and contains $\mu^{-1}[m]$ it follows that α_n is a supporting half-complex for $\mu^{-1}[m]$ and π is a supporting panel. Hence by Theorem 5.5, $\mu^{-1}[m]$ is convex. $\square$

Theorem 6.4 (A. Borovik and S. Roberts) *Let $\mu : W \to W$ be a W-matroid on an infinite Coxeter group W. Fix an element $m \in \mu[W]$. Then the full preimage $\mu^{-1}[m]$ of the element $m \in \mu[W]$ is infinite.*

Proof: Let $\{\alpha_k, k \in K\}$ be the set of half-complexes supporting $\mu^{-1}[m]$. By Theorem 6.3 $\mu^{-1}[m]$ is convex and by Theorem 5.5 $\mu^{-1}[m] = \cap_{k \in K} \alpha_k$. Then Theorem 6.1 implies that $m \in \cap_{k \in K} - \alpha_k$.

Now suppose that $X = \mu^{-1}[m]$ is finite. Let $x \in X$ be a chamber in X at maximal distance from m. Also let $\{\alpha_k, k \in K'\}$ W be the set of all half-complexes containing x but not m. Then since $x \in \cap_{k \in K} \alpha_k$ and $m \in \cap_{k \in K} - \alpha_k$,

$$X' = \bigcap_{k \in K'} \alpha_k \subset \bigcap_{k \in K} \alpha_k \subseteq X.$$

We claim that $X' = \cap_{k \in K'} \alpha_k = \{x\}$.

Assume the contrary. Since X' is convex by Fact 5.3(1), we can find a chamber $y \in X'$ adjacent to and distinct from x. By Fact 5.1 $d(m, y) = d(m, x) \pm 1$. Since $d(m, y) = d(m, x) - 1$ by our maximal choice of x then y lies on a geodesic gallery from m to x. Let σ be the wall containing the panel $\{x, y\}$. Now m and y lie on on one side of this wall and x lies on another. This is a contradiction to our choice of y.

So we have the following situation: m and x are chambers in W, $\{\alpha_k, \ k \in K'\}$ is the set of all half-complexes containing x but not m and $\cap_{k \in K'} \alpha_k = \{x\}$. Suppose $y \neq x$ is any adjacent chamber to x. Then $d(m, y) = d(m, x) \pm 1$. We want to prove $d(m, y) < d(m, x)$, then by Fact 5.13 W is finite.

If $y \in \mu^{-1}[m]$, then by our maximal choice of x, $d(m, y) \leq d(m, x)$ and so $d(m, y) = d(m, x) - 1$. So we can assume without loss that $y \notin \mu^{-1}[m]$. Let σ be the wall containing the panel $\pi = \{x, y\}$, α and $-\alpha$ its half-complexes containing x and y, correspondingly. By Theorem 6.1 m lies in $-\alpha$. Now by Fact 5.3(2) $d(m, y) < d(m, x)$. $\square$

Proof of the Main Theorem. Now the theorem stated in the Introduction easily follows from Theorems 6.1, 6.3, 6.4, and Lemma 5.15.

Acknowledgements

The first author would like to thank the organizers of the Lake Como Conference whose invitation to present a talk was a stimulus for his work on the problems discussed in the present paper.

The preparation of the paper has been completed when the first author was visiting the Discrete Mathematics Group at the University of Technology, Eindhoven. The author thanks Arjeh Cohen and all members of the Group for their warm hospitality. Special thanks to Peter Johnson whose advices helped to improve the text.

The authors thank the referee for several corrections in the text.

References

[BG1]   A. V. Borovik and I. M. Gelfand, *WP-matroids and thin Schubert cells on Tits systems*, to appear in Adv. Math.

[Bou]   N. Bourbaki, **Groupes et algebras de Lie, Chap. 4, 5, et 6**, Hermann, Paris, 1968.

[CoS] A. Cohen and R. Cushman, *Gröbner bases and standard monomial theory*, in **Computational Algebraic Geometry**, Birkhaüser, 1993.

[Deo1] V. V. Deodhar, *Some characterizations of Coxeter groups*, Enseignments Math. **32** (1986) 111–120.

[Deo2] V. V. Deodhar, *A splitting criterion for the Bruhat ordering on Coxeter groups*, Communications in Algebra **15** (1987) 1889–1894.

[Gal] D. Gale, *Optimal assignments in an ordered set: an application of matroid theory*, J. Combinatorial Theory **4** (1968) 1073–1082.

[GGMS] I. M. Gelfand, M. Goresky, R. D. MacPherson and V. V. Serganova, *Combinatorial Geometries, convex polyhedra, and Schubert cells*, Adv. Math. **63** (1987) 301–316.

[GS1] I. M. Gelfand and V. V. Serganova, *On a general definition of a matroid and a greedoid*, Soviet Math. Dokl. **35** (1987) 6–10.

[GS2] I. M. Gelfand and V. V. Serganova, *Combinatorial geometries and torus strata on homogeneous compact manifolds*, Russian Math. Surveys **42** (1987) 133-168; *also* I. M. Gelfand, **Collected papers, vol. III**, Springer-Verlag, New York a.o., 1989, 926–958.

[Grü] B. Grünbaum, **Convex Polytopes**, Interscience Publishers, New York a.o., 1967.

[Ron] M. Ronan, **Lectures on Buildings**, Academic Press, Boston, 1989.

[Ti1] J. Tits, **Buildings of Spherical Type and Finite BN-pairs**, Lecture Notes in Math. **386** Springer-Verlag, 1974.

[Ti2] J. Tits, *A local approach to buildings* in **The Geometric Vein (Coxeter Festschrift)**, Springer-Verlag, New York a.o., 1981, 317–322.

[Whi] H. Whitney, *On the abstract properties of linear dependence*, Amer. J. Math. **57** (1935) 509–533.

FINITE GROUPS AND GEOMETRIES: A VIEW ON THE PRESENT STATE AND ON THE FUTURE

Francis Buekenhout
Université Libre de Bruxelles
Campus Plaine C.P.216
Bd du Triomphe
B-1050 Bruxelles

e-mail: fbueken@ulb.ac.be

1 Introduction

1.1 The model: groups of Lie-Chevalley type and buildings

This paper is not the presentation of a completed theory but rather a report on a search progressing as in the natural sciences in order to better understand the relationship between groups and incidence geometry, in some future sought-after theory $\mathcal{T}$. The search is based on assumptions and on wishes some of which are time-dependent, variations being forced, in particular, by the search itself.

A major historical reference for this subject is, needless to say, Klein's Erlangen Programme. Klein's views were raised to a powerful theory thanks to the geometric interpretation of the simple Lie groups due to Tits (see for instance [12]), particularly his theory of buildings and of groups with a BN-pair (or Tits systems). Let us briefly recall some striking features of this.

Let G be a group of Lie-Chevalley type of rank r, defined over $GF(q)$, $q = p^n$, p prime. Let X_r denote the Dynkin diagram of G. To these data corresponds a unique thick building $B(G)$ of rank r over the Coxeter diagram X_r (assuming we forget arrows provided by the Dynkin diagram). It turns out that $B(G)$ can be constructed in a uniform way for all G, from a fixed p-Sylow subgroup U of G, its normalizer $N_G(U)$ and the r maximal subgroups of G containing

$N_G(U)$.

If H is an abstract group isomorphic to a group of Lie-Chevalley type X_r over $GF(q)$, then X_r and q are uniquely determined from H except for a short list of well known cases, such as $PSL(3,2) \cong PSL(2,7)$ $O^-(6,2) \cong U(4,2) \cong Sp(4,3) \cong O(5,3)$. Observe that in the latter example there are nevertheless no more than two distinct buildings related to H.

A most striking and useful fact is that every thick finite building of rank $r \geq 3$ is isomorphic to some $B(G)$ with G as above. For $r \leq 2$, characterizations of these buildings $B(G)$ exist assuming a flag-transitive automorphism group together with suitable additional conditions (see Buekenhout-Van Maldeghem [4]). The classification of all finite simple groups brings us back to a situation which is quite similar to Tits' starting point in the 1950's. At that time, his goal was to give a geometric interpretation of the five exceptional simple complex Lie groups. At present, the sporadic groups and the alternating groups are playing the role of exceptions.

1.2   Rules of experimentation

The expected outcome of a future theory $\mathcal{T}$ is to define a class $C(t)$ of finite groups, at some time t and, for each group $G \in C(t)$, a class of geometries $\Gamma(G,t)$ that would satisfy (at least) the following wishes.

(W1) Every finite almost simple group is in $C(t)$. Recall that G is almost simple if there is a simple nonabelian group S such that $S \leq G \leq Aut\ S$.

(W2) Every finite group of Lie-Chevalley type is in $C(t)$. This includes groups of rank one. This wish does not add very much to (W1): only Sym (4), Alt (4), Sym (3).

(W3) If $G \in C(t)$ and $\Gamma \in \Gamma(G,t)$ then Γ is a geometry, Γ is firm (each residue of rank one has at least two elements), Γ is residually connected and G is a flag-transitive type preserving automorphism group of Γ.

Here and at other places of the paper we refer to Buekenhout [3] (chapter 3) for the terminology used without explicit definitions.

(W4) If G is of Lie-Chevalley type and $B(G)$ is the corresponding building geometry then $B(G) \in \Gamma(G,t)$.

(W5) If $G \in C(t)$, $\Gamma \in \Gamma(G,t)$ and F is a flag of Γ, then the residue geometry Γ_F must be in $\Gamma(H,t)$ where H is the group induced on Γ_F by the stabilizer of F in G.

(W6) For each $G \in C(t)$, the class $\Gamma(G,t)$ must be a "reasonably small" nonempty set that we can classify up to isomorphism.

We could consider these wishes as a set of axioms if (W6) was not present. It is conceivable that there are several solutions satisfying (W1) to (W6). As a matter of fact, more wishes can be added if the present ones are satisfied already. Also, we cannot be sure that the problem has a solution. Optimism relies on 1.1. If we drop (W1) and if $C(t)$ consists only of the groups in (W2) then a solution is to choose the buildings in order to define $\Gamma(G,t)$ and then $\Gamma(G,t)$ contains at most two members for all G, which is a good answer to (W6).

1.3 Some good candidates

At this time our knowledge of geometries for sporadic groups rests on lots of data that we have been collecting since 1975 on the basis of work involving many authors. For some time, it was hoped that the restriction to geometries defined over a Coxeter diagram could be a reasonable approach. A remarkable theory grew from the efforts of Timmesfeld and of several colleagues (see Meixner [11] for a survey). However, that theory does not include most of the sporadic groups.

Being forced to consider at least somewhat more general diagrams, a good supply of examples indicates that we should allow at the very least rank 2 residues over the diagrams that follow.

We refer to [3] (chapter 22 by Buekenhout and Pasini) for a survey of classifications that rely on these diagrams. Actually there is evidence in favor of other rank 2 residues (like the tilda geometry which is a 3-cover of a generalized quadrangle).

Allowing the entrance of "circle" and "Petersen" means at once that standard properties of Lie-Chevalley groups and their buildings must be given up in $\mathcal{T}$. For instance:

1) a member of $\Gamma(G,t)$ need not be thick;

2) if Γ is in $\Gamma(G,t)$, if x is an element of Γ and if G_x is the stabilizer of x in G, then G_x need not be a maximal subgroup of G.

This has dramatic consequences. How are we going to reduce the wilderness of all subgroups of a random group G in the class $C(t)$? Are we going to consider thin geometries and to admit them without restrictions?

Another consequence of our choice and the wish of uniformity is to restrict the rank 2 residues to *quasi-generalized polygons* (Buekenhout [1]). This choice amounts to introducing primarily the linear spaces (together with the Petersen graph and the Hoffman-Singleton graph), hence the affine spaces. Then the door opens for the affine groups, perhaps for more primitive groups, etc.

2   Conditions at time $t_0 = 1991$

2.1   The groups

We require that $C(t_0)$ satisfies (W1), (W2) and we furthermore allow for all groups isomorphic to a primitive permutation group of affine type.

2.2   The geometries

For G in $C(t_0)$ and $\Gamma \in \Gamma(G, t_0)$ we require the following conditions:
(FT) G acts as a flag-transitive type-preserving automorphism group of Γ. We do not assume that the action is faithful;
(F) Γ is firm;
(RC) Γ is residually connected.

2.3   Experimentation

Experimentation with small groups G in $C(t_0)$, using CAYLEY [7] in order to find all Γ of rank ≥ 3 satisfying the above conditions, shows that their number tends to grow rather wildly and that (W6) is unlikely to be satisfied with such weak hypotheses.

However, these conditions lead to algorithms and to a possible, though modest, implementation on CAYLEY.

2.4   The Tits algorithms

Let X be a geometry in $\Gamma(G, t_0)$ and let n be the rank of Γ. Hence $G \leq Aut\ X$ and G acts transitively on the chambers of X. Fix a chamber $c_1, \ldots, c_n$ of X where c_i is an element of type i. Let G_i be the stabilizer of c_i in G, $i = 1, \ldots, n$. Then the structure of X can be entirely described in terms of G and the subgroups G_i, $i = 1, \ldots, n$. First, the elements of X of type i can be identified with the left cosets of the subgroups G_i in G. Next, two elements are incident if and only if the corresponding cosets have a nonempty intersection in G. Finally, G acts on the elements by left translation.

Therefore, the process can be reversed. No longer do we start with a geometry. We start with a group G and a collection $G_1, \ldots, G_n$ of subgroups. Then we define an incidence structure $X(G; G_1, \ldots, G_n)$ in the obvious way. Necessary and sufficient conditions on the G_i are known in order to ensure that conditions (FT), (F) and (RC) hold.

The isomorphism problem essentially reduces to conjugacy under G (actually under $Aut\ G$).

In Brussels, an efficient CAYLEY-library has been developed by Hermand [9] in order to automate this search of all geometries for a given group G. This was then greatly modified and further developed by Dehon [8]. These systems have been used effectively by their authors and by some students.

2.5   The wilderness

All geometries with (FT), (F), (RC) have explicitly been listed for small groups such as $Alt(5)$, $Sym(5)$, $PGL(2,7)$. In view of the complexity of the algorithms a group such as M_{11} is presently out of reach. Actually, there is no need to pursue this. For $PGL(2,7)$, hundreds of distinct non-isomorphic geometries were obtained. This is too much for (W6).

2.6   Maximal subgroups only

Let us consider an additional condition.
(PRI) Every subgroup G_i is maximal in G. This amounts to saying that G acts primitively on the elements of each given type in the geometry.
Of course, this may appear as a somewhat desperate move in view of observation 2) in 1.3 but in this game, we are not forced to respect all parts of 1.3 forever. Moreover, (PRI) may provide some better insight. In view of (W5) we also know that ultimately (for $t = \infty$) condition (PRI) if made official should actually become "Residually PRI" and hence hold on all residues.
A search along such lines, for amalgams rather than geometries, was made by Kommissartchik-Tsaranov [10], for the simple group $U(4,2)$. An independent search was made by Dehon [8] for the group $Aut\ U(4,2)$. For the ranks 3, 4, 5, 6, ≥ 7 Dehon obtained respectively 77, 87, 20, 4, 0 geometries. The system has been applied over the last years to various groups such as $AGL(2,q)$, $q = 3,4,5$, $AGL(3,2)$, $PSL(2,q) \leq G \leq P\Gamma L(2,q)$, $q = 5,7,8,9,11$ with rather similar results. For the Mathieu group M_{11}, the number of geometries of rank $3,4,\geq 5$ is respectively 78, 30, 0.

2.7   Restrictions that fail

The lists of geometries arising from the preceding experiments call for some restrictions. Here are some attempted restrictions that fail to rule out a "satisfactory" number of geometries.
(1) A connected diagram.
(2) The geometry is not a proper truncation of a geometry of higher rank in $\Gamma(G, t_0)$.
For this restriction, with $G = Aut\ U(4,2)$, Dehon still obtained 2, 41, 15, 4, 0 geometries in ranks 3,4,5,6, ≥ 7 respectively.

3   Conditions at time $t_1 = 1992$

3.1. We take $C(t_1) = C(t_0)$ and require first of all (FT), (F), (RC) and (PRI).

3.2. Observing the long lists of geometries thus obtained we notice the dominant presence of rank 2 residues having gonality 2 without being generalized digons. This leads to an interesting restriction which amounts to requiring that any two points are incident with at most one line in every rank 2 residue other than a generalized digon. Let us call this condition (LL) (as in Tits [13]) or (IP)$_2$ for the "rank 2 intersection property".
Hence we add (LL) to our conditions at $t = t_1$.

3.3. The case of *Aut* $U(4,2)$.
Applying these rules to Dehon's lists for *Aut* $U(4,2)$ under the additional condition (2) of 2.7 gives us respectively 0,0,1,4,0 geometries of rank 3,4,5,6,$\geq$ 7, which is encouraging. Of course, the full experiment ought to be done without condition (2). We believe that the outcome would be the same. Let us observe for the sake of completeness that there are 5 rank one and 5 rank two geometries. The latter are the two expected buildings and three generalized digons.
We must say also that some other smaller groups do not provide such a short list. Should this observation incline us to optimism or to pessimism?

3.4. The case of M_{11} .
With $G = M_{11}$, the number of geometries with (FT), (F), (RC), (PRI), (IP)$_2$ of rank 1,2,3,4,$\geq$ 5 is respectively 5,8,10,10,0.
As a side benefit, several previously unknown good-looking geometries have been unearthed.
If we want only thick geometries, the above numbers go down to 5,6,5,5,0. If we drop the condition (IP)$_2$, more than a hundred geometries arise.

4   More subgroups

4.1. As mentioned earlier, (PRI) is too strong if we intend to require it also on all residues and if we want to admit rank 2 residues such as circles, the Petersen graph and affine planes. Such examples made me think already in 1984 [2] that it may be interesting to replace (PRI) by:
(QPRI) all G_i are quasi-maximal subgroups of G, which means that there is a unique chain of overgroups from G_i to G.

4.2. Applying (FT), (F), (RC), (QPRI) and (IP)$_2$ to the groups $AGL(2,2)$, $AGL(2,3)$, $AGL(2,4)$ and $AGL(2,5)$ in joint work with M.Dehon and I.De Schutter, we found out that the number of geometries of rank at least 3 is respectively 3,15,13,11, which is quite encouraging.

Coming back to $G = Aut\ U(4,2)$ of order 51840, it has 5 conjugacy classes of maximal subgroups and, to my surprise, 32 more classes of quasimaximal subgroups were found by Dehon. These data are too large to be treated on our CAYLEY system.

4.3. A side benefit of the preceding study is that the lists of geometries of rank 3 obtained with $G = AGL(2,q)$, $q = 2,3,4,5$, has shown us new extensions of generalized quadrangles. These have been generalized to an unexpected extent in purely theoretical terms [6]. Similar benefits are likely to occur from the lists obtained for small groups $PSL(2,q)$.

5   The near future

5.1. In order to take all earlier observations into account we will work with a variation of (PRI).

(RPRI) At least one G_i is maximal and this holds true in each residue.

As a consequence, all rank 1 residues are endowed with a primitive permutation group.

This condition can be implemented in the spirit of the preceding experiments but it can also be pushed further in purely theoretical terms.

5.2. We shall keep an eye on another fairly natural condition, namely that the rank 1 residues be equipped with a split BN-pair.

5.3. Special attention has to be paid also to the new rank 2 geometries that have appeared in the above experiments. We now have a collection of more than thirty "exotic" (g, d_p, d_ℓ)-gons with $3 \leq g \leq d_p - 1$, $g \leq d_\ell - 1$. These have to be analyzed in depth with the help of CAYLEY and they need to be compared with the ideas in [5].

5.4. The present work has some influence on another front of the research, namely the study and classification of all flag-transitive geometries over specific diagrams. Experimentation provides examples or an absence of examples and so it gives insight on the paths to take.

5.5. On those fronts rather strong developments may be predicted for the coming years.

REFERENCES

[1] F.Buekenhout. (g, d^*, d)-gons. In "Finite geometries" Edited by N.Johnson, M.Kallaher, C.Long. Marcel Dekker, 1983, 93-111.

[2] F.Buekenhout. The geometry of the finite simple groups. In L.Rosati, ed. Buildings and the Geometry diagrams. Lect. Notes 1981. Springer. 1-78.

[3] F.Buekenhout (editor). Handbook of Incidence Geometry. Elsevier. Amsterdam 1994.

[4] F.Buekenhout, H.Van Maldeghem. Finite distance transitive generalized polygons. Geo. Ded. To appear.

[5] F.Buekenhout, H.Van Maldeghem. A characterization of some rank 2 incidence geometries by their automorphism group. Preprint.

[6] F.Buekenhout, M.Dehon, I.De Schutter. Projective injections of geometries and their affine extensions. Preprint.

[7] J.Cannon. A language for group theory. Univ. Sydney. 1982.

[8] M.Dehon. Classifying geometries with CAYLEY. Journ. Symbol. Comput. To appear.

[9] M.Hermand. Géométries, langage CAYLEY et groupe de Hall-Janko. Thèse. Université de Bruxelles, 1991.

[10] E.A.Komissartschik, S.V.Tsaranov. Construction of finite groups, amalgams and geometries. Geometries of the group $U_4(2)$. Comm. Alg. 18 (1990) 1071-1117.

[11] T.Meixner. Groups acting transitively on locally finite classical Tits chamber systems. In "Finite geometries, buildings and related topics", edited by W.M.Kantor, R.A.Liebler, S.E.Payne. E.E.Shult. Clarendon Press Oxford. 45-65.

[12] J.Tits. Buildings and Buekenhout geometries. In M.Collins, ed. Finite Simple Groups II. Acad. Press New-York, 1981, 309-320.

[13] J.Tits. A local approach to buildings. In "The Geometric Vein". Springer, 1981, 519-547.

Groups acting simply transitively on the vertices of a building of type $\tilde{A}_n$

Donald I. Cartwright[1]

Abstract. If a group Γ acts simply transitively on the vertices of a thick building of type $\tilde{A}_n$, Γ must have a presentation of a simple type. In the case $n = 1$, when Δ is a tree, the possible groups Γ are well understood. Recently, the case $n = 2$ was studied ([1], [2]). We now consider the case $n \geq 3$, and are lead to combinatorial objects $\mathcal{T}$ which we call $\tilde{A}_n$-*triangle presentations*. Associated to any $\tilde{A}_n$-triangle presentation $\mathcal{T}$ there is a group $\Gamma_{\mathcal{T}}$. We show that the Cayley graph of any group $\Gamma_{\mathcal{T}}$ is the 1-skeleton of a building $\Delta_{\mathcal{T}}$ of type $\tilde{A}_n$. For $n = 3$ and $n = 4$, and for any prime power q, we exhibit an $\tilde{A}_n$-triangle presentations $\mathcal{T}$, and an embedding of $\Gamma_{\mathcal{T}}$ into $PGL(n+1, \mathbf{F}_q(X))$. In these cases, the building $\Delta_{\mathcal{T}}$ is isomorphic to the building associated to $SL(n+1, \mathbf{F}_q((X)))$.

§1. Introduction.

It was shown in [1] that if Δ is an affine building with connected diagram, and if there is a group Γ of automorphisms of Δ acting transitively on the set $\mathcal{V}_\Delta$ of vertices of Δ, then the diagram of Δ must be $\tilde{A}_n$ for some $n \geq 1$. Now let Δ be a thick building of type $\tilde{A}_n$. Let Γ be a group of automorphisms of Δ which acts *simply* transitively on $\mathcal{V}_\Delta$. The case $n = 1$ is well understood (see [3] or [6]), and the case $n = 2$ was studied in [1] and [2], and so we shall be mainly concerned with the case $n \geq 3$ in this paper.

Let $d_\mathcal{V}(u, v)$ denote the natural graph distance on $\mathcal{V}_\Delta$. As we saw in [1], if we fix a vertex v_0 of Δ, and if we let $\Pi(v_0) = \{v \in \mathcal{V}_\Delta : d_\mathcal{V}(v_0, v) = 1\}$ be the set of vertices in the residue of v_0, then for each $v \in \Pi(v_0)$, there is a unique $g_v \in \Gamma$ such that $g_v v_0 = v$. Moreover, if $v \in \Pi(v_0)$, then $g_v^{-1} v_0$ is in $\Pi(v_0)$ too, and $\lambda(v) = g_v^{-1} v_0$ defines an involution $\lambda : \Pi(v_0) \to \Pi(v_0)$ such that $g_{\lambda(v)} = g_v^{-1}$. We also saw [1, Proposition 2.2] that if $\mathcal{T} = \{(u, v, w) \in \Pi(v_0)^3 : g_u g_v g_w = 1\}$, then the set $\{g_v : v \in \Pi(v_0)\}$, together with the relations $g_v g_{\lambda(v)} = 1$, $v \in \Pi(v_0)$, and $g_u g_v g_w = 1$, $(u, v, w) \in \mathcal{T}$, give a presentation of Γ.

There is a *type map* $\tau : \mathcal{V}_\Delta \to \{0, 1, \ldots, n\}$ such that each chamber of Δ contains one vertex of each type. We may assume that v_0 has type 0.

1 School of Mathematics and Statistics, University of Sydney, N.S.W. 2006, Australia.
1991 *Mathematics Subject Classification.* Primary 20G25, 51E24. Secondary 20F05, 51E15.
Key words and phrases. Affine buildings, local fields.

Now $\Pi(v_0)$ has a natural incidence structure: if $u, v \in \Pi(v_0)$ are distinct, we call u and v *incident* if u, v and v_0 lie on a common chamber of Δ. When $n = 2$, $\Pi(v_0)$ is a projective plane, the vertices of type 1 being the "points", and the vertices of type 2 being the "lines" (or vice versa). When $n \geq 3$, $\Pi(v_0)$ is a projective geometry (see [9, p. 105] or [5, p. 24], for example), and therefore isomorphic ([9, p. 203] or [5, pp. 27–28]) to the projective geometry $\Pi(\mathbf{V})$, the flag complex of an $n + 1$ dimensional vector space $\mathbf{V}$ over a not necessarily commutative field k. Also, the type map τ can be chosen so that the type $\tau(x_u)$ of the vertex $x_u \in \Pi(v_0)$ corresponding to $u \subset \mathbf{V}$ is just $\dim(u)$. When $n \geq 3$, any thick building of type $\widetilde{A_n}$ is isomorphic to the building $\widetilde{A_n}(K, v)$ associated with a not necessarily commutative field K with a discrete valuation v (see [8, §9.2 and Theorem 10.22]), but we shall not be really using this here. When $\Delta = \widetilde{A_n}(K, v)$, an isomorphism $\Pi(v_0) \to \Pi(\mathbf{V})$ is apparent: if v_0 is the lattice class $[L_0]$, then $\Pi(v_0)$ consists of the classes $[L]$ of lattices L satisfying $L_0 \pi \underset{\neq}{\subseteq} L \underset{\neq}{\subseteq} L_0$, where $\pi \in K$ and $v(\pi) = 1$, and we can associate to $[L] \in \Pi(v_0)$ the subspace $L/L_0\pi$ of $\mathbf{V} = L_0/L_0\pi$ (which is an $n + 1$-dimensional vector space over the residual field k of K). For $g \in GL(n + 1, K)$, we set $\tau([gL_0])$ equal to the $i \in \{0, 1, \ldots, n\}$ satisfying $i = -v(\det(g)) \pmod{n + 1}$, and then $\tau([L])$ equals the dimension of $L/L_0\pi$ over k for each $[L] \in \Pi(v_0)$.

An automorphism g of Δ is called *type-rotating* if there is an integer c such that $\tau(gx) = \tau(x) + c \pmod{n + 1}$ for each vertex x of Δ. Such automorphisms form a subgroup $\mathrm{Aut}_{\mathrm{tr}}(\Delta)$ of index at most 2 in $\mathrm{Aut}(\Delta)$, and $g \mapsto c \bmod n + 1$ is a homomorphism $\mathrm{Aut}_{\mathrm{tr}}(\Delta) \to \mathbf{Z}/(n + 1)\mathbf{Z}$. When $\Delta = \widetilde{A_n}(K, v)$, any $g \in PGL(n + 1, K)$ induces a type-rotating automorphism of Δ. We shall always assume that our group Γ is a subgroup of $\mathrm{Aut}_{\mathrm{tr}}(\Delta)$.

In the next section, given any projective geometry Π of dimension n, and any involution $\lambda : \Pi \to \Pi$ such that $\dim(\lambda(u)) = n + 1 - \dim(u)$ for each $u \in \Pi$, we define an $\widetilde{A_n}$-*triangle presentation* T compatible with λ. We associate to T a group Γ_T, generated by elements a_u, $u \in \Pi$, and show that the Cayley graph of Γ_T with respect to these generators is the 1-skeleton of a thick building Δ_T of type $\widetilde{A_n}$ on the vertices of which Γ_T acts simply transitively. When $n \geq 3$, Δ_T must be isomorphic to $\widetilde{A_n}(K, v)$ for some field K with a discrete valuation v, but it is not apparent what K should be, except that the order q of the residual field of K must be the order of Π. When $n = 2$, Δ_T may be isomorphic to $\widetilde{A_2}(K, v)$. In [2], by enumerating all $\widetilde{A_2}$-triangle presentations when $q = 2$ and $q = 3$, we saw that for those cases, only $K = \mathbf{F}_q((X))$ and $K = \mathbf{Q}_q$ could occur (assuming that K is complete with respect to v, as we may [8, p. 130]), and that when $q = 3$, some Δ_T are not isomorphic to any $\widetilde{A_2}(K, v)$.

In the subsequent sections, $\widetilde{A_n}$-triangle presentations T are exhibited for

$n = 3$ and $n = 4$, and any prime power q. Let us briefly indicate how these were found. Suppose that $\mathbf{V} = \mathbf{F}_{q^d}$, where $d = n + 1$. Then $\mathbf{F}_{q^d}^{\times}$ acts on $\Pi(\mathbf{V})$ by multiplication. In [1], when $n = 2$, all $\widetilde{A_n}$-triangle presentations $\mathcal{T}$ were found satisfying the further property

$$(u, v, w) \in \mathcal{T}, \quad t \in \mathbf{F}_{q^d}^{\times} \;\Rightarrow\; (tu, tv, tw) \in \mathcal{T} \tag{1.1}$$

Each such $\mathcal{T}$ was found to also satisfy

$$(u, v, w) \in \mathcal{T} \;\Rightarrow\; (\varphi(u), \varphi(v), \varphi(w)) \in \mathcal{T} \tag{1.2}$$

where $\varphi : x \mapsto x^q$ is the Frobenius automorphism of $\mathbf{F}_{q^d}$. Moreover, for each prime power q, an $\widetilde{A_2}$-triangle presentation $\mathcal{T}_0$ was exhibited such that $\Gamma_{\mathcal{T}_0}$ embeds as a subgroup of $PGL(3, \mathbf{F}_q(X))$ acting simply transitively on the vertices of $\widetilde{A_2}(\mathbf{F}_q(X), v)$ (or of $\widetilde{A_2}(\mathbf{F}_q((X)), v)$).

In this paper, for $n = 3$ and $n = 4$, $\widetilde{A_n}$-triangle presentations $\mathcal{T}$ were found satisfying (1.1) and (1.2), and an embedding of $\Gamma_{\mathcal{T}}$ into $PGL(n + 1, \mathbf{F}_q(X))$ was found in both cases. It remains an open question whether $\widetilde{A_n}$-triangle presentations exist for each integer $n \geq 2$.

§2. $\widetilde{A_n}$-triangle presentations $\mathcal{T}$, and the associated groups $\Gamma_{\mathcal{T}}$ and buildings $\Delta_{\mathcal{T}}$.

Let Δ be a thick building of type $\widetilde{A_n}$. We fix $v_0 \in \mathcal{V}_{\Delta}$. As discussed in the introduction, we may assume that $\tau(v_0) = 0$, and that (when $n \geq 3$) there is an isomorphism of $\Pi(v_0)$ onto $\Pi(\mathbf{V})$, the flag complex of an $n+1$ dimensional vector space $\mathbf{V}$ over a not necessarily commutative field k, so that the type $\tau(v)$ of $v \in \Pi(v_0)$ equals the dimension the corresponding subspace of $\mathbf{V}$. For $v \in \Pi(v_0)$, we shall write $\dim(v)$ in place of $\tau(v)$, even if $n = 2$. We shall write $u \subset v$ if $u, v \in \Pi(v_0)$ are incident and $\dim(u) \leq \dim(v)$.

Lemma 2.1. *Assume that* $\Gamma \leq \mathrm{Aut}_{\mathrm{tr}}(\Delta)$ *acts simply transitively on the set* $\mathcal{V}_{\Delta}$ *of vertices of* Δ. *Then the involution* $\lambda : \Pi(v_0) \to \Pi(v_0)$ *defined above (recall that* $\lambda(v) = g_v^{-1} v_0$, *where* $g_v \in \Gamma$ *and* $g_v v_0 = v$*) satisfies* $\dim(\lambda(u)) = n + 1 - \dim(u)$ *for each* $u \in \Pi$.

Proof. If $u \in \Pi(v_0)$ and $\dim(u) = i \in \{1, 2, \ldots, n\}$, then $\tau(v_0) + i = i = \tau(u) = \tau(g_u v_0)$. Hence $\tau(g_u v) = \tau(v) + i \pmod{n + 1}$ for every $v \in \mathcal{V}_{\Delta}$, because g_u is type-rotating. Thus $\tau(g_u^{-1} v) = \tau(v) - i \pmod{n+1}$ for every v, and so, in particular, $\dim(\lambda(u)) = \tau(g_u^{-1} v_0) = \tau(v_0) - i = -i = n + 1 - i \pmod{n + 1}$.

Let $\mathcal{T}$ be the set of triples (u, v, w), $u, v, w \in \Pi(v_0)$, associated to $\Gamma \leq \mathrm{Aut}_{\mathrm{tr}}(\Delta)$ as described in the introduction: $\mathcal{T} = \{(u, v, w) \in \Pi(v_0)^3 : g_u g_v g_w = 1\}$. Then, writing Π in place of $\Pi(v_0)$,

(A) given $u, v \in \Pi$, then $(u, v, w) \in \mathcal{T}$ for some $w \in \Pi$ if and only if $\lambda(u)$ and v are distinct and incident;

(B) if $(u, v, w) \in \mathcal{T}$, then $(v, w, u) \in \mathcal{T}$;

(C) if $(u, v, w_1) \in \mathcal{T}$ and $(u, v, w_2) \in \mathcal{T}$, then $w_1 = w_2$;

(D) if $(u, v, w) \in \mathcal{T}$, then $(\lambda(w), \lambda(v), \lambda(u)) \in \mathcal{T}$;

(E) if $(u, v, w) \in \mathcal{T}$, then $\dim(u) + \dim(v) + \dim(w) \equiv 0 \bmod n + 1$.

Indeed, properties (A)–(C) were derived in [1]. Property (D) is immediate from the fact that $g_{\lambda(u)} = g_u^{-1}$. As for (E), let the types of u, v and w be i, j and k, respectively. Then $\tau(g_u x) = \tau(x) + i \bmod n + 1$ for every vertex x, and similarly for g_v and g_w. Hence $0 = \tau(v_0) = \tau(g_u g_v g_w v_0) = i + j + k \bmod n + 1$.

Notice that if $(u, v, w) \in \mathcal{T}$ and $(\dim(u), \dim(v), \dim(w)) = (i, j, k)$, then $i + j + k$ is either $n + 1$ or $2(n + 1)$. If $v \subset \lambda(u)$, then $j < n + 1 - i$, and so $i + j + k = n + 1$ must hold (and $w \subset \lambda(v)$ and $u \subset \lambda(w)$). If $v \supset \lambda(u)$, then $j > n + 1 - i$, and so $i + j + k = 2(n + 1)$ must hold (and $w \supset \lambda(v)$ and $u \supset \lambda(w)$). Let us denote by $\mathcal{T}'$, respectively $\mathcal{T}''$, the set of triples $(u, v, w) \in \mathcal{T}$ such that $\dim(u) + \dim(v) + \dim(w)$ equals $n + 1$, respectively $2(n + 1)$. Axiom (D) shows that $(u, v, w) \mapsto (\lambda(w), \lambda(v), \lambda(u))$ is an involution of $\mathcal{T}$, interchanging the two subsets $\mathcal{T}'$ and $\mathcal{T}''$. Notice the following further property:

(F) if $(u, v, w) \in \mathcal{T}''$, and $(u, v', w') \in \mathcal{T}'$ then there is an $x \in \Pi$ such that $(\lambda(v), v', x) \in \mathcal{T}'$ and $(w, \lambda(w'), x) \in \mathcal{T}''$.

To see this, just notice that $v' \subset \lambda(u) \subset v$. Hence $(\lambda(v), v', x) \in \mathcal{T}'$ for some $x \in \Pi$, by Property (A). Also, $\lambda(w) \subset u \subset \lambda(w')$, and so $(w, \lambda(w'), x') \in \mathcal{T}''$ for some $x' \in \Pi$, again by Property (A). Thus $g_{\lambda(v)} g_{v'} g_x = 1 = g_w g_{\lambda(w')} g_{x'}$. But also $g_u g_v g_w = 1 = g_u g_{v'} g_{w'}$. Thus $g_v g_w = g_{v'} g_{w'}$, and so $g_{\lambda(v)} g_{v'} = g_w g_{\lambda(w')}$. Hence $g_x = g_{x'}$, so that $x = x'$.

We are led to the following definition:

Definition. Let Π be any projective geometry of dimension $n \geq 2$. For $i = 1, \ldots, n$, let $\Pi_i = \{u \in \Pi : \dim(u) = i\}$. Let $\lambda : \Pi \to \Pi$ be an involution such that $\lambda(\Pi_i)) = \Pi_{n+1-i}$ for each i. An $\widetilde{A}_n$-*triangle presentation compatible with* λ is a set $\mathcal{T}$ of triples (u, v, w), where $u, v, w \in \Pi$, satisfying properties (A)–(F) above.

When $n = 2$, the possible triples $(\dim(u), \dim(v), \dim(w))$, $(u, v, w) \in \mathcal{T}$, are just $(1, 1, 1)$ and $(2, 2, 2)$. So Axiom (F) holds vacuously. There is a slight difference between the above definition in the case $n = 2$ and the definition of $(\widetilde{A}_2$-)triangle presentation given in [1]. There, a triangle presentation is half of what we are now calling a triangle presentation; it is just what we are denoting $\mathcal{T}'$ here.

In view of property (D), an $\widetilde{A_n}$-triangle presentation is completely deter-mined by T'. Moreover, property (F) can be stated purely in terms of T':

(F') if $(x, y, u) \in T'$ and $(x', y', \lambda(u)) \in T'$, then for some $w \in \Pi$ we have $(y', x, w) \in T'$ and $(y, x', \lambda(w)) \in T'$.

So we could have defined an $\widetilde{A_n}$-triangle presentation to be an object like T'. This corresponds to the definition in [1], but seems less natural here.

In general, the possible triples $(\dim(u), \dim(v), \dim(w))$, $(u, v, w) \in T'$, are just the possible partitions of $n + 1$ into three positive integers. For example, if $n = 3$, the possible triples are just $(1, 1, 2)$ and its cyclic permu-tations, while if $n = 4$, the possibilities are $(1, 1, 3)$ and $(1, 2, 2)$, and their cyclic permutations.

An example will be given below of a set T of triples, where $n = 3$, satisfying (A)–(E), but not (F).

Given an $\widetilde{A_n}$-triangle presentation T, we can form the associated group Γ_T with a generating set indexed by Π:

$$\Gamma_T = \langle \{a_v\}_{v \in \Pi} \mid (1) \; a_{\lambda(v)} = a_v^{-1} \text{ for all } v \in \Pi$$

$$(2) \; a_u a_v a_w = 1 \text{ for all } (u, v, w) \in T \rangle.$$

The notation of the next theorem assumes that $\Pi = \Pi(\mathbf{V})$ of an $n + 1$ dimensional vector space $\mathbf{V}$ over a field k. However, the statement and the proof of the theorem are valid for an arbitrary n-dimensional projective geometry Π, provided that we interpret "$\lambda(u_i) + u_{i+1} = \mathbf{V}$" as meaning "there is no $v \in \Pi$ such that $\lambda(u_i) \subset v$ and $u_{i+1} \subset v$". In particular, the statement and proof generalize those of Proposition 3.2 in [1].

Theorem 2.2. Let T be an $\widetilde{A_n}$-triangle presentation, and let $\xi \in \Gamma_T$. Then we can write

$$\xi = a_{u_1} a_{u_2} \cdots a_{u_\ell} \tag{2.1}$$

for some integer $\ell \geq 0$ and for some $u_1, \ldots, u_\ell \in \Pi$ such that

$$\lambda(u_i) + u_{i+1} = \mathbf{V} \quad \text{for } i = 1, \ldots, \ell - 1 \tag{2.2}$$

(this implies that $\dim(u_1) \leq \dim(u_2) \leq \cdots \leq \dim(u_\ell)$). Moreover, this way of writing ξ as in (2.1) so that (2.2) holds is unique, and is of minimal length amongst all words equal to ξ.

Proof. Using the relations $a_v^{-1} = a_{\lambda(v)}$ for $v \in \Pi$, we see that any $\xi \in \Gamma_T$ is expressible as a word (2.1). To see that (2.2) can be arranged, choose a word $a_{u_1} \cdots a_{u_\ell}$ equal to ξ with ℓ minimal, and with $(\dim(u_1), \ldots, \dim(u_\ell))$ minimal for the lexicographic order. The first condition implies that $u_{i+1} \neq \lambda(u_i)$ (otherwise we could delete $a_{u_i} a_{u_{i+1}}$ from the word) and that u_{i+1} is not incident with $\lambda(u_i)$ (otherwise $(u_i, u_{i+1}, w) \in \mathcal{T}$ would hold for some $w \in \Pi$, and we could replace $a_{u_i} a_{u_{i+1}}$ by $a_{\lambda(w)}$, obtaining a shorter word). The second condition now implies that (2.2) holds. Indeed, it is enough to show that if $u, v \in \Pi$ with $v \neq \lambda(u)$, v not incident with $\lambda(u)$, and $\lambda(u) + v \neq \mathbf{V}$, then we can write $a_u a_v = a_{u'} a_{v'}$, where $\dim(u') < \dim(u)$ (and $\dim(v') > \dim(v)$). To see this, we write $\lambda(u) + v = \lambda(w)$ for some $w \in \Pi$. Because $\lambda(u) \subsetneq \lambda(w)$, by Property (A) of an $\widetilde{A}_n$-triangle presentation we can find $u' \in \Pi$ such that $(w, \lambda(u), u') \in \mathcal{T}'$. Also, $v \subsetneq \lambda(w)$, so we can find $v' \in \Pi$ such that $(w, v, \lambda(v')) \in \mathcal{T}'$. Then $a_w a_{\lambda(u)} a_{u'} = 1 = a_w a_v a_{\lambda(v')}$, so that $a_{\lambda(u)} a_{u'} = a_v a_{\lambda(v')}$ and $a_u a_v = a_{u'} a_{v'}$. Now $\dim(w) + \dim(\lambda(u)) + \dim(u') = n + 1$, so that $\dim(u') = \dim(u) - \dim(w) < \dim(u)$, and similarly, $\dim(v') = \dim(v) + \dim(w) > \dim(v)$. Note that by (2.2), $n + 1 = \dim(\lambda(u_i) + u_{i+1}) \leq \dim(\lambda(u_i)) + \dim(u_{i+1}) = n + 1 - \dim(u_i) + \dim(u_{i+1})$, so that $\dim(u_i) \leq \dim(u_{i+1})$.

The proof of the uniqueness of the word (2.1) for ξ satisfying (2.2) is more difficult, though quite similar to the proof of Proposition 3.2 in [1]. We form the set $\mathcal{W}$ of "words" (2.1) satisfying (2.2) in the letters $\{a_v : v \in \Pi\}$, writing 1 for the empty word. For each $u \in \Pi$, we define a map $T_u : \mathcal{W} \to \mathcal{W}$ (corresponding to left multiplication by a_u). Let $\mathcal{F}$ denote the free group on $\mathrm{Card}(\Pi)$ distinct letters g_u, $u \in \Pi$, and let N be the normal subgroup of $\mathcal{F}$ generated by $g_u g_{\lambda(u)}$, $u \in \Pi$, and $g_u g_v g_w$, $(u, v, w) \in \mathcal{T}$. Lemma 2.3 below shows that each T_u is a permutation of $\mathcal{W}$, and that the homomorphism φ from $\mathcal{F}$ into the group of permutations of $\mathcal{W}$ determined by $g_u \mapsto T_u$ factors through $\mathcal{F}/N \cong \Gamma_{\mathcal{T}}$. Then, as in the proof of Proposition 3.2 in [1], $gN \mapsto \varphi(g)(1)$ defines a bijection $\Gamma_{\mathcal{T}} \to \mathcal{W}$, and the theorem is proved.

Here now are the details: We write $|\xi|$ for ℓ if $\xi \in \mathcal{W}$ is as in (2.1). Now for $u \in \Pi$, we define a map $T_u : \mathcal{W} \to \mathcal{W}$ as follows: Firstly, we set $T_u(1) = a_u$ for each $u \in \Pi$. Now suppose that $\ell \geq 1$, and that $T_u(\eta) \in \mathcal{W}$ has been defined, and satisfies $|T_u(\eta)| \leq |\eta| + 1$, for each $u \in \Pi$ and for each $\eta \in \mathcal{W}$ satisfying $|\eta| < \ell$. Let $\xi \in \mathcal{W}$ be as in (2.1), and let $\xi' = a_{u_2} \cdots a_{u_\ell}$ ($\in \mathcal{W}$). We define $T_u(\xi)$ as follows:

(i) If $u_1 = \lambda(u)$, set $T_u(\xi) = \xi'$.

(ii) If $u_1 \neq \lambda(u)$, but u_1 is incident with $\lambda(u)$, let w be the unique element of Π such that $(u, u_1, w) \in \mathcal{T}$. We set $T_u(\xi) = T_{\lambda(w)}(\xi')$ (which is defined, because $|\xi'| = \ell - 1$).

(iii) If $\lambda(u) + u_1 = \mathbf{V}$ (which implies that $u_1 \neq \lambda(u)$ and that u_1 is not incident with $\lambda(u)$), then $a_u a_{u_1} \cdots a_{u_\ell} \in \mathcal{W}$, and we set $T_u(\xi)$ equal to this word.

If $\dim(u) = 1$, then one of the mutually exclusive possibilities (i)–(iii) must hold, because $\dim(\lambda(u)) = n$, and so $T_u(\xi)$ is defined. Suppose now that $1 < i < n+1$, and that $T_u(\eta)$ has been defined, and satisfies $|T_u(\eta)| \leq |\eta| + 1$, for all $u \in \Pi$ with $\dim(u) < i$ and for all $\eta \in W$ with $|\eta| \leq \ell$. Now let $u \in \Pi$, with $\dim(u) = i$, and let $\xi \in W$ be as in (2.1). We define $T_u(\xi)$ exactly as in (i)–(iii) above, except that there is now another possibility:

(iv) $u_1 \neq \lambda(u)$, and u_1 is not incident with $\lambda(u)$, but $\lambda(u) + u_1 \neq \mathbf{V}$. We then write $\lambda(u) + u_1 = \lambda(w')$. We know that $(w', \lambda(u), u') \in T'$ and $(w', u_1, v') \in T'$ for unique $u', v' \in \Pi$. Also, (see the beginning of this proof), $\dim(u') = \dim(u) - \dim(w') < \dim(u)$. We set $T_u(\xi) = T_{u'}(T_{\lambda(v')}(\xi'))$ (which is defined because $|\xi'| < \ell$, $|T_{\lambda(v')}(\xi')| \leq \ell$ and $\dim(u') < i$).

Lemma 2.3. *Let $\xi \in W$. Then*

(a) $T_u\big(T_{\lambda(u)}(\xi)\big) = \xi$ for all $u \in \Pi$;

(b) $T_u\big(T_v(\xi)\big) = T_{\lambda(w)}(\xi)$ if $(u, v, w) \in T$;

(c) $T_{\lambda(u)}\big(T_{u'}(\xi)\big) = T_v\big(T_{\lambda(v')}(\xi)\big)$ if $(s, u, v) \in T$ and $(s, u', v') \in T$ for some $s \in \Pi$.

Proof. Let us denote by (a_ℓ) assertion (a) in the case when $|\xi| = \ell$, and similarly for (b_ℓ) and (c_ℓ). We prove assertions (a_ℓ)–(c_ℓ) by induction on ℓ. The proof is divided into several steps.

STEP 1: $\ell = 0$. Now (a_0) and (b_0) are trivial to verify, but for (c_0), there are four cases to check.

1. If $(s, u, v) \in T''$ and $(s, u', v') \in T'$, then by Axiom (F) of an $\widetilde{A}_n$-triangle presentation, there exists a unique $x \in \Pi$ such that $(\lambda(u), u', x) \in T'$ and $(v, \lambda(v'), x) \in T''$. Thus (by part (ii) of the above definition)

$$T_{\lambda(u)}\big(T_{u'}(1)\big) = T_{\lambda(u)}(a_{u'}) = T_{\lambda(x)}(1) = T_v\big(a_{\lambda(v')}\big) = T_v\big(T_{\lambda(v')}(1)\big).$$

2. If $(s, u, v) \in T'$ and $(s, u', v') \in T''$, then $(\lambda(s), \lambda(v), \lambda(u)) \in T''$ and $(\lambda(s), \lambda(v'), \lambda(u')) \in T'$, and this case reduces to Case 1.

3. If $(s, u, v) \in T'$ and $(s, u', v') \in T'$, then we must compare $T_{\lambda(u)}\big(T_{u'}(1)\big) = T_{\lambda(u)}(a_{u'})$ and $T_v\big(T_{\lambda(v')}(1)\big) = T_v\big(a_{\lambda(v')}\big)$. We consider the four cases of the definition of the first of these.

(i). $u' = u$. Then $v = v'$ by Axiom (C), and $T_{\lambda(u)}\big(T_{u'}(1)\big) = 1 = T_v\big(T_{\lambda(v')}(1)\big)$.

(ii). $u' \neq u$, but u' is incident with u. Then $(\lambda(u), u', w) \in T$ for some $w \in \Pi$, and $T_{\lambda(u)}(a_{u'}) = a_{\lambda(w)}$. If $(\lambda(u), u', w) \in T'$, then $(u, \lambda(w), \lambda(u')) \in T''$, $(u, v, s) \in T'$ and Axiom (F) imply that $(w, v, x) \in T'$ and $(\lambda(u'), \lambda(s), x) \in T''$ for some $x \in \Pi$. The second of these shows that x must be $\lambda(v')$, while the first one now implies that $T_v\big(a_{\lambda(v')}\big) = a_{\lambda(w)}$ too. If $(\lambda(u), u', w) \in T''$, however, then $(u', w, \lambda(u)) \in T''$, $(u', v', s) \in T'$ and Axiom (F) show that $(\lambda(w), v', x) \in T'$ and $(\lambda(u), \lambda(s), x) \in T''$ for some x. This time x must be $\lambda(v)$, and again $T_v\big(a_{\lambda(v')}\big) = a_{\lambda(w)}$.

(iii). $u + u' = \mathbf{V}$. This cannot happen here, because $u, u' \subset \lambda(s)$ by the hypothesis of Case 3.

(iv). $u' \neq u$, u' is not incident with u, and $u + u' \neq \mathbf{V}$. Write $u + u' = \lambda(\tilde{s})$. Then $(\tilde{s}, u, \tilde{v}) \in T'$ and $(\tilde{s}, u', \tilde{v}') \in T'$ for some $\tilde{v}, \tilde{v}' \in \Pi$, and (by part (iv) of the above definition)

$$T_{\lambda(u)}(T_{u'}(1)) = T_{\lambda(u)}(a_{u'}) = T_{\tilde{v}}(T_{\lambda(\tilde{v}')}(1)). \qquad (2.3)$$

Now $u, u' \subset \lambda(s)$, and so $\lambda(\tilde{s}) = u + u' \subset \lambda(s)$. If $\lambda(\tilde{s}) = \lambda(s)$, then $\tilde{s} = s$, and so $\tilde{v} = v$ and $\tilde{v}' = v'$, and we are done. If $\lambda(\tilde{s}) \neq \lambda(s)$, then $(s, \lambda(\tilde{s}), y) \in T'$ for some y. Now $(\tilde{s}, \lambda(s), \lambda(y)) \in T''$, $(\tilde{s}, u, \tilde{v}) \in T'$ and Axiom (F) imply that $(\lambda(y), \lambda(\tilde{v}), v) \in T''$, so that $(y, \lambda(v), \tilde{v}) \in T'$. Similarly $(y, \lambda(v'), \tilde{v}') \in T'$. If (c_0) were to fail in Case 3, we could choose $(s, u, v) \in T'$ and $(s, u', v') \in T'$ with $T_{\lambda(u)}(T_{u'}(1)) \neq T_v(T_{\lambda(v')}(1))$ and $\dim(u)$ as large as possible. Now $(y, \lambda(v), \tilde{v}) \in T'$ and $(y, \lambda(v'), \tilde{v}') \in T'$, and (because $(s, u, v) \in T'$) $\dim(\lambda(v)) > \dim(u)$. Thus

$$T_v(T_{\lambda(v')}(1)) = T_{\tilde{v}}(T_{\lambda(\tilde{v}')}(1)). \qquad (2.4)$$

Comparing (2.3) and (2.4), we see that the proof of Case 3(iv) is complete.

4. If $(s, u, v) \in T''$ and $(s, u', v') \in T''$, then $(\lambda(s), \lambda(v), \lambda(u)) \in T'$ and $(\lambda(s), \lambda(v'), \lambda(u')) \in T'$, and this case reduces to Case 3.

We now assume that $\ell \geq 1$, and that $(a_{\ell-1})$, $(b_{\ell-1})$ and $(c_{\ell-1})$ have been proved. Let ξ be as in (2.1), and let $\xi' = a_{u_2} \cdots a_{u_\ell}$, as before.

STEP 2: We next prove (b_ℓ) in the case $(u, v, w) \in T'$. We shall denote this case (b'_ℓ) below. We consider the four cases of the definition of $T_v(\xi)$:

(i). If $u_1 = \lambda(v)$, then $T_v(\xi) = \xi'$. So

$$T_u(T_v(\xi)) = T_u(\xi') \overset{1}{=} T_{\lambda(w)}(T_{\lambda(v)}(\xi')) = T_{\lambda(w)}(T_{u_1}(\xi')) = T_{\lambda(w)}(\xi),$$

where the equation marked 1 holds by $(b_{\ell-1})$, because $(\lambda(w), \lambda(v), \lambda(u)) \in T$.

(ii). If $u_1 \neq \lambda(v)$, but u_1 is incident with $\lambda(v)$, then $(v, u_1, w_1) \in T$ for some $w_1 \in \Pi$, and $T_v(\xi) = T_{\lambda(w_1)}(\xi')$, by definition. Thus

$$T_u(T_v(\xi)) = T_u(T_{\lambda(w_1)}(\xi')) \overset{1}{=} T_{\lambda(w)}(T_{u_1}(\xi')) = T_{\lambda(w)}(\xi),$$

where for the equation marked 1, we have used the fact that $(v, w, u) \in T$, $(v, u_1, w_1) \in T$ and $(c_{\ell-1})$.

(iii). If $\lambda(v) + u_1 = \mathbf{V}$, then

$$T_u(T_v(\xi)) = T_u(a_v a_{u_1} \cdots a_{u_\ell}) = T_{\lambda(w)}(a_{u_1} \cdots a_{u_\ell}) = T_{\lambda(w)}(\xi).$$

(iv). If $u_1 \neq \lambda(v)$, u_1 is not incident with $\lambda(v)$, and $\lambda(v) + u_1 \neq \mathbf{V}$, write $\lambda(v) + u_1 = \lambda(w_1)$. Then $(w_1, \lambda(v), v'), (w_1, u_1, u_1') \in T'$ for some $v', u_1' \in \Pi$, and $T_v(\xi) = T_{v'}(T_{\lambda(u_1')}(\xi'))$, by definition. If (b_ℓ') were to fail, we could choose $(u, v, w) \in T'$ and $\xi \in W$ with $|\xi| = \ell$, $T_u(T_v(\xi)) \neq T_{\lambda(w)}(\xi)$ and $\dim(v)$ as small as possible. Knowing that $(\lambda(v), \lambda(u), \lambda(w)) \in T''$ and $(\lambda(v), v', w_1) \in T'$, Axiom (F) implies that $(u, v', x) \in T'$ and $(\lambda(w), \lambda(w_1), x) \in T''$ for some $x \in \Pi$. Thus $T_u(T_v(\xi))$ equals

$$T_u\big(T_{v'}(T_{\lambda(u_1')}(\xi'))\big) \overset{1}{=} T_{\lambda(x)}(T_{\lambda(u_1')}(\xi')) \overset{2}{=} T_{\lambda(w)}(T_{u_1}(\xi')) = T_{\lambda(w)}(\xi),$$

where the equation marked 1 holds because $(u, v', x) \in T'$ and $\dim(v') < \dim(v)$, and the equation marked 2 holds because $(\lambda(w_1), x, \lambda(w))$, $(\lambda(w_1), \lambda(u_1'), \lambda(u_1)) \in T$ and $(c_{\ell-1})$ holds. Thus (b_ℓ') is proved.

STEP 3: We next prove (a_ℓ). We consider the four cases of the definition of $T_{\lambda(u)}(\xi)$:

(i). If $u_1 = u$, then using part (iii) of the definition of T_u,

$$T_u(T_{\lambda(u)}(\xi)) = T_u(\xi') = \xi.$$

(ii). If $u_1 \neq u$, but u_1 is incident with u, then $(\lambda(u), u_1, v) \in T$ for some $v \in \Pi$, and $T_{\lambda(u)}(\xi) = T_{\lambda(v)}(\xi')$, by definition. Thus

$$T_u(T_{\lambda(u)}(\xi)) = T_u(T_{\lambda(v)}(\xi')) \overset{1}{=} T_{u_1}(\xi') = \xi,$$

where the equation marked 1 holds by $(b_{\ell-1})$ because $(u, \lambda(v), \lambda(u_1)) \in T$.

(iii). If $u_1 \neq u$, u_1 is not incident with u, and $u + u_1 = \mathbf{V}$, then using part (i) of the definition of T_u,

$$T_u(T_{\lambda(u)}(\xi)) = T_u(a_{\lambda(u)} a_{u_1} \cdots a_{u_\ell}) = \xi.$$

(iv). Finally, if $u_1 \neq u$, u_1 is not incident with u, and $u + u_1 \neq \mathbf{V}$, write $u + u_1 = \lambda(v)$. Then $(v, u, w), (v, u_1, w_1) \in T'$ for some $w, w_1 \in \Pi$. Then $T_{\lambda(u)}(\xi) = T_w(T_{\lambda(w_1)}(\xi'))$ by definition, and

$$T_u(T_{\lambda(u)}(\xi)) = T_u(T_w(T_{\lambda(w_1)}(\xi'))) \overset{1}{=} T_{\lambda(v)}(T_{\lambda(w_1)}(\xi')) \overset{2}{=} T_{u_1}(\xi') = \xi,$$

where the equation marked 1 holds by (b_ℓ') because $(u, w, v) \in T'$, and the equation marked 2 holds by $(b_{\ell-1})$ because $(\lambda(v), \lambda(w_1), \lambda(u_1)) \in T$.

STEP 4: We next prove (c_ℓ) in the case $(s, u, v), (s, u', v') \in T'$. We shall denote this case (c_ℓ') below.

If (c'_ℓ) were to fail, we could choose $(s, u, v), (s, u', v') \in T'$ and $\xi \in W$ with $|\xi| = \ell$ and $T_{\lambda(u)}(T_{u'}(\xi)) \neq T_v(T_{\lambda(v')}(\xi))$ and $\dim(u)$ as large as possible. Amongst such possible data with $\dim(u)$ maximal, we could choose $(s, u, v), (s, u', v') \in T'$ with $\dim(u')$ minimal. We know that $(s, u, v) \neq (s, u', v')$, as otherwise (a_ℓ) would give $T_{\lambda(u)}(T_{u'}(\xi)) = \xi = T_v(T_{\lambda(v')}(\xi))$.

We consider the four cases of the definition of $T_{u'}(\xi)$, in each case arriving at a contradiction to the above hypotheses:

1. If $u_1 = \lambda(u')$, then $(\lambda(v'), u_1, \lambda(s)) \in T''$, and so $T_{\lambda(v')}(\xi) = T_s(\xi')$ by definition. Thus

$$T_v(T_{\lambda(v')}(\xi)) = T_v(T_s(\xi')) \overset{1}{=} T_{\lambda(u)}(\xi') = T_{\lambda(u)}(T_{\lambda(u_1)}(\xi)) = T_{\lambda(u)}(T_{u'}(\xi)),$$

where the equation marked 1 holds by $(b_{\ell-1})$ because $(v, s, u) \in T$.

2. If $u_1 \neq \lambda(u')$, but u_1 is incident with $\lambda(u')$. Then $(u', u_1, w) \in T$ for some $w \in \Pi$. Thus $T_{\lambda(u)}(T_{u'}(\xi)) = T_{\lambda(u)}(T_{\lambda(w)}(\xi'))$, by definition. Hence

$$T_v(T_{\lambda(v')}(\xi)) = T_v(T_{\lambda(v')}(T_{u_1}(\xi'))) \overset{1}{=} T_v(T_s(T_{\lambda(w)}(\xi')))$$
$$\overset{2}{=} T_{\lambda(u)}(T_{\lambda(w)}(\xi')) = T_{\lambda(u)}(T_{u'}(\xi)),$$

where the equation marked 1 holds by $(c_{\ell-1})$ because $(u', v', s), (u', u_1, w) \in T$, and the equation marked 2 holds by (b'_ℓ) because $(v, s, u) \in T'$.

3. If $\lambda(u') + u_1 = V$, we have $T_{u'}(\xi) = a_{u'} a_{u_1} \cdots a_{u_\ell}$. We must now consider the four cases of the definition of $T_{\lambda(u)}(a_{u'} a_{u_1} \cdots a_{u_\ell})$.

(3.1). The first case, $u' = u$, is not allowed, as $(s, u, v) \neq (s, u', v')$.

(3.2). If $u' \neq u$, but u' is incident with u, then $(\lambda(u), u', w) \in T$ for some $w \in \Pi$, and $T_{\lambda(u)}(a_{u'} a_{u_1} \cdots a_{u_\ell}) = T_{\lambda(w)}(\xi)$, by definition.
Suppose first that $(\lambda(u), u', w) \in T'$. This, $(\lambda(u), \lambda(s), \lambda(v)) \in T''$ and Axiom (F) imply that $(\lambda(v), \lambda(w), v') \in T''$. Thus $T_v(T_{\lambda(v')}(\xi)) = T_{\lambda(w)}(\xi)$ by (b'_ℓ), as $(v, \lambda(v'), w) \in T'$.
So suppose that $(\lambda(u), u', w) \in T''$. This, $(u', v', s) \in T'$ and Axiom (F) now imply that $(\lambda(w), v', \lambda(v)) \in T'$. We must now consider four subcases according to the four cases of the definition of $T_{\lambda(v')}(\xi)$.

(3.2.1). The case $u_1 = v'$ is excluded. For, because $(s, u', v') \in T$, it would imply that u_1 is incident with $\lambda(u')$, contrary to the hypothesis of Case (3).

(3.2.2). If $u_1 \neq v'$, but u_1 is incident with v', then $(\lambda(v'), u_1, w_1) \in T$ for some w_1, and $T_{\lambda(v')}(\xi) = T_{\lambda(w_1)}(\xi')$ by definition. Thus

$$T_v(T_{\lambda(v')}(\xi)) = T_v(T_{\lambda(w_1)}(\xi')) \overset{1}{=} T_{\lambda(w)}(T_{u_1}(\xi'))$$
$$= T_{\lambda(w)}(\xi) = T_{\lambda(u)}(T_{u'}(\xi)),$$

where the equation marked 1 holds because $(\lambda(v'), w, v), (\lambda(v'), u_1, w_1) \in \mathcal{T}$, by $(c_{\ell-1})$, and the last equation is valid for the case (3.2).

(3.2.3). If $u_1 \neq v'$, u_1 is not incident with v', and $v' + u_1 = \mathbf{V}$, then

$$T_v\left(T_{\lambda(v')}(\xi)\right) = T_v(a_{\lambda(v')}a_{u_1}\cdots a_{u_\ell}) \overset{1}{=} T_{\lambda(w)}(a_{u_1}\cdots a_{u_\ell}) = T_{\lambda(u)}\left(T_{u'}(\xi)\right),$$

where the equation marked 1 holds by definition of T_v, because $(v, \lambda(v'), w) \in \mathcal{T}$.

(3.2.4). If $u_1 \neq v'$, u_1 is not incident with v', and $v' + u_1 \neq \mathbf{V}$, write $v' + u_1 = \lambda(w')$. Then $(w', v', z), (w', u_1, z') \in \mathcal{T}'$ for some $z, z' \in \Pi$, and $T_{\lambda(v')}(\xi) = T_z\left(T_{\lambda(z')}(\xi')\right)$ by definition. Thus

$$T_v\left(T_{\lambda(v')}(\xi)\right) = T_v\left(T_z\left(T_{\lambda(z')}(\xi')\right)\right) \overset{1}{=} T_{\lambda(w)}\left(T_{\lambda(w')}\left(T_{\lambda(z')}(\xi')\right)\right)$$
$$\overset{2}{=} T_{\lambda(w)}\left(T_{u_1}(\xi')\right) = T_{\lambda(w)}(\xi) = T_{\lambda(u)}\left(T_{u'}(\xi)\right).$$

Here the equation marked 1 holds because $(v', \lambda(v), \lambda(w)), (v', z, w') \in \mathcal{T}'$ and $\dim(\lambda(v)) > \dim(u)$ (as $(s, u, v) \in \mathcal{T}'$), by the hypotheses made at the beginning of Step 4. The equation marked 2 holds by $(b_{\ell-1})$, because $(\lambda(w'), \lambda(z'), \lambda(u_1)) \in \mathcal{T}$.

(3.3). The case $u + u' = \mathbf{V}$ cannot happen here, because $(s, u, v), (s, u', v') \in \mathcal{T}'$ implies that $u, u' \subset \lambda(s)$.

(3.4). If $u' \neq u$, u' is not incident with u, and $u + u' \neq \mathbf{V}$, the proof is exactly as in Case 3(iv) of Step 1, except that we replace the 1 there by ξ.

This completes the proof in Case 3 of Step 4.

4. If $u_1 \neq \lambda(u')$, u_1 is not incident with $\lambda(u')$, and $\lambda(u') + u_1 \neq \mathbf{V}$, write $\lambda(u') + u_1 = \lambda(s')$. Then $(s', \lambda(u'), z), (s', u_1, z') \in \mathcal{T}'$ for some $z, z' \in \Pi$, and $T_{u'}(\xi) = T_z\left(T_{\lambda(z')}(\xi')\right)$ by definition. Now $(u', \lambda(s'), \lambda(z)) \in \mathcal{T}'', (u', v', s) \in \mathcal{T}'$ and Axiom (F) imply that $(s', v', x) \in \mathcal{T}'$ and $(\lambda(z), \lambda(s), x) \in \mathcal{T}''$ for some $x \in \Pi$. Thus because $(s, u, v), (s, z, \lambda(x)) \in \mathcal{T}'$, $\dim(z) < \dim(u')$ and the hypotheses made at the beginning of Step 4, we have

$$T_{\lambda(u)}\left(T_z\left(T_{\lambda(z')}(\xi')\right)\right) = T_v\left(T_x\left(T_{\lambda(z')}(\xi')\right)\right).$$

Now by $(c_{\ell-1})$, $(s', v', x), (s', u_1, z') \in \mathcal{T}$ implies that $T_x\left(T_{\lambda(z')}(\xi')\right) = T_{\lambda(v')}\left(T_{u_1}(\xi')\right) = T_{\lambda(v')}(\xi)$. Combining the above facts, we obtain $T_{\lambda(u)}\left(T_{u'}(\xi)\right) = T_v\left(T_{\lambda(v')}(\xi)\right)$ again, and Step 4 is complete.

STEP 5: We next complete the proof of (b_ℓ), proving it for the case when $(u, v, w) \in \mathcal{T}''$.

The proof follows exactly as for the case when $(u, v, w) \in \mathcal{T}'$ dealt with in Step 2 until we get to the fourth case of the definition of $T_v(\xi)$. Suppose then that $u_1 \neq \lambda(v)$, u_1 is not incident with $\lambda(v)$, and $\lambda(v) + u_1 \neq \mathbf{V}$, and write $\lambda(v) + u_1 = \lambda(w_1)$. Then $(w_1, \lambda(v), v'), (w_1, u_1, u_1') \in T'$ for some $v', u_1' \in \Pi$, and $T_v(\xi) = T_{v'}(T_{\lambda(u_1')}(\xi'))$, by definition. Then

$$T_u(T_v(\xi)) = T_u(T_{v'}(T_{\lambda(u_1')}(\xi'))) \overset{1}{=} T_{\lambda(w)}(T_{\lambda(w_1)}(T_{\lambda(u_1')}(\xi')))$$

$$\overset{2}{=} T_{\lambda(w)}(T_{u_1}(\xi')) = T_{\lambda(w)}(\xi),$$

where the equation marked 1 holds by (c_ℓ'), because $(\lambda(v), \lambda(u), \lambda(w))$, $(\lambda(v), v', w_1) \in T'$, and the equation marked 2 holds by $(b_{\ell-1})$, because $(\lambda(w_1), \lambda(u_1'), \lambda(u_1)) \in T$.

STEP 6: The final step in the proof is to complete the proof of (c_ℓ).

If $(s, u, v), (s, u', v') \in T''$, then $(\lambda(s), \lambda(v), \lambda(u)), (\lambda(s), \lambda(v'), \lambda(u')), \in T'$, and so $T_v(T_{\lambda(v')}(\xi)) = T_{\lambda(u)}(T_{u'}(\xi))$ holds if $|\xi| \leq \ell$ because of (c_ℓ'). Suppose that $(s, u, v) \in T''$ and $(s, u', v') \in T'$. Then by Axiom (F), $(\lambda(u), u', x) \in T'$ and $(v, \lambda(v'), x) \in T''$ for some $x \in \Pi$. Now (b_ℓ) implies that

$$T_{\lambda(u)}(T_{u'}(\xi)) = T_{\lambda(x)}(\xi) = T_v(T_{\lambda(v')}(\xi)).$$

Finally, if $(s, u, v) \in T'$ and $(s, u', v') \in T''$, then $(\lambda(s), \lambda(v), \lambda(u)) \in T''$ and $(\lambda(s), \lambda(v'), \lambda(u')) \in T'$, and we are reduced to the previous case.

Lemma 2.4. *With the notation as in Theorem 2.2, let $u, v, w \in \Pi$, and suppose that $a_u a_v a_w = 1$ in Γ_T. Then $(u, v, w) \in T$.*

Proof. If not, choose a counterexample with $\dim(u)$ minimal. If $v = \lambda(u)$, then $a_u a_v = 1$, so that $a_w = 1$, contrary to Theorem 2.2. If v and $\lambda(u)$ are distinct and incident, then $(u, v, x) \in T$ for some $x \in \Pi$. Thus $a_u a_v a_x = 1$ too, so that $a_x = a_w$, and hence $x = w$ by the theorem, and we are done. If $\lambda(u) + v = \mathbf{V}$, then $a_u a_v = a_{\lambda(w)}$ would contradict the theorem. Finally, if v and $\lambda(u)$ are distinct and not incident, with $\lambda(u) + v \neq \mathbf{V}$, write $\lambda(u) + v = \lambda(s)$. We have $(s, \lambda(u), u'), (s, v, \lambda(v')) \in T'$ for some u', v', and $a_u a_v$ can be re-written $a_{u'} a_{v'}$. Also, $\dim(u') < \dim(u)$, and so by the choice at the beginning, we have $(u', v', w) \in T$. Treating the cases $(u', v', w) \in T'$ and $(u', v', w) \in T''$ separately, and using Axiom (F), we get that $(u, v, w) \in T$, contrary to hypothesis.

Theorem 2.5 *Let T be any $\widetilde{A}_n$-triangle presentation. Then the Cayley graph of Γ_T with respect to the generators a_u, $u \in \Pi$, is the 1-skeleton of a thick building Δ_T of type $\widetilde{A}_n$, on which Γ_T acts simply transitively, as a group of type rotating automorphisms.*

Proof. The proof is very similar to that of Theorem 3.4 in [1], so we shall be brief. We show first that Γ_T is a geometry of type $\widetilde{A}_n$ (see [10]). To begin with, Axiom (E) implies that there is a homomorphism (the type map) $\tau : \Gamma_T \to \mathbf{Z}/(n+1)\mathbf{Z}$ such that $\tau(a_u) = \dim(u)$. Next, we define an incidence relation $*$ on Γ_T, calling $g, g' \in \Gamma_T$ *incident* if $g' = g$ or if $g' = ga_u$ for some $u \in \Pi$. Clearly, if g and g' are incident and $\tau(g) = \tau(g')$, then $g = g'$. So $(\Gamma_T, \tau, *)$ is a geometry. The associated graph is just the Cayley graph of Γ_T (relative to the generators a_u). This graph is connected because the a_u's generate Γ_T. Theorem 2.2 implies in particular that, given $g \in \Gamma_T$, the vertices ga_u, $u \in \Pi$, are distinct and different from g. The (vertex set of the) residue of $g \in \Gamma_T$ is $\{ga_u : u \in \Pi\}$. The map $u \mapsto ga_u$, is an isomorphism of Π onto this residue. To check this, we must verify that if $u, v \in \Pi$ are distinct, then ga_u and ga_v are incident if and only if u, v are incident in Π. If u, v are incident in Π, then $(\lambda(u), v, w) \in T$ for some $w \in \Pi$. Thus $a_u = a_v a_w$, and so $ga_u = (ga_v)a_w$ is incident with ga_v in the residue of g. Conversely, if $ga_u = (ga_v)a_w$, then $a_u = a_v a_w$, and so $(\lambda(u), v, w) \in T$ by Lemma 2.4. Hence u and v are incident in Π.

Thus $(\Gamma_T, \tau, *)$ is a geometry of type $\widetilde{A}_n$. The rest of the proof follows closely that of Theorem 3.4 in [1], and we omit it.

The next proposition (a generalization of Theorem 3.5 in [1]) is useful when we are verifying that a homomorphism $\varphi : \Gamma_T \to \text{Aut}_{\text{tr}}(\Delta)$ is an isomorphism onto a subgroup of $\text{Aut}_{\text{tr}}(\Delta)$ which acts simply transitively on $\mathcal{V}_\Delta$. It is also useful for verifying that a set T of triples satisfying axioms (A)–(E) of an $\widetilde{A}_n$-triangle presentation also satisfies (F). We can form the group Γ_T even when (F) is not assumed.

Proposition 2.6. *Let Π be a projective geometry of order $q < \infty$ and dimension $n \geq 2$. Let λ be an involution of Π such that $\lambda(\Pi_i) = \Pi_{n+1-i}$ for $i = 1, \ldots, n$. Let T be a family of triples $(u, v, w) \in \Pi^3$ satisfying axioms (A)–(E) of an $\widetilde{A}_n$-triangle presentation compatible with λ, but perhaps not axiom (F). Let Δ be a thick building of type $\widetilde{A}_n$, and let $v_0 \in \mathcal{V}_\Delta$ have type 0. Write $\Pi_i(v_0)$ for the set of neighbours of v_0 having type i. Suppose that $\psi : \Gamma_T \to \text{Aut}_{\text{tr}}(\Delta)$ is a homomorphism such that*

(1) $\psi(a_u)v_0 \in \Pi_i(v_0)$ for each $u \in \Pi_i$;

(2) $u \mapsto \psi(a_u)v_0$ is a bijection $\Pi_1 \to \Pi_1(v_0)$.

Then T does satisfy axiom (F), $u \mapsto \psi(a_u)v_0$ is an isomorphism from Π onto the residue $\Pi(v_0)$ of v_0, ψ is an isomorphism onto a subgroup of $\text{Aut}_{\text{tr}}(\Delta)$ which acts simply transitively on $\mathcal{V}_\Delta$, and $g \mapsto \psi(g)v_0$ is an isomorphism of Δ_T onto Δ.

Proof. We first show that for $i = 1, \ldots, n$, $u \mapsto \psi(a_u)v_0$ is a bijection $\Pi_i \to \Pi_i(v_0)$. Hypothesis (2) implies that the projective geometry $\Pi(v_0)$ is of order q. Suppose that $i > 1$, and that for each $j < i$, $u \mapsto \psi(a_u)v_0$ is a bijection $\Pi_j \to \Pi_j(v_0)$. Let $u, u' \in \Pi_i$ with $\psi(a_u)v_0 = \psi(a_{u'})v_0$. If $u \neq u'$, there exists $v \in \Pi_{i-1}$ such that $v \subset u$ but $v \not\subset u'$. Then $(\lambda(u), v, w) \in \mathcal{T}$ for some $w \in \Pi$, so that $a_u = a_v a_w$ in $\Gamma_{\mathcal{T}}$. Thus

$$d(\psi(a_u)v_0, \psi(a_v)v_0) = d(\psi(a_v)(\psi(a_w)v_0), \psi(a_v)v_0) = d(\psi(a_w)v_0, v_0) = 1 \tag{2.5}$$

So $\psi(a_v)v_0$ is a type $i-1$ neighbour of $\psi(a_u)v_0 = \psi(a_{u'})v_0$. But $\psi(a_{v'})v_0$ is also a type $i-1$ neighbour of $\psi(a_{u'})v_0$ for each of the $(q^i - 1)/(q-1)$ elements $v' \in \Pi_{i-1}$ such that $v' \subset u'$. By the induction hypothesis, $\psi(a_v)v_0$ is different from each such $\psi(a_{v'})v_0$. This is a contradiction, as $\psi(a_{u'})v_0$ has only $(q^i - 1)/(q-1)$ neighbours of type $i-1$. Thus $u = u'$.

Thus $u \mapsto \psi(a_u)v_0$ is a bijection $\Pi \to \Pi(v_0)$. If $u, v \in \Pi$ are distinct and incident, then $(\lambda(u), v, w) \in \mathcal{T}$ for some $w \in \Pi$, and the calculation (2.5) shows that $\psi(a_u)v_0$ and $\psi(a_v)v_0$ are incident in $\Pi(v_0)$. Conversely, suppose that $\psi(a_u)v_0$ and $\psi(a_v)v_0$ are incident in $\Pi(v_0)$, where $u \in \Pi_i$ and $v \in \Pi_j$, say. For each $u' \in \Pi_i$ incident with v, we have $\psi(a_{u'})v_0 \in \Pi_i(v_0)$ incident with $\psi(a_v)v_0$. If u and v were not incident, then $\psi(a_u)v_0$ would be different from each such $\psi(a_{u'})v_0$, and $\psi(a_v)v_0$ would have one type i neighbour too many. So u and v are incident, and $u \mapsto \psi(a_u)v_0$ is an isomorphism from Π onto $\Pi(v_0)$.

Next, we check that $\mathcal{T}$ satisfies axiom (F). Let $(u, v, w) \in \mathcal{T}''$ and $(u, v', w') \in \mathcal{T}'$. Now $v' \subset \lambda(u) \subset v$. Hence $(\lambda(v), v', x) \in \mathcal{T}'$ for some $x \in \Pi$. Also, $\lambda(w) \subset u \subset \lambda(w')$, and so $(w, \lambda(w'), x') \in \mathcal{T}''$ for some $x' \in \Pi$. Thus $a_{\lambda(v)} a_{v'} a_x = 1 = a_w a_{\lambda(w')} a_{x'}$ in $\Gamma_{\mathcal{T}}$. But also $a_u a_v a_w = 1 = a_u a_{v'} a_{w'}$. Thus $a_v a_w = a_{v'} a_{w'}$, and so $a_{\lambda(v)} a_{v'} = a_w a_{\lambda(w')}$. Hence $a_x = a_{x'}$, so that $\psi(a_x)v_0 = \psi(a_{x'})v_0$. This implies that $x = x'$, by what we have proved so far. Thus axiom (F) is satisfied.

Finally, as in the proof of Theorem 3.5 in [1], we see that $g \mapsto \psi(g)v_0$ is a covering $\Delta_{\mathcal{T}} \to \Delta$, and hence an isomorphism, because Δ is simply connected. This implies the last assertion of our proposition.

Let us now state as a lemma two ways of getting new $\widetilde{A_n}$-triangle presentations from a given one $\mathcal{T}$. The proof is as in [2, Lemma 2.1], and we omit it. Recall that a *collineation* of Π is an incidence-preserving bijection of Π mapping each Π_i to Π_i, and a *correlation* of Π is an incidence-preserving bijection of Π mapping each Π_i to Π_{n+1-i}.

Lemma 2.7 (a) *If $\mathcal{T}$ is compatible with an involution $\lambda : \Pi \to \Pi$, and if $h : \Pi \to \Pi$ is a collineation, then $h(\mathcal{T}) = \{(h(u), h(v), h(w)) : (u, v, w) \in \mathcal{T}\}$ is an $\widetilde{A_n}$-triangle presentation compatible with $h \circ \lambda \circ h^{-1}$.*

(b) *If $\mathcal{T}$ is compatible with an involution $\lambda : \Pi \to \Pi$, and if $\alpha : \Pi \to \Pi$ is a correlation, then $\alpha\lambda(\mathcal{T}^{rev}) = \{(\alpha\lambda(w), \alpha\lambda(v), \alpha\lambda(u)) : (u, v, w) \in \mathcal{T}\}$ is an $\widetilde{A_n}$-triangle presentation compatible with $\alpha \circ \lambda \circ \alpha^{-1}$.*

When h is a collineation and $h(\mathcal{T}) = \mathcal{T}$, we call $\mathcal{T}$ *invariant under* h. It is easy to see that an $\widetilde{A_n}$-triangle presentation $\mathcal{T}$ cannot be compatible with two different λ's. Thus the λ of a $\mathcal{T}$ invariant under h must satisfy $\lambda(h(u)) = h(\lambda(u))$ for each $u \in \Pi$.

We conclude this section with the following fact, which we use later:

Lemma 2.8. *Let $\mathcal{T}$ be an $\widetilde{A_n}$-triangle presentation. Then $\Gamma_{\mathcal{T}}$ is generated by the elements a_u, $u \in \Pi_1$.*

Proof. Let Γ' denote the subgroup of $\Gamma_{\mathcal{T}}$ generated by the elements a_u, $u \in \Pi_1$. Using the relations $a_{\lambda(v)} = a_v^{-1}$ for $v \in \Pi$, and Lemma 2.1, we see that $a_v \in \Gamma'$ if $v \in \Pi_1 \cup \Pi_n$. Suppose that $i \geq 1$, and that we know that $a_v \in \Gamma'$ if $v \in \Pi_1 \cup \cdots \Pi_i \cup \Pi_{n-i+1} \cup \cdots \cup \Pi_n$ (where $2i < n + 1$). Let $w \in \Pi_{n-i}$. Then $w \subset v'$ for some $v' \in \Pi_{n-i+1}$. Let $v = \lambda(v')$. As $w \subset \lambda(v)$, there exists $u \in \Pi$ such that $(u, v, w) \in \mathcal{T}'$. We must have $u \in \Pi_1$. Then $a_w = (a_u a_v)^{-1} \in \Gamma'$. Using the relations $a_{\lambda(v)} = a_v^{-1}$ for $v \in \Pi$ again, we see that $a_v \in \Gamma'$ for any $v \in \Pi_1 \cup \cdots \Pi_{i+1} \cup \Pi_{n-i} \cup \cdots \cup \Pi_n$. This completes the proof.

§3. Invariant $\widetilde{A_n}$-triangle presentations $\mathcal{T}$.

Consider a prime power q, and an integer $n \geq 2$. Let $\mathbf{V}$ be a vector space of dimension $d = n + 1$ over $\mathbf{F}_q$, and let $\Pi = \Pi(\mathbf{V})$. It is well known (see, e.g., [5, p. 28]) that for $i = 1, \ldots, d - 1 = n$,

$$|\Pi_i| = \frac{(q^d - 1)(q^{d-1} - 1) \cdots (q - 1)}{(q^i - 1)(q^{i-1} - 1) \cdots (q - 1)(q^{d-i} - 1)(q^{d-i-1} - 1) \cdots (q - 1)}.$$

In particular,

$$|\Pi_1| = |\Pi_{d-1}| = \frac{q^d - 1}{q - 1} = q^{d-1} + \cdots + q + 1$$

and

$$|\Pi_2| = |\Pi_{d-2}| = \frac{(q^d - 1)(q^{d-1} - 1)}{(q^2 - 1)(q - 1)}.$$

In what follows, it is convenient to choose $\mathbf{V}$ equal to the field $\mathbf{F}_{q^d}$. Then $\mathbf{F}_{q^d}^{\times}$ acts on Π by multiplication: for each $t \in \mathbf{F}_{q^d}^{\times}$, $M_t : U \mapsto tU$ is a collineation of Π. As each $t \in \mathbf{F}_q^{\times}$ fixes each $U \in \Pi$, this induces an action of $\mathbf{F}_{q^d}^{\times}/\mathbf{F}_q^{\times}$ on Π. Also, the Frobenius automorphism $\varphi : x \mapsto x^q$ of $\mathbf{F}_{q^d}$ induces a collineation of Π.

The $\widetilde{A_n}$-triangle presentations $\mathcal{T}$ found in this paper satisfy (1.1) and (1.2), and so are invariant under each collineation in the group of order $d(q^d -$

$1)/(q-1)$ generated by the M_t, $t \in \mathbf{F}_{q^d}^\times$, and by φ. By the remarks after Lemma 2.7 above, the λ of any $\mathcal{T}$ satisfying (1.1) and (1.2) must satisfy

$$\lambda(tU) = t\lambda(U) \quad \text{for each } t \in \mathbf{F}_{q^d}^\times \text{ and } U \in \Pi \qquad (3.1)$$

and

$$\lambda(\varphi(U)) = \varphi(\lambda(U)) \quad \text{for each } U \in \Pi. \qquad (3.2)$$

Let Tr denote the trace function $\text{Tr} : \mathbf{F}_{q^d} \to \mathbf{F}_q$. For $S \subset \mathbf{F}_{q^d}$, let

$$S^\perp = \{y \in \mathbf{F}_{q^d} : \text{Tr}(xy) = 0 \text{ for all } x \in S\}.$$

Then $\alpha : U \mapsto U^\perp$ is a polarity (i.e., an involutive correlation) of Π, because $(x, y) \mapsto \text{Tr}(xy)$ is a nondegenerate bilinear form on $\mathbf{F}_{q^d}$. This is useful when applying Lemma 2.7(b).

Clearly the action of $\mathbf{F}_{q^d}^\times/\mathbf{F}_q^\times$ on Π_1 is simply transitive, and we can identify Π_1 and $\mathbf{F}_{q^d}^\times/\mathbf{F}_q^\times$, with $\mathbf{F}_q x_0 = x_0.\mathbf{F}_q \in \Pi_1$ and $x_0\mathbf{F}_q^\times \in \mathbf{F}_{q^d}^\times/\mathbf{F}_q^\times$ being identified. In particular, $\mathbf{F}_q \in \Pi_1$ is identified with $1 \in \mathbf{F}_{q^d}^\times/\mathbf{F}_q^\times$. Using $(tU)^\perp = t^{-1}U^\perp$, we see that $\mathbf{F}_{q^d}^\times/\mathbf{F}_q^\times$ also acts simply transitively on Π_{d-1}.

Suppose that an $\widetilde{A}_n$-triangle presentation $\mathcal{T}$ compatible with $\lambda : \Pi \to \Pi$, satisfies (1.1) and (1.2). Then as $\mathbf{F}_{q^d}^\times$ acts transitively on $\Pi_1 \cong \mathbf{F}_{q^d}^\times/\mathbf{F}_q^\times$ and on Π_{d-1}, $\lambda_{|\Pi_1}$ is determined by $\lambda(1)$, and $\lambda(1) = \gamma\{1\}^\perp$ must hold for some $\gamma \in \mathbf{F}_{q^d}^\times/\mathbf{F}_q^\times$. Also, (3.2) implies that

$$\gamma^q\{1\}^\perp = \varphi(\gamma\{1\}^\perp) = \varphi(\lambda(1)) = \lambda(\varphi(1)) = \lambda(1) = \gamma\{1\}^\perp$$

and so $\gamma^{q-1} = 1$ in $\mathbf{F}_{q^d}^\times/\mathbf{F}_q^\times$. So the order, m, say, of γ in $\mathbf{F}_{q^d}^\times/\mathbf{F}_q^\times$ must divide both $q-1$ and $|\mathbf{F}_{q^d}^\times/\mathbf{F}_q^\times| = (q^d-1)/(q-1)$. Now $(q^d-1)/(q-1) \equiv d$ (mod $q-1$), and so m must divide d (and $q-1$). In particular, $\gamma^d = 1$.

The next lemma gives another way of forming a new $\widetilde{A}_n$-triangle presentation from a given one $\mathcal{T}$, provided that $\mathcal{T}$ satisfies (1.1) and (1.2). The proof is routine, and we omit it.

Lemma 3.1 *Suppose that $\mathcal{T}$ is an $\widetilde{A}_n$-triangle presentation compatible with $\lambda : \Pi \to \Pi$, and satisfying (1.1). Let $\beta \in \mathbf{F}_{q^d}^\times/\mathbf{F}_q^\times$ satisfy $\beta^d = 1$. Then*

$$\mathcal{T}^\beta = \{(u, \beta^{\dim(u)}v, \beta^{\dim(u)+\dim(v)}w) : (u, v, w) \in \mathcal{T}\}$$

is an $\widetilde{A}_n$-triangle presentation satisfying (1.1) compatible with $\lambda^\beta : \Pi \to \Pi$, where $\lambda^\beta(u) = \beta^{\dim(u)}\lambda(u)$. If $\mathcal{T}$ also satisfies (1.2), and if $\beta^{q-1} = 1$, then $\mathcal{T}^\beta$ also satisfies (1.2).

The next corollary is immediate from the last lemma and the remarks preceding it.

Corollary 3.2 *Suppose that there is an $\widetilde{A}_n$-triangle presentation T compatible with $\lambda : \Pi \to \Pi$, and satisfying (1.1) and (1.2). Then either $\lambda(1) = \{1\}^{\perp}$, or there is some other $\widetilde{A}_n$-triangle presentation satisfying (1.1) and (1.2) which is compatible with some other λ satisfying $\lambda(1) = \{1\}^{\perp}$.*

For general $i \in \{1, \ldots, d-1\}$, we can count the number of orbits in Π_i under the action of $\mathbf{F}_{q^d}^{\times}$ by considering $\mathrm{Stab}(U) = \{t \in \mathbf{F}_{q^d}^{\times} : tU = U\}$ for $U \in \Pi_i$. It is easy to check that $\mathrm{Stab}(U) \cup \{0\} = \{t \in \mathbf{F}_{q^d} : tU \subset U\}$ is a subfield of $\mathbf{F}_{q^d}$, and so equals $\mathbf{F}_{q^s}$ for some divisor s of d. Moreover, U is a vector space over $\mathbf{F}_{q^s}$. If its dimension as such is j, then $i = js$, and so s divides i too.

In particular, taking $i = 2$, we see that when d is even, Π_2 consists of the orbit $\mathcal{O}_0$ of $\mathbf{F}_{q^2}$, which has order $|\mathbf{F}_{q^d}^{\times}/\mathbf{F}_{q^2}^{\times}|$, and other orbits, each of order $|\mathbf{F}_{q^d}^{\times}/\mathbf{F}_q^{\times}|$. When d is odd, Π_2 divides into orbits of order $|\mathbf{F}_{q^d}^{\times}/\mathbf{F}_q^{\times}|$.

Let us state the above observations as a lemma:

Lemma 3.3. *Suppose that d is even. Then under the action of $\mathbf{F}_{q^d}^{\times}$, Π_2 divides into exactly $N + 1$ orbits, where $N = q(q^{d-2} - 1)/(q^2 - 1)$:*

$$\Pi_2 = \mathcal{O}_0 \cup \mathcal{O}_1 \cup \cdots \cup \mathcal{O}_N .$$

Moreover, $|\mathcal{O}_0| = |\mathbf{F}_{q^d}^{\times}/\mathbf{F}_{q^2}^{\times}| = (q^d - 1)/(q^2 - 1)$, and $|\mathcal{O}_i| = |\mathbf{F}_{q^d}^{\times}/\mathbf{F}_q^{\times}| = (q^d - 1)/(q-1)$ for $i = 1, \ldots, N$. When d is odd, Π_2 divides into $(q^{d-1} - 1)/(q^2 - 1)$ orbits, each of order $|\mathbf{F}_{q^d}^{\times}/\mathbf{F}_q^{\times}| = (q^d - 1)/(q - 1)$.

For example, if $d = 4$, then $N = q$, $|\mathcal{O}_0| = q^2 + 1$, and $|\mathcal{O}_i| = (q^4 - 1)/(q - 1) = (q^2 + 1)(q + 1)$ for $i = 1, \ldots, q$.

Let $V \in \Pi_r$, and suppose that $v_1, \ldots, v_r \in \mathbf{F}_{q^d}$ form a basis for V. For distinct integers $m_1, \ldots, m_r \in \{0, \ldots, d - 1\}$, form the determinant of the matrix whose (i, j)-th entry is $v_j^{q^{m_i}}$. If a different basis for V is used, the determinant is multiplied by the determinant of the change of basis matrix, and is thus determined, mod $\mathbf{F}_q^{\times}$, by V and the m_j's. Moreover, the ratio of two such determinants for V, if the denominator is nonzero, is determined as an element of $\mathbf{F}_{q^d}$. For example, if $V \in \Pi_2$, $m_1 = 0$ and $m_2 = 1$, the determinant is $v_1 v_2^q - v_1^q v_2 \neq 0$, and we can set

$$\delta(V) = \text{ the image of } v_1^q v_2 - v_1 v_2^q \text{ in } \mathbf{F}_{q^d}^{\times}/\mathbf{F}_q^{\times} . \tag{3.3}$$

Notice that $\delta(tV) = t^{q+1}\delta(V)$ for $t \in \mathbf{F}_{q^d}^{\times}$ (using t on the right to also denote its image in $\mathbf{F}_{q^d}^{\times}/\mathbf{F}_q^{\times}$). Also, $\delta(\varphi(V)) = \delta(V)^q$. We can also form

$v_1^{q^2} v_2 - v_1 v_2^{q^2}$, and assuming that it is nonzero (which holds unless d is even and $V \in \mathcal{O}_0$), we set

$$\delta_2(V) = \text{ the image of } v_1^{q^2} v_2 - v_1 v_2^{q^2} \text{ in } \mathbf{F}_{q^d}^\times / \mathbf{F}_q^\times . \qquad (3.4)$$

Notice that $\delta_2(tV) = t^{q^2+1} \delta_2(V)$ for $t \in \mathbf{F}_{q^d}^\times$. Also, $\delta_2(\varphi(V)) = \delta_2(V)^q$.

A method of embedding $\Gamma_\mathcal{T}$ in $\mathrm{Aut}_{\mathrm{tr}}(\Delta)$.

We apply Proposition 2.6 to verify that certain homomorphisms $\Gamma_\mathcal{T} \to PGL(n+1,K)$ we find are embeddings onto subgroups of $PGL(n+1,K)$ which act simply transitively on the set of vertices of the building $\widetilde{A_n}(K,v)$. The field K here is $\mathbf{F}_q(X)$, where X is an indeterminate, and v is the usual valuation (so that $v(X) = 1$). It will be convenient to work with the building $\widetilde{A_n}(\hat{K},v)$, where $\hat{K} = \mathbf{F}_q((X))$ is the completion of K with respect to v, which is isomorphic to $\widetilde{A_n}(K,v)$ as a chamber system [8, p. 130]. Embeddings of this type were discussed in [1, Section 4], and we refer the reader for more details, and to [4, Chapter 7] for generalities on cyclic simple algebras. We start with an indeterminate Y. It will be convenient to also use $Z = 1 + Y$. Let $K_0 = \mathbf{F}_q(Y) = \mathbf{F}_q(Z)$ and $L_0 = \mathbf{F}_{q^d}(Y) = \mathbf{F}_{q^d}(Z)$, where $d = n + 1$, as usual. We form the cyclic simple algebra $\mathcal{A} = L_0[\sigma]$, whose elements may be written uniquely

$$x_0 + x_1\sigma + \cdots + x_{d-1}\sigma^{d-1}, \quad \text{where } x_0, \ldots, x_{d-1} \in L_0, \qquad (3.5)$$

and where multiplication is determined by the rules $\sigma^d = Z$ and $\sigma x \sigma^{-1} = \varphi(x)$ for $x \in L_0$, where φ is the automorphism of L_0 determined by $\varphi(t) = t^q$ (for $t \in \mathbf{F}_{q^d}$) and $\varphi(Z) = Z$. Now $L_0[\sigma]$ is actually a division algebra, because Z^d is the smallest power of Z which is a norm $N_{L_0/K_0}(\xi)$ of some $\xi \in L_0$ (see, for example, [7, p. 84]), though this fact is not used below. Now consider an indeterminate X, and let $L = \mathbf{F}_{q^d}(X)$ and $K = \mathbf{F}_q(X)$. Choose any $\beta \in \mathbf{F}_{q^d}$ such that $\mathrm{Tr}_{\mathbf{F}_{q^d}/\mathbf{F}_q}(\beta) = t_1 \neq 0$. Then the norm $N_{L/K}(1 + \beta X)$ equals $1 + t_1 X + t_2 X^2 + \cdots + t_d X^d$ for certain other elements $t_2, \ldots, t_d$ of $\mathbf{F}_q$. We embed $K_0 = \mathbf{F}_q(Y)$ into $K = \mathbf{F}_q(X)$ and $L_0 = \mathbf{F}_{q^d}(Y)$ into $L = \mathbf{F}_{q^d}(X)$ by mapping Y to $t_1 X + t_2 X^2 + \cdots + t_d X^d$. The algebra $L[\sigma]$, whose elements are as in (3.5), but with the x_j's now in L, splits, i.e., is isomorphic to the algebra $M_{d \times d}(K)$ of $d \times d$ matrices over K. To see this, as in [1, Section 4] one first defines an embedding $\Psi : L[\sigma] \to M_{d \times d}(L)$ via

$$x \mapsto \Psi(x) = \begin{pmatrix} x & 0 & \cdots & 0 \\ 0 & \varphi(x) & \cdots & 0 \\ \vdots & \vdots & \ddots & \vdots \\ 0 & 0 & \cdots & \varphi^{d-1}(x) \end{pmatrix} \qquad \text{for } x \in L,$$

and

$$
\sigma \mapsto \Psi(\sigma) = \begin{pmatrix}
0 & 1+\beta X & 0 & \cdots & 0 \\
0 & 0 & 1+\beta^q X & \cdots & 0 \\
\vdots & \vdots & \vdots & \ddots & \vdots \\
0 & 0 & 0 & \cdots & 1+\beta^{q^{d-2}} X \\
1+\beta^{q^{d-1}} X & 0 & 0 & \cdots & 0
\end{pmatrix}.
$$

Let $\xi_0, \ldots, \xi_{n-1}$ be a basis for $\mathbf{F}_{q^d}$ over $\mathbf{F}_q$, and let $A \in GL(d, \mathbf{F}_{q^d})$ have (i,j)-th entry $\varphi^j(\xi_i)$ for $i, j = 0, \ldots, n-1$. The conjugation $M \mapsto AMA^{-1}$ maps the image of Ψ into $M_{d \times d}(K)$. The map $\xi \mapsto A\Psi(\xi)A^{-1}$ is an isomorphism $L[\sigma] \to M_{d \times d}(K)$.

Suppose now that we have a projective geometry Π of dimension n, an involution λ of Π such that $\lambda(\Pi_i) = \Pi_{n+1-i}$ for each i, and a family $\mathcal{T}$ of triples $(u, v, w) \in \Pi^3$ satisfying (A)-(E), but perhaps not (F). We can still form $\Gamma_{\mathcal{T}}$, and suppose that for each of its generators a_u, $u \in \Pi$, we can find $b_u \in L_0[\sigma]$ such that

(i) For each $u \in \Pi$ and for each $(u, v, w) \in \mathcal{T}$, $b_u b_{\lambda(u)}$ and $b_u b_v b_w$ are nonzero scalars, i.e., in the centre K_0 of $L_0[\sigma]$;

(ii) when b_u is written as in (3.5), the coefficients are in $\mathbf{F}_{q^d}[Z]$;

(iii) for each $u \in \Pi$, the reduced norm $N(b_u)$ (which is the determinant of the image g_u of b_u in $M_{d \times d}(K)$ under the isomorphism defined above) has valuation $d - \dim(u)$.

Then (i) means that the assignment $a_u \mapsto b_u$, $u \in \Pi$, and the isomorphism $L[\sigma] \to M_{d \times d}(K)$ described above, induce a homomorphism ψ of $\Gamma_{\mathcal{T}}$ into $PGL(d, K) \subset PGL(d, \mathbf{F}_q((X)))$. Condition (ii) implies that g_u has entries in $\mathbf{F}_q[X] \subset \mathbf{F}_q[[X]]$. Consider the building $\widetilde{A}_n(\mathbf{F}_q((X)), v)$. Let v_0 be the class of the lattice $V_0 = \{a_1 e_1 + \cdots + a_d e_d \mid a_1, \ldots, a_d \in \mathbf{F}_q[[X]]\}$, where $\{e_1, \ldots, e_d\}$ is the usual basis of column vectors of length d over $\mathbf{F}_q((X))$. Recall that the neighbours of v_0 are the classes of lattices V satisfying $XV_0 \subsetneq V \subsetneq V_0$. Thus $\Pi_i(v_0)$ consists of the lattice classes $[gV_0]$, where $g \in GL(d, \mathbf{F}_q((X)))$ has entries in $\mathbf{F}_q[[X]]$, Xg^{-1} has entries in $\mathbf{F}_q[[X]]$, and $v(\det(g)) = d - i$. For the first condition implies that $gV_0 \subset V_0$, the second that $XV_0 \subset gV_0$, and the three conditions imply that for some $k_1, k_2 \in GL(d, \mathbf{F}_q[[X]])$, $k_1 g k_2$ is a diagonal matrix, with i diagonal entries 1 and $d-i$ diagonal entries X. This implies that gV_0/XV_0 is an i-dimensional subspace of $V_0/XV_0 \cong \mathbf{F}_q^d$. Now let $u \in \Pi_i$. If we write $g_u g_{\lambda(u)} = c_u I$, then taking determinants and then valuations, we get $dv(c_u) = d - i + (d - (d-i)) = d$, so that $v(c_u) = 1$. Thus $Xg_u^{-1} = Xc_u^{-1} g_{\lambda(u)}$ has entries in $\mathbf{F}_q[[X]]$. Thus $[g_u v_0] \in \Pi_i(v_0)$ for each $u \in \Pi_i$. Thus condition (1) of Proposition 2.6 is satisfied.

We now suppose that, in addition, that $\mathcal{T}$ satisfies (1.1) and (1.2). Moreover, suppose that there are $a_0, \ldots, a_{d-1} \in \mathbf{F}_q[Z]$ so that the b_u, for $u \in \Pi_1$, are

$$b_u = u_0(a_0 + a_1\sigma + \cdots + a_{d-1}\sigma^{d-1})u_0^{-1} \qquad (3.6)$$

where $u_0 \in \mathbf{F}_{q^d}^\times$ and $u = u_0\mathbf{F}_q^\times \in \Pi_1$. Then (see the proof of Lemma 4.2 in [1]) to check condition (2) of Proposition 2.6, all we have to do is show that when $t \in \mathbf{F}_{q^d}^\times$, an equation $TA = AM_0$, where

$$T = \begin{pmatrix} t & 0 & \cdots & 0 \\ 0 & \varphi(t) & \cdots & 0 \\ \vdots & \vdots & \ddots & \vdots \\ 0 & 0 & \cdots & \varphi^{d-1}(t) \end{pmatrix} \quad \text{and} \quad A = \begin{pmatrix} \bar{a}_0 & \bar{a}_1 & \bar{a}_2 & \cdots & \bar{a}_{d-1} \\ \bar{a}_{d-1} & \bar{a}_0 & \bar{a}_1 & \cdots & \bar{a}_{d-2} \\ \vdots & \vdots & \ddots & \vdots & \vdots \\ \bar{a}_2 & \bar{a}_3 & \cdots & \bar{a}_0 & \bar{a}_1 \\ \bar{a}_1 & \bar{a}_2 & \bar{a}_3 & \cdots & \bar{a}_0 \end{pmatrix}$$

and where M_0 is a $d \times d$ matrix with entries in $\mathbf{F}_{q^d}$, can only happen if $t \in \mathbf{F}_q^\times$ (here $\bar{a}_i$ denotes a_i, reduced mod X, i.e., mod $Y = Z - 1$). This is certainly the case if the $\bar{a}_i$ are all equal to some $a \in \mathbf{F}_q^\times$. For if $t \notin \mathbf{F}_q$, we can find $s \in \mathbf{F}_{q^d}$ such that $\mathrm{Tr}(s) = 0$, but $\mathrm{Tr}(st) \neq 0$. If we multiply both sides of the equation $TA = AM_0$ on the left by the row vector $(s, s^q, \ldots, s^{q^{d-1}})$, then the right hand side equals the zero vector, while the left hand side is the constant vector whose entries are all $a\mathrm{Tr}(st) \neq 0$. We have therefore proved the following result:

Corollary 3.4 *Suppose that $\Pi = \Pi(\mathbf{F}_{q^d})$, and that we have an involution λ of Π such that $\lambda(\Pi_i) = \Pi_{n+1-i}$ for each i, and a family $\mathcal{T}$ of triples $(u, v, w) \in \Pi^3$ satisfying (A)-(E), and also (1.1) and (1.2), but perhaps not (F). Suppose that for each $u \in \Pi$ an element $b_u \in \mathbf{F}_{q^d}(Z)[\sigma]$ is given, so that (i)-(iii) above hold. Suppose also that each b_u, $u \in \Pi_1$, is of the form (3.6), where each $a_j \in \mathbf{F}_q[Z]$ is congruent to 1 mod $Z - 1$. Then $\mathcal{T}$ does satisfy axiom (F), and $u \mapsto [g_uV_0]$ is an isomorphism from Π onto the residue $\Pi(v_0)$ of $v_0 = [V_0]$ in $\Delta = \tilde{A}_n(\mathbf{F}_q((X)), v)$. Also, $a_u \mapsto g_u$ induces an isomorphism ψ of $\Gamma_{\mathcal{T}}$ onto a subgroup of $PGL(d, \mathbf{F}_q(X)) \subset \mathrm{Aut}_{\mathrm{tr}}(\Delta)$ which acts simply transitively on $\mathcal{V}_\Delta$, and $g \mapsto \psi(g)v_0$ is an isomorphism of $\Delta_{\mathcal{T}}$ onto Δ.*

§4. A family of $\widetilde{A}_3$-triangle presentations.

In this section, let $\Pi = \Pi(\mathbf{F}_{q^4})$. To define an $\widetilde{A}_3$-triangle presentation satisfying (1.1) and (1.2), we must first define an involution $\lambda : \Pi \to \Pi$ satisfying $\lambda(\Pi_i) = \Pi_{4-i}$ for $i = 1, 2, 3$, and also (3.1) and (3.2) above. Defining λ on Π_1 (and thus on Π_3) is easy, for by Corollary 3.2, we may as well assume that $\lambda(1) = \{1\}^\perp$. It is not so clear how λ should be defined on Π_2, and we start this section with some results about Π_2. Recall that $\mathcal{O}_0$ denotes the $\mathbf{F}_{q^4}^\times$-orbit of $V_0 = \mathbf{F}_{q^2}$ in Π_2 (see Lemma 3.3).

Lemma 4.1. Let $V \in \Pi_2$. Then if $V \in \Pi_2 \setminus \mathcal{O}_0$ [resp. $V \in \mathcal{O}_0$] there exist unique $a_V, \gamma_V \in \mathbf{F}_{q^4}^\times / \mathbf{F}_q^\times$ [resp. $a_V, \gamma_V \in \mathbf{F}_{q^4}^\times / \mathbf{F}_{q^2}^\times$] such that $V^\perp = a_V V$ and $\varphi^2(V) = \gamma_V V$.

(a) $a_{tV} = t^{-2} a_V$ and $\gamma_{tV} = t^{q^2-1} \gamma_V$ for any $V \in \Pi_2$ (and $t \in \mathbf{F}_{q^4}^\times / \mathbf{F}_q^\times$ or $t \in \mathbf{F}_{q^4}^\times / \mathbf{F}_{q^2}^\times$, according as $V \in \Pi_2 \setminus \mathcal{O}_0$ or $V \in \mathcal{O}_0$).

(b) If $V \in \Pi_2 \setminus \mathcal{O}_0$, then

$$a_V = \delta(V)^{q+q^2} \delta_2(V)^{2q+1} \tag{4.1}$$

and

$$\gamma_V = \delta(V)^{q+q^2} \delta_2(V)^{q+1}, \tag{4.2}$$

where $\delta(V)$ and $\delta_2(V)$ are defined in (3.3) and (3.4) above.

(c) If $V = V_0$, then $\gamma_V = 1$, and a_V is either 1 (if q is even) or (if q is odd) a_V is the unique element in $\mathbf{F}_{q^4}^\times / \mathbf{F}_{q^2}^\times$ of order 2. In view of (a), this gives us a_V and γ_V for any $V \in \mathcal{O}_0$.

Proof. Suppose first that $V = V_\alpha$, the span of 1 and $\alpha \in \mathbf{F}_{q^4} \setminus \mathbf{F}_q$. To show that $V^\perp = tV$ for some $t \in \mathbf{F}_{q^4}^\times$, we must show that for some t, $\mathrm{Tr}(t(a + b\alpha)(c + d\alpha)) = 0$ for each $a, b, c, d \in \mathbf{F}_q$. But this amounts to finding a nonzero $t \in \{1, \alpha, \alpha^2\}^\perp$. Now 1, α and α^2 span a subspace U of $\mathbf{F}_{q^4}$ of dimension at most 3. Thus $U^\perp \neq \{0\}$, and so we just need to pick any nonzero $t \in U^\perp$. The existence and uniqueness of a_V is now clear, as well as the first part of (a).

To verify (4.1) and (4.2), notice that $\xi_1 = \alpha^q - \alpha^{q^3}$ and $\xi_2 = \alpha^{q^2} \xi_1$ are in $\{1, \alpha\}^\perp$. For example,

$$\mathrm{Tr}(\xi_2 \alpha) = \mathrm{Tr}(\alpha^{1+q+q^2}) - \mathrm{Tr}(\alpha^{1+q^2+q^3}) = 0$$

because $\alpha^{1+q^2+q^3} = \varphi^2(\alpha^{1+q+q^2})$. Thus $r\xi_1 + s\xi_2 \in \{1, \alpha, \alpha^2\}^\perp$ for $r = \mathrm{Tr}(\alpha^2 \xi_2)$ and $s = -\mathrm{Tr}(\alpha^2 \xi_1)$. One may verify that

$$\mathrm{Tr}(\alpha^2 \xi_2)\xi_1 - \mathrm{Tr}(\alpha^2 \xi_1)\xi_2 = (\alpha - \alpha^{q^2})(\alpha^q - \alpha^{q^2})(\alpha^{q^2} - \alpha^{q^3})(\alpha^q - \alpha^{q^3})^2$$

which is nonzero if $\alpha \notin \mathbf{F}_{q^2}$, and the image of this in $\mathbf{F}_{q^4}^\times / \mathbf{F}_q^\times$ is $\delta(V_\alpha)^{q+q^2} \cdot \delta_2(V_\alpha)^{2q+1}$. This proves (4.1) when $V = V_\alpha$, $\alpha \notin \mathbf{F}_{q^2}$. To verify (4.2) in this case, we must check that $\varphi^2(V_\alpha) = \gamma V_\alpha$ for $\gamma = (\alpha - \alpha^q)^{q+q^2}(\alpha - \alpha^{q^2})^{q+1}$. This amounts to verifying that $t\gamma^{-1}\varphi^2(V_\alpha) = V_\alpha^\perp$ for $t = (\alpha - \alpha^q)^{q+q^2}(\alpha -$

$\alpha^{q^2})^{2q+1}$. But $t\gamma^{-1} = \alpha^q - \alpha^{q^3} = \xi_1$ for this t, and we have seen that both ξ_1 and $\xi_2 = \xi_1\alpha^{q^2}$ are in $V_\alpha^\perp$. Thus (4.2) is proved when $V = V_\alpha$, $\alpha \notin \mathbf{F}_{q^2}$.

Also, $\varphi(V_0) = V_0$, and so $\gamma_{V_0} = 1$. The existence of γ_V for all $V \in \Pi_2$, and the second part of (a) is now clear.

Using $\delta(tV) = t^{q+1}\delta(V)$ and $\delta_2(tV) = t^{q^2+1}\delta_2(V)$, and the above, we see that (4.1) and (4.2) hold.

It remains to calculate $\tau = a_{V_0}$. When q is even, $\text{Tr}(x) = 0$ for all $x \in \mathbf{F}_{q^2}$, because $\mathbf{F}_{q^4}$ is of degree 2 over $\mathbf{F}_{q^2}$. Thus $V_0^\perp = V_0$, and $a_{V_0} = 1$. So suppose that q is odd. Applying φ to both sides of $V_0^\perp = \tau V_0$, we have $\tau^q V_0 = \varphi(\tau V_0) = \varphi(V_0^\perp) = \varphi(V_0)^\perp = V_0^\perp = \tau V_0$. Thus $\tau^{q-1}V_0 = V_0$. Hence $\tau \in \mathbf{F}_{q^4}^\times/\mathbf{F}_{q^2}^\times$ has order m dividing $q - 1$. But m must also divide $|\mathbf{F}_{q^4}^\times/\mathbf{F}_{q^2}^\times| = q^2 + 1$. The greatest common divisor of $q - 1$ and $q^2 + 1$ is 2. Thus $\tau^2 = 1$. Now $\tau \neq 1$. For otherwise $V_0^\perp = V_0$, which implies that $1 \in V_0^\perp$, so that $4 = \text{Tr}(1) = 0$, which is false, as q is odd. Thus $a_{V_0} = \tau$ is the unique element in $\mathbf{F}_{q^4}^\times/\mathbf{F}_{q^2}^\times$ of order 2.

If in the last lemma we have $V = V_\alpha$, where $\alpha \in \mathbf{F}_{q^4} \setminus \mathbf{F}_{q^2}$, then (mod $\mathbf{F}_q^\times$)

$$a_V = \frac{(\alpha - \alpha^{q^2})(\alpha^q - \alpha^{q^3})}{(\alpha - \alpha^q)(\alpha^{q^3} - \alpha)}(\alpha^q - \alpha^{q^3}) \quad \text{and} \quad \gamma_V = \frac{(\alpha - \alpha^{q^2})(\alpha^q - \alpha^{q^3})}{(\alpha - \alpha^q)(\alpha^{q^3} - \alpha)}$$

as we see by dividing the expressions (4.1) and (4.2) by the norm of $\alpha - \alpha^q$.

Lemma 4.2. For $V \in \Pi_2$, let $c_V = 1/\delta(V^\perp)$. Then

(a) $c_{tV} = t^{q+1}c_V$ for $t \in \mathbf{F}_{q^4}^\times$;

(b) $c_{\varphi(V)} = c_V^q$;

(c) $\text{Tr}(1/(c_V x^{1+q})) = 0$ for any nonzero $x \in V^\perp$ (where "c_V" here is interpreted as any $t \in \mathbf{F}_{q^4}^\times$ whose image in $\mathbf{F}_{q^4}^\times/\mathbf{F}_q^\times$ is c_V);

(d) $c_V^{q+1}a_V^{q^2+q}\gamma_V = 1$ (where, when $V \in \mathcal{O}_0$, "c_V" here is interpreted as its image in $\mathbf{F}_{q^4}^\times/\mathbf{F}_{q^2}^\times$).

Proof. The statements (a) and (b) are immediate from the properties of $\delta(V)$ noted above. To prove (c), let $x \in V^\perp$ be nonzero. Pick any $y \in V^\perp$ such that x and y span $V^\perp$. Then c_V^{-1} is the image in $\mathbf{F}_{q^4}^\times/\mathbf{F}_q^\times$ of $x^q y - xy^q$. Hence $\text{Tr}(1/(c_V x^{1+q})) = \text{Tr}((y/x) - (y/x)^q) = 0$.

To prove (d), suppose first that $V \in \Pi_2 \setminus \mathcal{O}_0$. Now $\delta(V^\perp) = \delta(a_V V) = a_V^{q+1}\delta(V)$. Notice that $\delta_2(V)^{q^2} = \delta_2(V)$, and that the norm of $\alpha - \alpha^{q^2}$ is $(\alpha - \alpha^{q^2})^{2(q+1)}$, and so $\delta_2(V)^{2(q+1)} = 1$ for all $V \in \Pi_2 \setminus \mathcal{O}_0$. Using (4.1) and (4.2) and these facts, we find that $c_V = \delta(V)^{-q^2}\delta_2(V)^{-q-1}$. For the same reasons, we obtain

$$c_V^{q+1}a_V^{q^2+q} = \delta(V)^{-q-q^2}\delta_2(V)^{-q-1} = 1/\gamma_V$$

Suppose now that $V \in \mathcal{O}_0$. Now $c_{tV}^{q+1} a_{tV}^{q^2+q} \gamma_{tV} = c_V^{q+1} a_V^{q^2+q} \gamma_V$, so we need only show that $c_{V_0}^{q+1} a_{V_0}^{q^2+q} \gamma_{V_0} = 1$ in $\mathbf{F}_{q^4}^{\times}/\mathbf{F}_{q^2}^{\times}$. Note that $\delta(V_0) \in V_0 = \mathbf{F}_{q^2}$. When q is even, we have $a_{V_0} = \gamma_{V_0} = 1$, and $c_{V_0} \in \mathbf{F}_{q^2}^{\times}/\mathbf{F}_q^{\times}$, and we are finished. When q is odd, we have $\gamma_{V_0} = 1$ and $a_{V_0}^2 = 1$. Thus $a_{V_0}^{q^2+q} = 1$. Also $c_{V_0} = 1/\delta(V_0^{\perp}) = 1/\delta(a_{V_0} V_0) = 1/(a_{V_0}^{q+1} \delta(V_0)) \in \mathbf{F}_{q^2}^{\times}/\mathbf{F}_q^{\times}$, and again (d) holds.

We are now finally ready to define an $\widetilde{A_3}$-triangle presentation $\mathcal{T}$ satisfying (1.1) and (1.2). We first define a λ satisfying (3.1) and (3.2) as follows:

1. To define λ on Π_1, for $x_0 \in \mathbf{F}_{q^4}^{\times}$ we set $\lambda(x_0 \mathbf{F}_q) = x_0 \mathbf{F}_q^{\perp} = (x_0^{-1} \mathbf{F}_q)^{\perp} = \{y_0 \in \mathbf{F}_{q^4} : \mathrm{Tr}(y_0/x_0) = 0\}$.

2. To define λ on Π_3, for $x_0 \in \mathbf{F}_{q^4}^{\times}$ we set $\lambda((x_0 \mathbf{F}_q)^{\perp}) = x_0^{-1} \mathbf{F}_q$.

3. For $V \in \Pi_2$, we set $\lambda(V) = c_V \varphi(V^{\perp})$, where $c_V \in \mathbf{F}_{q^4}^{\times}/\mathbf{F}_q^{\times}$ was defined in Lemma 4.2.

It is immediate that (3.1) and (3.2) hold for $U \in \Pi_1 \cup \Pi_3$, and they hold for $U \in \Pi_2$ by Lemma 4.2(a) and (b). To verify that λ is an involution, we need only check that λ is an involution on Π_2. Now for $V \in \Pi_2$, we have, using Lemma 4.2(d),

$$
\begin{aligned}
\lambda(\lambda(V)) = \lambda(c_V \varphi(V^{\perp})) &= c_V \lambda(\varphi(V^{\perp})) \\
&= c_V \varphi(\lambda(V^{\perp})) \\
&= c_V \varphi(\lambda(a_V V)) \\
&= c_V a_V^q \varphi(\lambda(V)) \\
&= c_V a_V^q \varphi(c_V \varphi(V^{\perp})) \\
&= c_V^{1+q} a_V^{q+q^2} \varphi(\varphi(V)) \\
&= c_V^{1+q} a_V^{q+q^2} \gamma_V V \\
&= V
\end{aligned}
$$

We claim that the set consisting of the triples

$$(c_V x^q, x^{-1}, V) \quad \text{where } V \in \Pi_2, x = x_0 \mathbf{F}_q^{\times} \in \mathbf{F}_{q^4}^{\times}/\mathbf{F}_q^{\times} \text{ and } x_0 \in V^{\perp},$$
$$(4.3)$$

together with their cyclic permutations, is the half $\mathcal{T}'$ of an $\widetilde{A_3}$-triangle presentation $\mathcal{T}$ compatible with λ. Note that we are identifying Π_1 and $\mathbf{F}_{q^4}^{\times}/\mathbf{F}_q^{\times}$ here, as usual.

Let us verify that $\mathcal{T}$ is an $\widetilde{A_3}$-triangle presentation. We need only check properties (A), (B) and (C) of an $\widetilde{A_3}$-triangle presentation for $\mathcal{T}'$, plus verify property (F), because properties (D) and (E) clearly hold. It is easiest to check (F) at the end of the section, after embedding Γ_T in $PGL(4, \mathbf{F}_q((X)))$, by applying Corollary 3.4. Indeed, the only direct proof we have that (F) holds is long and messy.

We first check the "only if" part of property (A). Notice that if $(c_V x^q, x^{-1}, V)$ is as in (4.3), then x^{-1} is incident with $\lambda(c_V x^q)$, for this just means that $\mathrm{Tr}(1/(c_V x_0^{1+q})) = 0$, which is Lemma 4.2(c). Also, V is incident with $\lambda(x^{-1}) = x_0^{-1}\{1\}^\perp$ because $x_0 \in V^\perp$ is equivalent to $x_0 V \subset \{1\}^\perp$. Finally $c_V x^q$ is incident with $\lambda(V)$ because $\lambda(V) = c_V \varphi(V^\perp)$.

Let us now prove that T satisfies the "if" part of property (A). Suppose that $x = x_0 \mathbf{F}_q \in \Pi_1$ and $V \in \Pi_2$ are given, with V and $\lambda(x)$ incident. Then $V \subset x_0\{1\}^\perp$, and so $x_0^{-1} \in V^\perp$. Thus $(x, V, c_V x^{-q}) \in T$. Suppose next that $y = y_0 \mathbf{F}_q \in \Pi_1$ and $V \in \Pi_2$ are given with y and $\lambda(V)$ incident. Thus $y_0 \in \lambda(V) = c_V \varphi(V^\perp)$, and so $y = c_V x^q$ for some $x \in V^\perp$, and $(V, y, x^{-1}) = (V, c_V x^q, x^{-1}) \in T$. Finally, suppose that $x = x_0 \mathbf{F}_q \in \Pi_1$ and $y = y_0 \mathbf{F}_q \in \Pi_1$ are given with x and $\lambda(y)$ incident. Then $x_0 \in \lambda(y) = y_0\{1\}^\perp$, and so $\mathrm{Tr}(x_0/y_0) = 0$. Thus $x_0/y_0 = u - u^q$ for some $u \in \mathbf{F}_{q^4}$ (because the map $u \mapsto u - u^q$ is $\mathbf{F}_q$-linear $\mathbf{F}_{q^4} \to \{v \in \mathbf{F}_{q^4} : \mathrm{Tr}(v) = 0\}$, has 1-dimensional kernel $\mathbf{F}_q$, and is therefore onto). Let $z_0 = u/x_0$ and $V = \{x_0^{-1}, z_0\}^\perp$. Then

$$\frac{1}{x_0^q} z_0 - \frac{1}{x_0} z_0^q = \frac{1}{y_0 x_0^q} \neq 0$$

which shows that $1/x_0$ and z_0 are linearly independent, so that $V \in \Pi_2$, $x_0^{-1} \in V^\perp$, and c_V is the image in $\mathbf{F}_{q^4}^\times/\mathbf{F}_q^\times$ of $y_0 x_0^q$. Hence $y = c_V x^{-q}$, and $(y, x, V) = (c_V x^{-q}, x, V) \in T$.

We next check property (C). Suppose that $(c_V x^q, x^{-1}, V) \in T$ and $(c_V x^q, x^{-1}, U) \in T$. We must check that $U = V$. We know that $c_U = c_V$ and that $x = x_0 \mathbf{F}_q$, where $x_0 \in U^\perp$ and $x_0 \in V^\perp$. Pick $y_0, z_0 \in \mathbf{F}_{q^4}$ such that $U^\perp$ and $V^\perp$ are the spans of $\{x_0, y_0\}$ and $\{x_0, z_0\}$, respectively. By assumption, $c_U = c_V$, and so $x_0^q y_0 - x_0 y_0^q = a(x_0^q z_0 - x_0 z_0^q)$ for some $a \in \mathbf{F}_q^\times$. Thus $x_0^q(y_0 - az_0) = x_0(y_0 - az_0)^q$, so that $y_0 - az_0 = bx_0$ for some $b \in \mathbf{F}_q$. Thus $y_0 = bx_0 + az_0 \in V^\perp$. Thus $U^\perp = V^\perp$, and so $U = V$. Suppose next that both $(V, y, x) \in T$ and $(V, y, x') \in T$ (where $V \in \Pi_2$ and $x, x', y \in \Pi_1 = \mathbf{F}_{q^4}^\times/\mathbf{F}_q^\times$). Then $y = c_V x^{-q}$ and also $y = c_V x'^{-q}$. Thus $x^q = x'^q$, so that $x = x'$. Finally, if $(x, V, y) \in T$ and also $(x, V, y') \in T$ (where $V \in \Pi_2$ and $x, x', y \in \mathbf{F}_{q^4}^\times/\mathbf{F}_q^\times$), then $y = c_V x^{-q} = y'$.

Embedding Γ_T in $PGL(4, \mathbf{F}_q(X))$.

Consider the cyclic simple algebra $\mathcal{A} = \mathbf{F}_{q^4}(Y)[\sigma]$ defined in Section 3, taking $d = 4$ here. Recall that each element of $\mathcal{A}$ can be written uniquely in the form (3.5). We check the conditions of Corollary 3.4 by mapping each

a_u $(u = u_0 \mathbf{F}_q \in \Pi_1)$ to

$$b_u = u_0(Z + \sigma + \sigma^2 + \sigma^3)u_0^{-1} \tag{4.4}$$

(note that this depends only on u, not u_0), and mapping each a_V $(V \in \Pi_2)$ to

$$b_V = Z + \left(\frac{x_1^{q^3} x_2^{q^2} - x_1^{q^2} x_2^{q^3}}{x_1 x_2^{q^3} - x_1^{q^3} x_2}\right)\sigma^2 + \left(\frac{x_1^{q^3} x_2^{q} - x_1^{q} x_2^{q^3}}{x_1^{q} x_2 - x_1 x_2^{q}}\right)\sigma^3 \tag{4.5}$$

where $\{x_1, x_2\}$ is any basis of $V^\perp$. As noted before (3.3), the coefficients of σ^2 and σ^3 in (4.5) depend only on V, not on the particular basis chosen for $V^\perp$.

One may easily verify that for $u = u_0 \mathbf{F}_q \in \Pi_1$,

$$b_u^{-1} = Z^{-1}(Z - 1)^{-1}u_0(Z - \sigma^3)u_0^{-1}.$$

We therefore also map a_W to $b_W = u_0(Z - \sigma^3)u_0^{-1}$ for each $W = \lambda(u) \in \Pi_3$. If $u = u_0 \mathbf{F}_q \in \Pi_1$ and $v = v_0 \mathbf{F}_q \in \Pi_1$, then a simple calculation gives

$$b_u^{-1}b_v^{-1} = Z^{-1}(Z - 1)^{-2}\left(Z + u_0^{1-q^3} v_0^{q^3-q^2}\sigma^2 - (u_0^{1-q^3} + v_0^{1-q^3})\sigma^3\right).$$

Suppose now that $x \in V^\perp$, and that $u = x^{-1}$ and $v = cvx^q$. Then pick any basis $\{x_1, x_2\}$ for $V^\perp$ with $x = x_1 \mathbf{F}_q$. We can take $u_0 = x_1^{-1}$ and $v_0 = x_1^q/(x_1^q x_2 - x_1 x_2^q)$. Some routine algebra yields

$$u_0^{1-q^3} + v_0^{1-q^3} = -\frac{x_1^{q^3} x_2^{q} - x_1^{q} x_2^{q^3}}{x_1^{q} x_2 - x_1 x_2^{q}}$$

and

$$u_0^{1-q^3} v_0^{q^3-q^2} = \frac{x_1^{q^3} x_2^{q^2} - x_1^{q^2} x_2^{q^3}}{x_1 x_2^{q^3} - x_1^{q^3} x_2}.$$

It follows that

$$b_v b_u b_V = Z(Z - 1)^2 \in Z(\mathcal{A}^\times). \tag{4.6}$$

Also,

$$\lambda(V)^\perp = \left(c_V \varphi(V^\perp)\right)^\perp = c_V^{-1}\varphi(V) = c_V^{-1} a_V^{-q}\varphi(V^\perp).$$

Using $c_V^{-1} = \delta(V^\perp)$ and (if $V \notin \mathcal{O}_0$) $a_V = a_{V^\perp}^{-1} = \delta(V^\perp)^{-q^2-q}\delta_2(V^\perp)^{-2q-1}$, we find that $c_V a_V^q = \delta(V^\perp)^q \delta_2(V^\perp)^q$. So if x_1 and x_2 span $V^\perp$, then y_1 and y_2 span $\lambda(V)^\perp$, where

$$y_j = \frac{x_j^{q}}{(x_1^{q^2} x_2^{q} - x_1^{q} x_2^{q^2})(x_1^{q^3} x_2^{q} - x_1^{q} x_2^{q^3})} \qquad \text{(for } j = 1, 2).$$

After a little algebra, we then find that

$$b_{\lambda(V)} = Z + \left(\frac{x_1^q x_2^{q^2} - x_1^{q^2} x_2^q}{x_1^q x_2 - x_1 x_2^q}\right)\sigma^2 + \left(\frac{x_1^q x_2^{q^3} - x_1^{q^3} x_2^q}{x_1^q x_2 - x_1 x_2^q}\right)\sigma^3 .$$

More routine calculations show that

$$b_V b_{\lambda(V)} = Z(Z-1) \in Z(\mathcal{A}^\times). \qquad (4.7)$$

The verification of (4.7) is a little different when $V \in \mathcal{O}_0$. For if $V = tV_0$, where $t \in \mathbf{F}_{q^4}^\times$, then $V^\perp = t^{-1}V_0^\perp = t^{-1}a_{V_0}V_0 = sV_0$, say, for some $s \in \mathbf{F}_{q^4}^\times$. We can take $x_1 = s$ and $x_2 = s\alpha$, for any $\alpha \in \mathbf{F}_{q^2} \setminus \mathbf{F}_q$. Then the coefficient of σ^3 in (4.5) is 0, and the coefficient of σ^2 in (4.5) is $-s^{q^2-1}$. Now $a_{V_0} \in \mathbf{F}_{q^4}^\times/\mathbf{F}_{q^2}^\times$ is either 1 (if q is even) or of order 2 (if q is odd), by Lemma 4.1(c). Thus $-s^{q^2-1} = -1/t^{q^2-1}$, whatever the parity of q. Thus

$$b_V = Z + \frac{1}{t^{q^2-1}}\sigma^2 .$$

Now $\lambda(V) = t\lambda(V_0) = tc_{V_0}\varphi(V_0)^\perp = tc_{V_0}V_0^\perp$. Thus $\lambda(V)^\perp = (tc_{V_0})^{-1}V_0 = s'V_0$, say. Then $y_1 = s'$ and $y_2 = s'\alpha$ span $\lambda(V)^\perp$. The coefficient of σ^2 in $b_{\lambda(V)}$ is thus $-(s')^{q^2-1}$. Now $1/c_{V_0} = \delta(V_0^\perp) = a_{V_0}^{q+1}\delta(V_0) = \delta(V_0)$. Using the fact that $\delta(V_0) \in \mathbf{F}_{q^2}^\times$, we see that $c_{V_0} = 1$. Hence $(s')^{q^2-1} = 1/t^{q^2-1}$, and so

$$b_{\lambda(V)} = Z - \frac{1}{t^{q^2-1}}\sigma^2 ,$$

and (4.7) for this case is easily verified.

It follows from (4.6) and (4.7) that condition (i) of Corollary 3.4 is satisfied. Clearly condition (ii) is satisfied. Now (4.4) implies that $N(b_u) = N(b_1)$ for all $u \in \Pi_1$. Thus (4.6) shows that $N(b_V)$ is independent of $V \in \Pi_2$. Then (4.7) implies that $N(b_V) = \pm Z^2(Z-1)^2$. Now (4.6) shows that $N(b_1)^2 = \pm Z^2(Z-1)^6$. Remembering that $Z-1 = Y = t_1 X + t_2 X^2 + t_3 X^3 + t_4 X^4$, where $t_1 \neq 0$, we see that each $N(b_u)$, $u \in \Pi_1$, has valuation 3, each $N(b_V)$, $V \in \Pi_2$, has valuation 2, and each $N(b_W)$, $W \in \Pi_3$, has valuation 1. So condition (iii) of Corollary 3.4 is satisfied. In fact, using the explicit embedding of $\mathcal{A}$ in $M_{4\times4}(\mathbf{F}_{q^4}(Z))$ described last section, we can calculate that $N(b_u) = Z(Z-1)^3$ for each $u \in \Pi_1$, and so $N(b_V) = Z^2(Z-1)^2$ and $N(b_W) = Z^3(Z-1)$ for each $V \in \Pi_2$ and $W \in \Pi_3$.

Finally, the coefficients Z, 1, 1 and 1 in (4.4) are all congruent to 1 mod $Z-1$. Thus Corollary 3.4 may be applied, showing that $\mathcal{T}$ satisfies (F), and that $\Delta_{\mathcal{T}} \cong \widetilde{A_3}(\mathbf{F}_q((X)), v)$, etc.

Arithmeticity of $\Gamma_{\mathcal{T}}$.

If $s \in \mathbf{F}_{q^4}$, and $j \in \{0, 1, 2, 3\}$, then

$$(Z + \sigma + \sigma^2 + \sigma^3)\, s\sigma^j (Z + \sigma + \sigma^2 + \sigma^3)^{-1}$$

$$= \frac{1}{Z - 1}\left((Z - s^q) + (s^q - s^{q^2})\sigma + (s^{q^2} - s^{q^3})\sigma^2 + (s^{q^3} - 1)\sigma^3\right)\sigma^j \ .$$

It follows that if $\{s_0, s_1, s_2, s_3\}$ is a basis of $\mathbf{F}_{q^4}$ over $\mathbf{F}_q$, so that $\{s_i\sigma^j : i, j = 0, 1, 2, 3\}$ is a basis for $\mathcal{A} = \mathbf{F}_{q^4}(Z)[\sigma]$ over $\mathbf{F}_q(Z)$, then the automorphisms $\xi \mapsto b_u \xi b_u^{-1}$ ($u \in \Pi_1$) of $\mathcal{A}$ have matrices with respect to this basis which have entries in $\mathbf{F}_q[1/Y]$ (recall that $Y = Z - 1$). By Lemma 2.8, the same is true for all $u \in \Pi$. Hence if $\mathbf{A} = \mathrm{Aut}(\mathcal{A})$, then the adjoint representation maps $\Gamma_{\mathcal{T}}$ into $\mathbf{A}(\mathbf{F}_q[1/Y]) = \mathbf{A} \cap GL(16, \mathbf{F}_q[1/Y])$. This exhibits the arithmeticity of $\Gamma_{\mathcal{T}}$ (see [1, §4] for more details).

Families satisfying (A)–(E), but not (F).

If we use the polarity $\alpha(U) = U^{\perp}$, $U \in \Pi$, and let $\mathcal{T}^*$ denote the $\widetilde{A_3}$-triangle presentation $\alpha\lambda(\mathcal{T}^{\mathrm{rev}})$ (see Lemma 2.7(b)), then for the $\mathcal{T}$ defined above, the half $(\mathcal{T}^*)'$ of $\mathcal{T}^*$ consists of triples

$$(c_V x, x^{-q}, \varphi(V)), \tag{4.8}$$

where $V \in \Pi_2$ and $x \in V^{\perp}$, together with their cyclic permutations. Moreover, $\mathcal{T}^*$ is compatible with λ^*, say, where λ^* agrees with λ on $\Pi_1 \cup \Pi_3$, and is given for $V \in \Pi_2$ by

$$\lambda^*(V) = c_{V^{\perp}}^{-1} \varphi(V^{\perp}).$$

Suppose that we take a subset Π_2' of Π_2 which satisfies (i): if $V \in \Pi_2'$, then $tV \in \Pi_2'$ for each $t \in \mathbf{F}_{q^4}^{\times}$; and (ii): if $V \in \Pi_2'$, then $\varphi(V) \in \Pi_2'$. Suppose we define an involution $\tilde{\lambda}$ of Π which agrees with λ on $\Pi_1 \cup \Pi_3$, and on Π_2 satisfies

$$\tilde{\lambda}(V) = \begin{cases} \lambda(V) & \text{if } V \in \Pi_2', \\ \lambda^*(V) & \text{if } V \in \Pi_2 \setminus \Pi_2'. \end{cases}$$

Form a family $\tilde{\mathcal{T}}$ of triples consisting of the triples (4.3), but only for $V \in \Pi_2'$, and the triples (4.8) for $V \in \Pi_2 \setminus \Pi_2'$, together with the triples $(\tilde{\lambda}(w), \tilde{\lambda}(v), \tilde{\lambda}(u))$, where (u, v, w) is one the triples already given, and the cyclic permutations of all these triples. Then one may check that $\tilde{\mathcal{T}}$ satisfies properties (A)–(E), and also the invariance properties (1.1) and (1.2). However, $\tilde{\mathcal{T}}$ does not satisfy (F), in general. To give a concrete example, take $q = 3$. Then $\mathbf{F}_{81} = \mathbf{F}_3(\theta)$, where $\theta^4 = -\theta + 1$. Then Π_2 divides into

4 orbits under the action of $\mathbf{F}_{81}^{\times}$ (cf. Lemma 3.3): the orbit $\mathcal{O}_0$ of $V_0 = \mathbf{F}_9 = \{0, \pm 1, \pm \theta^{10}, \pm \theta^{20}, \pm \theta^{30}\}$, the orbit $\mathcal{O}_1$ of $V_1 = \{0, \pm 1, \pm \theta, \pm \theta^4, \pm \theta^{13}\}$, the orbit $\mathcal{O}_2$ of $V_2 = \{0, \pm 1, \pm \theta^{15}, \pm \theta^{22}, \pm \theta^{38}\}$, and the orbit $\mathcal{O}_3$ of $V_3 = \varphi(V_2)$. Notice that $\varphi(V_0) = V_0$, $\varphi(V_1) = \theta^{-1} V_1$, $\varphi(V_2) = V_3$ and $\varphi(V_3) = V_2$. Also, $V_0^{\perp} = \theta^5 V_0$, $V_1^{\perp} = \theta V_1$, $V_2^{\perp} = \theta^5 V_2$ and $V_3^{\perp} = \theta^{15} V_3$, while $\lambda(V_0) = \theta^5 V_0$, $\lambda(V_1) = \theta^{20} V_1$, $\lambda(V_2) = V_3$ and $\lambda(V_3) = V_2$. Using the invariance properties (1.1) and (1.2), all the triples in T and T^* can be found from the lists

$$(\theta^5, \theta^{25}, V_0) \in T \qquad\qquad (\theta^5, \theta^{25}, V_0) \in T^*$$
$$(\theta^{20}, \theta^{26}, V_1) \in T \qquad\qquad (\theta^{20}, \theta^{38}, V_1) \in T^*$$
$$(\theta^{21}, \theta^{39}, V_1) \in T \qquad\qquad (\theta^{21}, \theta^{35}, V_1) \in T^*$$
$$(\theta^{24}, \theta^{38}, V_1) \in T \qquad\qquad (\theta^{24}, \theta^{26}, V_1) \in T^*$$
$$(\theta^{33}, \theta^{35}, V_1) \in T \quad\text{and}\quad (\theta^{33}, \theta^{39}, V_1) \in T^*$$
$$(1, \theta^{35}, V_2) \in T \qquad\qquad (\theta^4, \theta^{13}, V_2) \in T^*$$
$$(\theta^5, \theta^{20}, V_2) \in T \qquad\qquad (\theta^{10}, \theta^{35}, V_2) \in T^*$$
$$(\theta^{26}, \theta^{13}, V_2) \in T \qquad\qquad (\theta^{15}, \theta^{20}, V_2) \in T^*$$
$$(\theta^{34}, \theta^{37}, V_2) \in T \qquad\qquad (\theta^{36}, \theta^{37}, V_2) \in T^*$$

For brevity, we have written simply "θ^i" here to denote $\theta^i \mathbf{F}_3^{\times} \in \mathbf{F}_{81}^{\times}/\mathbf{F}_3^{\times}$, or equivalently, the 1-dimensional subspace $\{0, \theta^i, -\theta^i\}$ spanned by θ^i. These lists could be shortened slightly; for example, that $(\theta^{34}, \theta^{37}, V_2) \in T$ follows by applying φ twice to $(\theta^{26}, \theta^{13}, V_2)$.

If we take $\Pi_2' = \mathcal{O}_0 \cup \mathcal{O}_2 \cup \mathcal{O}_3$, then the set $\tilde{T}$ of triples will contain

$$(\theta^5, \theta^{25}, V_0) \qquad\qquad (1, \theta^{35}, V_2)$$
$$(\theta^{20}, \theta^{38}, V_1) \qquad\qquad (\theta^5, \theta^{20}, V_2)$$
$$(\theta^{21}, \theta^{35}, V_1) \quad\text{and}\quad (\theta^{26}, \theta^{13}, V_2)$$
$$(\theta^{24}, \theta^{26}, V_1) \qquad\qquad (\theta^{34}, \theta^{37}, V_2)$$
$$(\theta^{33}, \theta^{39}, V_1)$$

The family $\tilde{T}$ does not satisfy (F). For $\tilde{\lambda}(V_1) = \lambda^*(V_1) = \theta^{20} V_1 = \lambda(V_1)$. So $(\theta^{20}, \theta^{38}, V_1) \in \tilde{T}$ and $(\theta^1, \theta^{15}, \tilde{\lambda}(V_1)) \in \tilde{T}$. We have $(\theta^{15}, \theta^{20}, V_3) \in \tilde{T}$ and $(\theta^{38}, \theta^1, \theta^4 V_2) \in \tilde{T}$, but $\theta^4 V_2 \neq V_2 = \tilde{\lambda}(V_3)$, so property (F') fails.

Remark. By Lemmas 4.1 and 3.3, for general q, φ acts as an involution on the set $\{\mathcal{O}_0, \mathcal{O}_1, \ldots, \mathcal{O}_q\}$ of $\mathbf{F}_{q^d}^{\times}$-orbits in Π_2. Clearly it fixes $\mathcal{O}_0$. One can show that if q is even, then φ fixes no other $\mathcal{O}_i$, and that if q is odd, then φ fixes exactly one other $\mathcal{O}_i$, say $\mathcal{O}_1$. The parts of T and T^* involving $V \in \mathcal{O}_0$ coincide, so we may as well assume that $\mathcal{O}_0 \subset \Pi_2'$. There are thus

$2^{q/2}$ possible subsets Π_2' if q is even, and $2^{(q+1)/2}$ if q is odd, and there-fore this many families $\tilde{T}$. Calculations with the first few prime powers q suggests that (a): these families $\tilde{T}$ are the only families satisfying (A)–(E) as well as (1.1) and (1.2), and compatible with a λ satisfying $\lambda(1) = \{1\}^{\perp}$ (see Corollary 3.2), and (b): of these families $\tilde{T}$, only two, T and T^* also satisfy (F).

§5. A family of $\widetilde{A_4}$-triangle presentations.

Now let $\Pi = \Pi(\mathbf{F}_{q^5})$. We define an involution $\lambda : \Pi \to \Pi$ satisfying $\lambda(\Pi_i) = \Pi_{5-i}$ for $i = 1, 2, 3, 4$, and (3.1) and (3.2) as follows:

1. To define λ on Π_1, for $x_0 \in \mathbf{F}_{q^4}^{\times}$ we set

$$\lambda(x_0\mathbf{F}_q) = x_0\mathbf{F}_q^{\perp} = (x_0^{-1}\mathbf{F}_q)^{\perp} = \{y_0 \in \mathbf{F}_{q^5} : \mathrm{Tr}(y_0/x_0) = 0\}$$

2. For $U \in \Pi_2$, we set $\lambda(U) = k_U \varphi^3(U^{\perp})$ where k_U is defined as the image in $\mathbf{F}_{q^5}^{\times}/\mathbf{F}_q^{\times}$ of $u_1^{q^3}u_2 - u_1 u_2^{q^3}$, for any basis $\{u_1, u_2\}$ of U.

3. For $V \in \Pi_3$, we set $\lambda(V) = \ell_V \varphi^2(V^{\perp})$ where ℓ_V is defined as the image in $\mathbf{F}_{q^5}^{\times}/\mathbf{F}_q^{\times}$ of $1/(u_1^{q^2}u_2 - u_1 u_2^{q^2})$ for any basis $\{u_1, u_2\}$ of $U = V^{\perp}$.

4. To define λ on Π_4, for $x_0 \in \mathbf{F}_{q^4}^{\times}$ we set $\lambda((x_0\mathbf{F}_q)^{\perp}) = x_0^{-1}\mathbf{F}_q$.

In this case, it is routine to verify that $\lambda : \Pi \to \Pi$ is an involution sat-isfying (3.1) and (3.2)—just observe that $k_{tU} = t^{q^3+1}k_U$, $k_{\varphi(U)} = \varphi(k_U)$, $\ell_{tV} = t^{q^2+1}\ell_V$, $\ell_{\varphi(V)} = \varphi(\ell_V)$ and that $\ell_V = 1/\varphi^2(k_{V^{\perp}})$ for each $U \in \Pi_2$ and $V \in \Pi_3$.

We claim that the set consisting of the triples

$$(\ell_V x^{q^2}, x^{-1}, V) \quad \text{where } V \in \Pi_3, x = x_0\mathbf{F}_q^{\times} \in \mathbf{F}_{q^4}^{\times}/\mathbf{F}_q^{\times} \text{ and } x_0 \in V^{\perp} \quad (5.1)$$

and the triples

$$(\frac{1}{k_U^{q^4}k_W^{q^3}}, U, W) \quad \text{where } U, W \in \Pi_2, \text{ and } W \subset \lambda(U), \qquad (5.2)$$

together with their cyclic permutations, is the T' of an $\widetilde{A_4}$-triangle presen-tation T compatible with λ and satisfying (1.1) and (1.2).

Checking that (1.1) holds is easy, when we notice that $(t^{q^3+1})^{q^4+q^3}$ equals $N_{\mathbf{F}_{q^5}/\mathbf{F}_q}(t)/t$ for $t \in \mathbf{F}_{q^5}^{\times}$. Clearly (1.2) holds.

Checking (C).

For the triples of the form (5.1) and their cyclic permutations, (C) is checked in the same way as it was for the $\widetilde{A_3}$-triangle presentations of Sec-tion 4. For the triples of the form (5.2) and their cyclic permutations, we have to show three things: (i) if $(v, U, W), (v, U, W') \in T'$, where $v \in \Pi_1$ and $U, W, W' \in \Pi_2$, then $W = W'$; (ii) if $(v, U, W), (v, U', W) \in T'$, where $v \in \Pi_1$ and $U, U', W \in \Pi_2$, then $U = U'$; and (iii) if $(v, U, W), (v', U, W) \in T'$, where $v, v' \in \Pi_1$ and $U, W \in \Pi_2$, then $v = v'$. In (i), notice that W and W' are two 2-dimensional subspaces of the 3-dimensional subspace $\lambda(U)$. So $W \cap W' \neq \{0\}$. Choose a nonzero $x \in W \cap W'$, and then $y \in W$ and $y' \in W'$ so that W is spanned by $\{x, y\}$, and W' by $\{x, y'\}$. Now $(v, U, W), (v, U, W') \in T'$, and so $v = 1/k_U^{q^4} k_W^{q^3}$ and also $v = 1/k_U^{q^4} k_{W'}^{q^3}$. Thus $k_W = k_{W'}$. Hence $x^{q^3} y - x y^{q^3} = a(x^{q^3} y' - x y'^{q^3})$ for some $a \in \mathbf{F}_q^{\times}$. Thus $x^{q^3}(y - ay') = x(y - ay')^{q^3}$, so that $y - ay' = bx$ for some $b \in \mathbf{F}_q$. Thus $y \in W'$, so that $W = W'$, and (i) is proved. Similarly, the hypotheses in (ii) imply that $k_U = k_{U'}$, and that $W \subset \lambda(U) \cap \lambda(U')$. These then imply that U and U' are two 2-dimensional subspaces of the 3-dimensional subspace $\left(\varphi^2(k_U^{-1} W)\right)^{\perp}$. As before, this and $k_U = k_{U'}$ force $U = U'$ to hold. Finally, (iii) is clear, as $v = 1/k_U^{q^4} k_{W'}^{q^3} = v'$.

Checking (A).

To check the "only if" part of (A) for the triples (5.1) and their cyclic permutations, we must check that if $x = x_0 \mathbf{F}_q^{\times}$, where $x_0 \in V^{\perp}$, then $x_0^{-1} \in \lambda(\ell_V x^{q^2})$, $V \subset \lambda(x^{-1})$ and $\ell_V x^{q^2} \in \lambda(V)$. Only the first of these is nontrivial, and it is proved as in Lemma 4.2(c).

To check the "only if" part of (A) for the triples (5.2) and their cyclic permutations, we must check that if $U, W \in \Pi_2$, and $W \subset \lambda(U)$, then $k_U^{q^4} k_W^{q^3} \in U^{\perp}$ and $1/k_U^{q^4} k_W^{q^3} \in \lambda(W)$. Let u_1, u_2 span U. Then W is spanned by $w_j = k_U v_j^{q^3}$, $j = 1, 2$, for some $v_1, v_2 \in U^{\perp}$. Thus $k_W = k_U^{q^3+1}(v_1^q v_2^{q^3} - v_1^{q^3} v_2^q)$, and so $k_U^{q^4} k_W^{q^3} = k_U^{q+q^3+q^4}(v_1^{q^4} v_2^q - v_1^q v_2^{q^4})$. To show that $k_U^{q^4} k_W^{q^3} \in U^{\perp}$, we must check that $\mathrm{Tr}(x_j) = 0$ for $j = 1, 2$, where

$$x_j = u_j \left(u_1^{q^3} u_2 - u_1 u_2^{q^3}\right)^{q+q^3+q^4} \left(v_1^{q^4} v_2^q - v_1^q v_2^{q^4}\right).$$

When we write down $\mathrm{Tr}(x_j) = x_j + x_j^q + x_j^{q^2} + x_j^{q^3} + x_j^{q^4}$, we get a complicated expression involving $v_j^{q^k}$ for $j = 1, 2$ and $k = 0, 1, 2, 3, 4$. Whenever $v \in U^{\perp}$, by solving the equations $\mathrm{Tr}(u_j v) = 0$ for $j = 1, 2$, we see that

$$v^{q^3} = c_1 v + c_2 v^q + c_3 v^{q^2} \quad \text{and} \quad v^{q^4} = c_4 v + c_5 v^q + c_6 v^{q^2}$$

for

$$c_1 = \frac{u_2^{q^4} u_1 - u_1^{q^4} u_2}{u_1^{q^4} u_2^{q^3} - u_2^{q^4} u_1^{q^3}}, \quad c_2 = \frac{u_2^{q^4} u_1^q - u_1^{q^4} u_2^q}{u_1^{q^4} u_2^{q^3} - u_2^{q^4} u_1^{q^3}}, \quad c_3 = \frac{u_2^{q^4} u_1^{q^2} - u_1^{q^4} u_2^{q^2}}{u_1^{q^4} u_2^{q^3} - u_2^{q^4} u_1^{q^3}}$$

and

$$c_4 = \frac{u_1^{q^3} u_2 - u_2^{q^3} u_1}{u_1^{q^4} u_2^{q^3} - u_2^{q^4} u_1^{q^3}}, \quad c_5 = \frac{u_1^{q^3} u_2^q - u_2^{q^3} u_1^q}{u_1^{q^4} u_2^{q^3} - u_2^{q^4} u_1^{q^3}}, \quad c_6 = \frac{u_1^{q^3} u_2^{q^2} - u_2^{q^3} u_1^{q^2}}{u_1^{q^4} u_2^{q^3} - u_2^{q^4} u_1^{q^3}}$$

Applying these to $v = v_1$ and $v = v_2$, and substituting into the above mentioned complicated expression for $\text{Tr}(x_j)$, we obtain 0, as desired. In fact, $\text{Tr}(x_j)$ is a linear combination of $v_1^q v_2 - v_1 v_2^q$, $v_1^{q^2} v_2 - v_1 v_2^{q^2}$ and $v_1^{q^2} v_2^q - v_1^q v_2^{q^2}$, and the three coefficients are all 0. Thus $k_U^{q^4} k_W^{q^3} \in U^\perp$.

Next we must check that $1/k_U^{q^4} k_W^{q^3} \in \lambda(W)$ if $W \subset \lambda(U)$. This amounts to showing that $1/k_U^q k_W^{q^2+1} \in W^\perp$, and so we must show that $w_j/k_U^q k_W^{q^2+1}$ has trace 0 for $j = 1, 2$. Using the formula $k_W = k_U^{q^3+1}(v_1^q v_2^{q^3} - v_1^{q^3} v_2^q)$ derived above, we are reduced to showing that $\text{Tr}(y_j) = 0$ for $j = 1, 2$, where

$$y_j = v_j \left(u_1^{q^3} u_2 - u_1 u_2^{q^3}\right)^q \left(v_1^q v_2^{q^3} - v_1^{q^3} v_2^q\right)^{1+q+q^3}$$

But y_j is just x_j, as defined above, with the u's and v's interchanged.

The proof of the "if" part of (A) for the triples (5.1) and their cyclic permutations is similar to the proof of the corresponding fact for the triples (4.3). The only nontrivial thing is this: suppose that $x = x_0 \mathbf{F}_q \in \Pi_1$ and $y = y_0 \mathbf{F}_q \in \Pi_1$ are given with x and $\lambda(y)$ incident. Then $x_0 \in \lambda(y) = y_0\{1\}^\perp$, and so $\text{Tr}(x_0/y_0) = 0$. Thus $x_0/y_0 = u - u^{q^2}$ for some $u \in \mathbf{F}_{q^4}$. Let $z_0 = u/x_0$ and $V = \{x_0^{-1}, z_0\}^\perp$. Then

$$\frac{1}{x_0^{q^2}} z_0 - \frac{1}{x_0} z_0^{q^2} = \frac{1}{y_0 x_0^{q^2}} \neq 0$$

which shows that $1/x_0$ and z_0 are linearly independent, so that $V \in \Pi_3$, $x_0^{-1} \in V^\perp$, and ℓ_V is the image in $\mathbf{F}_{q^5}^\times/\mathbf{F}_q^\times$ of $y_0 x_0^{q^2}$. Hence $y = \ell_V x^{-q^2}$, and $(y, x, V) = (\ell_V x^{-q^2}, x, V) \in \mathcal{T}$.

As for the proof of the "if" part of (A) for the triples (5.2) and their cyclic permutations, we must show that (i) if $x \in \Pi_1$, $U \in \Pi_2$, and $U \subset \lambda(x)$, then there is a $W \in \Pi_2$ such that $W \subset \lambda(U)$ and $x = 1/k_U^{q^4} k_W^{q^3}$, and (ii) if $x \in \Pi_1$, $W \in \Pi_2$, and $x \subset \lambda(W)$, then there is a $U \in \Pi_2$ such that $W \subset \lambda(U)$ and $x = 1/k_U^{q^4} k_W^{q^3}$. These are most quickly verified "nonconstructively", using (C) and counting. For there are $|\Pi_2|(q^2 + q + 1)$ pairs (U, W) such that $U, W \in \Pi_2$ and $W \subset \lambda(U)$, since for each fixed $W \in \Pi_2$ there are $q^2 + q + 1$ 1-dimensional subspaces of $\mathbf{F}_{q^5}/W$. Hence the set of triples (5.2) has $|\Pi_2|(q^2 + q + 1)$ elements. By the "only if" part of (A) and by (C), $(1/k_U^{q^4} k_W^{q^3}, U, W) \mapsto (1/k_U^{q^4} k_W^{q^3}, U)$ is a 1–1 map of this set into the set of pairs (x, U), where $x \in \Pi_1$, $U \in \Pi_2$ and $U \subset \lambda(x)$. But this last set also has $|\Pi_2|(q^2 + q + 1)$ elements. So the map is onto, and (i) is proved. The proof of (ii) is similar.

Properties (B), (D) and (E) are clearly satisfied, while (F) will be verified by an application of Corollary 3.4, once Γ_T has been embedded in $PGL(5, \mathbf{F}_q((X)))$ as described in the next subsection.

Embedding Γ_T in $PGL(5, \mathbf{F}_q(X))$.

Consider the cyclic simple algebra $\mathcal{A} = \mathbf{F}_{q^5}(Z)[\sigma]$ defined in Section 3 above, taking $d = 5$ here. Recall that each element of $\mathcal{A}$ can be written uniquely in the form (3.5). We now check the conditions of Corollary 3.4. We map each a_u ($u = u_0 \mathbf{F}_q \in \Pi_1$) to

$$b_u = u_0(Z^2 + Z\sigma + \sigma^2 + Z\sigma^3 + \sigma^4)u_0^{-1}$$

(note that this depends only on u, not u_0). A simple calculation shows that b_u has norm $Z^2(Z^2 - 1)^4$, and inverse $b_u^{-1} = Z^{-1}(Z^2 - 1)^{-1}u_0(Z - \sigma^3)u_0^{-1}$. So if $T = \lambda(u) \in \Pi_4$, we map a_T to $b_T = u_0(Z - \sigma^3)u_0^{-1}$, which has norm $Z^3(Z^2 - 1)$. Then $b_u b_{\lambda(u)} = b_{\lambda(u)} b_u = Z(Z^2 - 1)$ for each $u \in \Pi_1$. We map each a_U ($U \in \Pi_2$) to

$$b_U = Z^2 + \frac{Z(u_1^{q^4} u_2 - u_1 u_2^{q^4})}{u_1^{q^4} u_2^q - u_1^q u_2^{q^4}}\sigma - \frac{Z(u_1^q u_2 - u_1 u_2^q)}{u_1^{q^3} u_2^q - u_1^q u_2^{q^3}}\sigma^3 - \frac{u_1^{q^2} u_2 - u_1 u_2^{q^2}}{u_1^{q^4} u_2^{q^2} - u_1^{q^2} u_2^{q^4}}\sigma^4$$

where $\{u_1, u_2\}$ is any basis of U. The norm of b_U is $Z^4(Z^2 - 1)^3$. Finally, we map a_V ($V \in \Pi_3$) to

$$b_V = Z - \frac{u_1^{q^3} u_2^q - u_1^q u_2^{q^3}}{u_1^{q^3} u_2 - u_1 u_2^{q^3}}\sigma + \frac{u_1^{q^3} u_2^{q^2} - u_1^{q^2} u_2^{q^3}}{u_1^{q^2} u_2 - u_1 u_2^{q^2}}\sigma^3$$

where $\{u_1, u_2\}$ is any basis of $V^\perp$. One can calculate that b_V has norm $Z(Z^2 - 1)^2$, and that $b_u b_v b_V = Z(Z^2 - 1)^2$ if $u = u_0 \mathbf{F}_q$ and $v = v_0 \mathbf{F}_q$, where $v_0 = u_1^{-1}$ and $u_0 = u_1^{q^2}/(u_1^{q^2} u_2 - u_1 u_2^{q^2})$. Also, if $U \in \Pi_2$ is spanned by u_1 and u_2, then $(\lambda(U))^\perp = k_U^{-1}\varphi^3(U)$ is spanned by $u_j^{q^3}/(u_1^{q^3} u_2 - u_1 u_2^{q^3})$ for $j = 1, 2$. We find that

$$b_{\lambda(U)} = Z - \frac{u_1^{q^3} u_2 - u_1 u_2^{q^3}}{u_1^{q^3} u_2^q - u_1 u_2^{q^3}}\sigma + \frac{u_1^q u_2 - u_1 u_2 2q}{u_1^{q^3} u_2^q - u_1 u_2^{q^3}}\sigma^3 \; ,$$

and that $b_u b_{\lambda(U)} = b_{\lambda(U)} b_u = Z(Z^2 - 1)$. If $U, W \in \Pi_2$ and $W \subset \lambda(U)$, and if $x = 1/k_U^{q^4} k_W^{q^3}$, then one finds that $b_x b_u b_W = Z^2(Z^2 - 1)^2$. So all the conditions of Corollary 3.4 are satisfied.

Another family of $\widetilde{A_4}$-triangle presentations.

There seems to be a second $\widetilde{A_4}$-triangle presentation with all the symmetry (1.1) and (1.2). It is defined as follows:

If $V \in \Pi_3$, we set $\lambda(V) = c_V \varphi(V^{\perp})$ for $c_V = 1/\delta(V^{\perp})$, where, as in Section 3, for $U \in \Pi_2$, we let $\delta(U)$ be the image in $\mathbf{F}_{q^5}^{\times}/\mathbf{F}^{\times}$ of $u_1^q u_2 - u_1 u_2^q$ for any basis $\{u_1, u_2\}$ of U over $\mathbf{F}_q$. In order for λ to be an involution on Π, this forces us to set $\lambda(U) = \varphi^{-1}(\delta(U)U^{\perp})$ for $U \in \Pi_2$. We also set $\lambda(x_0 \mathbf{F}_q) = x_0\{1\}^{\perp}$ and $\lambda(\{x_0\}^{\perp}) = x_0^{-1}\mathbf{F}_q$ if $x_0 \in \mathbf{F}_{q^5}^{\times}$. I believe that the set of triples

$$\{(c_V x^q, x^{-1}, V) : V \in \Pi_3, x = x_0 \mathbf{F}_q^{\times} \in \Pi_1 \text{ and } x_0 \in V^{\perp}\} \, ,$$

and the triples

$$\{\left(\frac{1}{\delta(U)^q \delta(W)^{q^3}}, U, W\right) : U, W \in \Pi_2 \text{ and } W \subset \lambda(U)\} \, ,$$

together with their cyclic permutations, is the $\mathcal{T}'$ of an $\widetilde{A_4}$-triangle presentation $\mathcal{T}$ compatible with λ. Certainly, (A)–(E) hold, but I have verified (F) only for $q = 2$, and in this case, my attempt to embed $\Gamma_{\mathcal{T}}$ into the cyclic simple algebra $\mathbf{F}_{q^5}(Z)[\sigma]$ failed.

References

[1] Cartwright, D.I., Mantero, A.M., Steger, T. and Zappa, A, Groups acting simply transitively on the vertices of a building of type $\tilde{A}_2$ I, *Geom. Ded.* **47** (1993), 143–166.

[2] Cartwright, D.I., Mantero, A.M., Steger, T. and Zappa, A, Groups acting simply transitively on the vertices of a building of type $\tilde{A}_2$ II: the cases $q = 2$ and $q = 3$. *Geom. Ded.* **47** (1993), 167–226.

[3] Choucroun, F.M., "Groupes opérant simplement transitivement sur un arbre homogène et plongments dans $PGL_2(k)$", *Comptes Rendus de l'Académie des Sciences* **298** (1984), 313–315.

[4] Cohn, P.M., "Algebra, Volume 3" (Second Edition), John Wiley & Sons, Chichester, 1991.

[5] Dembowski, P., "Finite Geometries", Ergebnisse der Mathematik und ihrer Grenzgebiete, Band **44**, Springer-Verlag, Berlin, Heidelberg, New York, 1968.

[6] Figá-Talamanca, A., Nebbia, C., "Harmonic Analysis and Representation Theory for groups Acting on Homogeneous Trees", London Mathematical Society Lecture Note Series **162**, Cambridge University Press 1991.

[7] Jacobson, N., "PI-Algebras, an introduction", Springer Lecture Notes in Mathematics **441**, Springer-Verlag, Berlin Heidelberg New York, 1975.

[8] Ronan, M., "Lectures on Buildings", Academic Press, New York, 1989.

[9] Tamaschke, O., "Projektive Geometrie, I", Bibliographisches Institut, Mannheim, Wein, Zürich, 1969.

[10] Tits, J., A local approach to buildings, pp. 519–547 in *The Geometric Vein. The Coxeter Festschrift.* Springer Verlag, New York-Heidelberg-Berlin, 1981.

Finite simple subgroups of semisimple complex Lie groups – a survey

Arjeh M. Cohen

Fac. Wisk. en Inf.
TUE
P.O. Box 513
5600 MB Eindhoven
The Netherlands

David B. Wales

Sloan Lab
Caltech
Pasadena
CA 91125
USA

Keywords: finite simple groups, Lie groups

Abstract

We survey recent results regarding embeddings of finite simple groups (and their nonsplit central extensions) in complex Lie groups, especially the Lie groups of exceptional type.

1. Introduction

Throughout this paper, L will be a finite group. Representation theory for L is usually understood to be the study of group morphisms $L \to GL(n, k)$ for distinguished collections of fields k (e.g., all overfields of a fixed field F) and positive integers n. The topic of this survey is motivated by the question as to what happens if $GL(n, \cdot)$ is replaced by another algebraic group $G(\cdot)$.

We shall mainly be concerned with the case where L is a finite simple group (that is, a finite nonabelian simple group) or a central extension thereof, and $G(k)$ is a connected simple algebraic group over a field k. A further restriction of our discussion concerns the field k. It will mostly be taken to be the complex numbers, in which case we will mainly study group morphisms from L to the complex Lie group $G(\mathbb{C})$. (See below for some exceptions in §3 and §5.)

For $G(\cdot)$ of classical type, the theory for representations $L \to G(\mathbb{C})$ differs little from the usual one for $GL(n, \mathbb{C})$. Indeed, a representation $L \to GL(n, \mathbb{C})$ decomposes into irreducible subrepresentations. The decomposition is well controlled by character theory. Given an irreducible representation $\rho : L \to GL(n, \mathbb{C})$, it can be checked whether it is conjugate to a symplectic representation $L \to Sp(n, \mathbb{C})$ or an orthogonal representation $L \to O(n, \mathbb{C})$ by verifying whether its Frobenius-Schur index (that is, $\sum_{g \in L} \rho(g^2)/|L|$) takes the value -1, or 1, respectively (cf. (Isaacs [1976])). Using the criterion for irreducible subrepresentations, it can also be successfully employed for arbitrary (reducible) representations $L \to GL(n, \mathbb{C})$.

Thus, the simple connected complex algebraic groups of exceptional type remain. There are five of them; their universal covers form a chain with respect to group embeddings:

$$G_2(\mathbb{C}) < F_4(\mathbb{C}) < 3 \cdot E_6(\mathbb{C}) < 2 \cdot E_7(\mathbb{C}) < E_8(\mathbb{C}).$$

Here, $3 \cdot E_6(\mathbb{C})$ denotes the universal covering group of type E_6, which has a center of order 3.

In §4 we indicate what is known about the occurrence of finite simple groups in each of these. In §2 we deal with some general theory, and in §3 with correspondences between ordinary (characteristic 0) and modular representations. §5 deals with related embedding problems, mainly focussing on finite maximal subgroups of the same overgroups, and finite simple subgroups of simple algebraic groups in positive characteristic. §6 is concerned with an overview of the calculations needed to establish the harder embeddings. In §7, we end by a discussion of the computational aspects of the constructive proofs outlined in the previous section.

We gratefully acknowledge comments by R.L. Griess, Jr. and J-P. Serre on earlier versions of this paper.

2. Finiteness results

The theorem below generalises a well-known fact known for $GL(n, \cdot)$. A good reference for a proof is (Slodowy [1993]); it is based on (Weil [1964]).

Consider the set of all maps from L to G, denoted by G^L, as an affine variety by viewing it as the product of $|L|$ copies of the affine variety G. Regard the set of all representations as the subvariety of G^L consisting of all points $\rho : L \to G$ satisfying the polynomial equations $\rho(g)\rho(h) = \rho(gh)$ for all $g, h \in L$. Note that G acts on X by conjugation:

$$(g \cdot \rho)(h) = g\rho(h)g^{-1} \quad (g \in G, \ \rho \in X, \ h \in L).$$

If k is a field, we denote by $\overline{k}$ its algebraic closure. If G is an algebraic group, we denote by G^0 its connected component containing the identity.

2.1. Theorem. *Suppose L is a finite group and G is an algebraic group. If k is a field such that $H^1(L, V) = 0$ for all finite-dimensional kL-modules V, then the number of conjugacy classes of representations $L \to G(\overline{k})$ is finite. In fact, in the variety X of all representations, each G^0-orbit of a representation is an irreducible component of X.*

The vanishing cohomology condition for L on kL-modules is satisfied if $|L|$ and char k are coprime (this includes the case char $k = 0$). At least some condition in this direction is necessary as, for any natural number i, the elementary abelian group L of order p^2 embeds into $SL(2, k)$, where $k = (\mathbf{Z}/p)(t)$, via

$$\phi_i : (a, b) \mapsto \begin{pmatrix} 1 & a + bt^i \\ 0 & 1 \end{pmatrix};$$

this gives an infinite set of representations $\{\phi_i\}_i$, no two of which are $G(k)$-conjugate.

There are more detailed results in this direction. For instance, for $p = $ char $k > 0$, Slodowy (Slodowy [1993]) proves that, when fixing a particular representation of a Sylow p-subgroup of L, the number of conjugacy classes of representations of L extending this representation is finite (thus answering a question of Külshammer).

A very useful consequence of Theorem 2.1 is the following result.

2.2. Corollary. *Let K be an algebraically closed overfield of F and suppose char F and $|L|$ are coprime. Then any finite subgroup L of $G(K)$ is conjugate to a subgroup of $G(k)$ where k is a finite extension of F inside K.*

PROOF. As before, let X be the subvariety of G^L consisting of all group morphisms $L \to G$. Then X is clearly defined over F. Let $\rho : L \to G(K)$ be the embedding afforded by the hypothesis. The above theorem yields that the $G(K)^0$-orbit of $\rho : L \to G(K)$ is the set of K-points of an irreducible component of X. Therefore, this orbit contains a point defined over a finite extension k of F, that is, there is $g \in G(K)$ such that $g \cdot \rho : L \to G(K)$ satisfies $(g \cdot \rho)(L) < G(k)$. The assertion follows as $(g \cdot \rho)(L) = g\rho(L)g^{-1}$. $\square$

Thus, for any given representation with specified ground field, one may ask for the minimal degree of an extension field of the ground field that

realizes it. In particular, it would be interesting to have an analogue of
Brauer's result (Brauer [1980]), which states that if F is a finite field, each
irreducible representation $\rho : L \to GL(n, F)$ is $GL(n, F)$-conjugate to a
representation $\sigma : L \to GL(n, k)$, where k is the smallest subfield of F
containing all traces of $\rho(g)$ for $g \in L$.

In order to give a meaning to such an extension from $GL(n, \cdot)$ to arbitrary reductive algebraic groups G, two notions need appropriate generalizations. The first is irreducibility of a representation: a good candidate for
simple algebraic groups might be that $\rho(L)$ is not contained in any parabolic
subgroup of G. The second is the extension field of the ground field: taking
k to be the smallest subfield of F containing all traces of elements of $\rho(L)$ on
any of the fundamental weight modules, we would regain Brauer's subfield
in case $G = GL(n, \cdot)$.

For example it is shown in (Testerman [1989]) that $G_2(q)$ embeds in
$E_6(q)$ as an irreducible group on a 27-dimensional module for $E_6(q)$ if and
only if $\sqrt{-7}$ is in $GF(q)$. It is shown in (Cohen & Wales [1993]) that a
certain embedding of $L(2, 13)$ into $E_6(q^2)$ is in $E_6(q)$ if and only if $\sqrt{-91}$ is
in $GF(q)$.

3. Relation with the finite groups of Lie type

In this section we review some of the folklore on the connection between
group embeddings in groups of Lie type defined over the complex numbers
and those over a finite field. See also (Griess [1991]), (Cohen, Griess &
Lisser [1993]) and (Cohen & Wales [1992]). We are indebted to Prasad,
Ramakrishnan, and others, for helpful discussions concerning the contents
of this section.

For the duration of this section, let G be a semi-simple algebraic group
scheme. Denote by Δ its Dynkin diagram, by r its number of nodes (i.e.,
the rank of G), and by $\tilde{\Delta}$ the extended Dynkin diagram of Δ. Furthermore,
fix a prime number p. We let $\mathbf{Q}_p$ be the p-adic field and K the p-adic
completion of the algebraic closure of $\mathbf{Q}_p$. As is well known, K and $\mathbb{C}$ are
isomorphic as fields. Pick an isomorphism which identifies these fields.

For k a finite extension of $\mathbf{Q}_p$, let o denote the ring of integers in k and
p the maximal ideal of o. Then $\mathsf{o}/\mathsf{p} \cong \mathbf{F}_q$ where q is a power of p and $\mathbf{F}_q$
is a finite field with q elements. The groups $G(\mathbb{C}) \cong G(K)$, $G(k)$, $G(\mathbf{F}_q)$,
and the subgroup $G_\Gamma(\mathsf{o})$ of $G(K)$, where Γ is a rank r subdiagram of $\tilde{\Delta}$, are
now defined by the group scheme.

Reduction modulo p is a homomorphism from $G_\Gamma(\mathsf{o})$ onto $G_\Gamma(\mathbf{F}_q)$. The
kernel of this map is a profinite p-group. The quotient $G_\Gamma(\mathbf{F}_q)$ is a finite
group of Lie type Γ. The quotient $G_\Delta(\mathbf{F}_q)$ coincides with $G(\mathbf{F}_q)$. Reduction
modulo p works because of the following key result.

3.1. Theorem. *Suppose k is a finite extension of $\mathbf{Q}_p$. Then the group $G(k)$ is a locally compact group, when endowed with the p-adic topology. Any finite subgroup of $G(k)$ is compact and is contained in a maximal compact subgroup of $G(k)$. Given a maximal compact subgroup M of $G(k)$ there is a rank r subdiagram Γ of $\tilde{\Delta}$ and an algebraic subgroup G_Γ of G defined over k such that M is conjugate within $G(k)$ to $G_\Gamma(\mathsf{o})$, where o denotes the ring of integers in k.*

PROOF. See (Bruhat & Tits [1972]; Tits [1979]).   □

Actually, we have been informed (Serre [1994]) that one can do better: there is a totally ramified extension k_e of k such that $G_\Gamma(\mathsf{o})$ embeds in $G_\Delta(\mathsf{o}_e)$, where o_e is the ring of integers of k_e. It follows that each finite subgroup of $G(k)$ is conjugate in $G(\bar{k})$ to a subgroup of $G_\Delta(\mathsf{o}_e)$.

3.2. Theorem. *Suppose L is a finite subgroup of $G(K)$. Then there is a finite extension field k of $\mathbf{Q}_p$ in K, a rank r subdiagram of $\tilde{\Delta}$ and a subgroup L_1 of $G(k)$ conjugate to L such that L_1 is a subgroup of $G_\Gamma(\mathsf{o})$, where o is the ring of integers of k. Reduction modulo the maximal ideal p of o is a homomorphism from L_1 onto a subgroup of $G_\Gamma(\mathbf{F}_q)$ for some power q of p. The kernel is a p-group.*

PROOF. Choose k as in Corollary 2.2, so that, identifying K and $\mathbf{C}$, the subgroup L of $G(\mathbf{C})$ is conjugate to a subgroup of $G(k)$.

By Theorem 3.1 there is a conjugate L_2 of L_1 in $G(k)$ which is a subgroup of $G_\Gamma(\mathsf{o})$ for some rank r subdiagram Γ of Δ. The kernel in $G_\Gamma(\mathsf{o})$ of reduction modulo p being a profinite p-group, the kernel of its restriction to L_2 is a finite p-group. The image of L_2 under reduction modulo p is a subgroup of $G_\Gamma(\mathbf{F}_q)$.   □

In particular, if L has no normal p-subgroup, we find that L embeds in $G(\mathbf{F}_q)$. For, take k as in Theorem 3.2. Then the residue field of o_e as above is again $\mathbf{F}_q$, so by the second assertion of the theorem, there is a homomorphism from a conjugate L_1 of L to $G_\Delta(\mathbf{F}_q) = G(\mathbf{F}_q)$. By the last assertion and the hypothesis on normal subgroups of L, the homomorphism is injective.

Now, conversely, given an embedding of L in $G(\mathbf{F}_q)$, we can lift the embedding to one of L in $G(\mathbf{C})$ under a familiar condition, which finds its origin in the following well-known result.

3.3. Lemma. *Suppose H is a finite group which contains a normal subgroup P of order a power of the prime p and H/P is of order prime to p. Then H contains a subgroup isomorphic to H/P, and all such subgroups are conjugate in H.*

PROOF. This is the Schur-Zassenhaus theorem. See, e.g., (Suzuki [1982]), Theorem 8.10 of Chapter 2.   □

3.4. Theorem. *Suppose q is a power of p and L is a subgroup of $G(\mathbf{F}_q)$ whose order is not divisible by p. Choose a finite extension k of $\mathbf{Q}_p$ with ring of integers $\mathfrak{o}$ and maximal ideal $\mathfrak{p}$ of $\mathfrak{o}$ such that $\mathfrak{o}/\mathfrak{p} \cong \mathbf{F}_q$. Then there is a subgroup of $G_\Delta(\mathfrak{o})$ which reduces modulo $\mathfrak{p}$ to L. For a fixed isomorphism between the p-adic completion K of the algebraic closure of k and $\mathbb{C}$, there is a unique conjugacy class of subgroups of $G(\mathbb{C})$ for which some conjugate in $G_\Delta(\mathfrak{o})$ reduces to L.*

PROOF. Denote by $\widetilde{L}$ the inverse image in $G_\Delta(\mathfrak{o})$ of L under reduction modulo $\mathfrak{p}$. If N_i is the kernel of reduction modulo $\mathfrak{p}^i$, then N_i is normal in $\widetilde{L}$ and N_1/N_i is a finite p-group. Since the latter quotients are p-groups and p does not divide $|L|$, the lemma above and induction on i yield that there exists a unique (up to conjugacy) subgroup L_i in $\widetilde{L}$ isomorphic to L and mapping onto L_{i-1} under the natural quotient map $\widetilde{L}/N_i \to \widetilde{L}/N_{i-1}$. Thus, we can find a complement of N unique up to conjugacy in $\widetilde{L}$ isomorphic to L. This complement provides an embedding of L in $G_\Delta(\mathfrak{o})$, which by the isomorphism $K \cong \mathbb{C}$ as above leads to an embedding of L in $G(\mathbb{C})$. The conjugacy condition follows from the uniqueness of L in $\widetilde{L}$.   □

Caution with the uniqueness statement in the theorem is needed, as, changing the isomorphism between K and $\mathbb{C}$, the conjugacy class of L may change by a Galois conjugation.

4. Established embeddings and open cases

The existence of finite simple subgroups of complex Lie groups has been of interest for some time. Systematic searches for such embeddings received an impetus by *Kostant's conjecture*, formulated in 1983. It asserts that every simple complex Lie group $G(\mathbb{C})$ with a Coxeter number h such that $2h+1$ is a prime power, has a subgroup isomorphic to $L(2, 2h+1)$. For $G(\mathbb{C})$ of classical type, this is readily checked using ordinary representation theory and the Frobenius-Schur index. For $G(\mathbb{C})$ of exceptional type the theorem below and the knowledge that $h = 6, 12, 12, 18, 30$ for the five respective exceptional types give an affirmative case-by-case answer.

A quick overview of the state of the art is supplied by Table 1.

X_n	L
Table 1. Nonabelian simple groups L a central extension of which embeds in a complex Lie group of exceptional type X_n	
G_2	Alt_5, Alt_6, $L(2,7)$, $L(2,8)$, $L(2,13)$, $U(3,3)$
F_4	Alt_7, Alt_8, Alt_9, $L(2,25)$, $L(2,27)$, $L(3,3)$, $^3D_4(2)$, $U(4,2)$, $O(7,2)$, $O^+(8,2)$
E_6	Alt_{10}, Alt_{11}, $L(2,11)$, $L(2,17)$, $L(2,19)$, $L(3,4)$, $U(4,3)$, $^2F_4(2)'$, M_{11}, J_2
E_7	Alt_{12}, Alt_{13}, $L(2,29)^?$, $L(2,37)$, $U(3,8)$, M_{12}
E_8	Alt_{14}, Alt_{15}, Alt_{16}, Alt_{17}, $L(2,16)$, $L(2,31)$, $L(2,41)^?$, $L(2,32)^?$, $L(2,61)$, $L(3,5)$, $Sp(4,5)$, $G_2(3)$, $Sz(8)^?$

There are two meanings to be attached to this table:

4.1. Theorem. *Let L be a finite simple group and let G be a simple algebraic group of exceptional type X_n.*

(i) *If L occurs on a line corresponding to X_n in Table 1, then a central extension of it embeds in $G(\mathbb{C})$, with a possible exception for the four groups marked with a "?".*

(ii) *If X_n is as in some line of Table 1 and L appears neither in the line corresponding to X_n nor in a line above it, then no central extension of L embeds in $G(\mathbb{C})$.*

Warnings. To simplify the presentation,

a. we have deliberately neglected questions of conjugacy classes of embeddings, and

b. we have not specified the particular nonsplit central extensions of the simple groups involved.

Ad a. An example where the conjugacy class question is more subtle than suggested by the table is provided by $L(2,13)$. By (Cohen & Wales [1993]), it is isomorphic to a subgroup of $F_4(\mathbb{C})$ whose normalizer is a finite maximal closed Lie subgroup of $F_4(\mathbb{C})$, whereas the table only hints at the existence of embeddings via a closed Lie subgroup of $F_4(\mathbb{C})$ of type G_2.

Ad b. For instance, the simple group $L(2,37)$ listed embeds into a group of type E_7 but not in a group of type E_8 because each embedding in an adjoint group of type E_7 lifts to an embedding of $SL(2,37)$ into the universal covering group $2 \cdot E_7(\mathbb{C})$. Of course, the double cover $SL(2,37)$ of $L(2,37)$ embeds in the universal Lie group of type E_7, whence in a Lie group of type E_8.

Another warning concerning Table 1 is perhaps in order: The main theorems in (Cohen & Wales [1992]) and (Cohen & Griess [1987]) only concern subgroups not contained in closed Lie subgroups of positive dimension whereas Table 1 lists all finite simple subgroups (whether in a closed Lie subgroup of positive dimension or not).

Remarks.

i. The choice of central extensions of simple groups rather than just simple groups is important because they are the ones needed for the generalized Fitting subgroup.

ii. The table does not account for all groups that are involved in $E_8(\mathbb{C})$. For instance, no central extension of $L(5,2)$ is embeddable in $E_8(\mathbb{C})$, but a nonsplit extension $2^{\{5+10\}} \cdot L(5,2)$ does embed (cf. (Alekseevskii [1974])).

iii. The group $L(2,29)$ appears in a Lie group of type B_7, whence in one of type E_8. So, if the question whether a central cover of $L(2,29)$ embeds in $E_7(\mathbb{C})$ has a negative answer, the group should appear at the bottom line of Table 1.

iv. Unlike the $GL(n,\cdot)$ case, knowledge of the classes of the individual elements of an embedded group L does not suffice to determine the conjugacy class of L in G. This has been observed by Borovik for the alternating group Alt_6 in $E_8(\mathbb{C})$. The problem of how many conjugacy classes of embeddings of L exist only has a partial solution. See (Griess [1994]) for the full solution concerning G_2.

v. The group $L(2,41)$ does not appear as a possible subgroup of $E_8(\mathbb{C})$ in (Cohen & Griess [1987]), but neither does the argument ruling it out. Also, the group $Sz(8)$ does not appear as a possible subgroup of $E_8(\mathbb{C})$ in [loc. cit.], whereas the argument ruling it out is erroneous.

vi. Another error in [loc. cit.] concerns the character given for $L(2,31)$. The restriction of the adjoint character for $E_8(\mathbb{C})$ to the subgroup isomorphic to $L(2,31)$ constructed by Serre (see below) has a different character.

PROOF OF THEOREM 4.1(i). In some cases where the subgroup to be embedded is "big enough", L can be shown to embed by use of character theoretic arguments, without explicit constructions. Two useful examples are the following two criteria, valid for both finite and algebraically closed fields F:

I. If $\rho : L \rightarrow GL(7, F)$ is an irreducible representation and $\rho(L)$ leaves fixed a 3-linear alternating form on F^7, then there is a subgroup of $GL(7, F)$ isomorphic to $G_2(F)$ such that $\rho(L) < G_2(F)$ (cf. (Cohen & Helminck [1988])).

II. If, for some positive integer n, the map $\rho : L \rightarrow GL(n, F)$ is an irreducible representation such that $\rho(L)$ fixes a nonzero symmetric bilinear form and a nonzero alternating trilinear form on F^n, but no nonzero alternating quadrilinear form, then $\rho(L)$ preserves a non-trivial Lie algebra product on F^n (cf. (Norton [1988]); earlier, in (Griess [1977]), similar conditions were given).

In both cases, the conditions can be verified using character tables and power maps only.

We now deal with the individual groups occurring in the table.

G_2: Alt_5, Alt_6, $L(2, 7)$ have 3-dimensional projective representations, so a central extension occurs in a group of type A_2; the latter diagram occurs in $\widetilde{G}_2$, and so, by (Borel & Siebenthal [1949]), there is a closed Lie subgroup in $G_2(\mathbb{C})$ of type A_2, via which central covers of the three simple groups embed in $G_2(\mathbb{C})$. The above argument I. applies to $L(2, 8)$, $L(2, 13)$. The group $U(3, 3)$ is an index 2 subgroup of the full automorphism group of the Cayley integers, and as such is known to embed in G_2, see (Coxeter [1946]). For more details, see (Cohen & Wales [1983]). Recently, a new approach to this classification appeared in (Griess [1994]).

F_4: $Alt_7 < Alt_8 < Alt_9$, and the latter has an orthogonal representation of degree 8, so embeds in a Lie group of type D_4, whence a central cover embeds in a Lie group of type F_4. Similarly for $O(7, 2)$ and $O^+(8, 2)$. A central cover of the group $U(4, 2)$ has a 4-dimensional orthogonal representation, and so embeds in a group of type A_3, whence in $F_4(\mathbb{C})$. The group $L(2, 25)$ occurs as a subgroup of $^2F_4(2)$, which can easily be seen to embed in $E_6(\mathbb{C})$ (see below); restriction of the E_6-character on a 27-dimensional high-weight module to the subgroup $L(2, 25)$ shows that a vector is left fixed; the stabilizer of this vector must then be a Lie group of type F_4, whence $L(2, 25) < F_4(\mathbb{C})$. The group $^3D_4(2)$ can be seen to embed in $F_4(\mathbb{C})$ by argument II. above. The group $L(3, 3)$ occurs in the split extension $3^3 : L(3, 3)$ found by (Alekseevskii [1974]). For an embedding of $L(2, 27)$ and more details, see (Cohen & Wales [1992]).

E_6: $Alt_{10} < Alt_{11}$ embeds in a Lie group of type D_5. The groups $L(2, 11)$, $L(3, 4)$, $U(4, 3)$, and J_2 have nontrivial projective representations of dimension less than or equal to 6, and so central extensions embed in $A_5(\mathbb{C})$, whence in $E_6(\mathbb{C})$. Similarly M_{11} has an orthogonal representation of degree

10, so embeds in a group of type D_5, whence in one of type E_6. By arguments II., it can be established that $^2F_4(2)'$ embeds in $E_6(\mathbb{C})$. For $L(2,17)$ and $L(2,19)$, and more details, see (Cohen & Wales [1992]).

E_7: $Alt_{12} < Alt_{13}$ is in $D_6(\mathbb{C})$ whence in $E_7(\mathbb{C})$. An embedding for $U(3,8)$ is given in (Griess & Ryba [1991]). Finally, M_{12} embeds in Alt_{12}. For more details on maximality and characters, see (Cohen & Griess [1987]).

E_8: $Alt_{14} < Alt_{15} < Alt_{16} < Alt_{17}$ embed in $D_8(\mathbb{C})$. The group $L(2,16)$ embeds in Alt_{17}. The group $L(3,5)$ occurs in a split extension $5^3 : SL(3,5)$ (cf. (Alekseevskii [1974])). The groups $Sp(4,5)$ and $G_2(3)$ embed in a group of type D_7, whence in $E_8(\mathbb{C})$. Finally, $L(2,61)$ is constructed in (Cohen, Griess & Lisser [1993]). For more details, see (Cohen & Griess [1987]). Using a more elaborate lifting criterion than the one of Theorem 3.4 Serre recently proved (Serre [1994]), at least for groups G of exceptional type, the existence of a subgroup of $G(\mathbb{C})$ isomorphic to $PGL(2, h + 1)$ where h is the Coxeter number (starting from the existence of a particular subgroup of type A_1 in $G(\mathbf{F}_{h+1})$). This establishes the existence of a subgroup of $E_8(\mathbb{C})$ isomorphic to $PGL(2, 31)$.

SKETCH OF PROOF OF THEOREM 4.1(ii). This part of the theorem uses the classification of finite simple groups. The proof can be found in (Cohen & Griess [1987]) for E_7 and E_8, (Cohen & Wales [1992]) for F_4 and E_6, and (Cohen & Wales [1983]) for G_2. Some of the main techniques are discussed below.

First, due mainly to (Landazuri & Seitz [1974]), for any given finite simple L there is an explicitly known number r_L such that each nontrivial projective representation of L has degree at least r_L. If a central cover of L embeds in $G(\mathbb{C})$, then the smallest high weight representation has dimension at least r_L. For $G = G_2, F_4, E_6, E_7, E_8$, this gives $r_L \le 7, 26, 27, 56, 248$. This leads to an explicitly known finite list of simple groups for which the existence of an embedding needs to be checked.

In most cases, the list resulting from the Landazuri-Seitz bound r_L is still too big for a detailed analysis. An extremely useful result that helps to trim down the list further is due to (Borel & Serre [1953]). It states that every supersolvable subgroup of $G(\mathbb{C})$ is embeddable in the normalizer N of a maximal torus T of $G(\mathbb{C})$. Its use lies in the fact that the structure of N is completely determined: $T \cong \mathbb{C}^r$ where r is the rank of G, and N/T is the Weyl group of G. Thus necessary conditions for the existence of an embedding of L in G can be derived in terms of the structure of all supersolvable subgroups of L. For example, the rank of a maximal abelian p-subgroup of L is at most $r + 1$ when L is embedded in $G(\mathbb{C})$ (cf. (Cohen & Seitz [1987]; Griess [1991])).

Another useful criterion comes from the limited number of classes of elements of given order in $G(\mathbb{C})$ and knowledge of their centralizers. For instance, the possible traces on small G-modules can be readily computed (it is fully automated in LiE, cf. (Leeuwen, Cohen & Lisser [1992])). This gives rise to necessary conditions on the characters of L for them to be restrictions of characters of the ambient Lie group $G(\mathbb{C})$ on a given small-dimensional high-weight module.

There are other conditions on the characters of L that must hold for them to be restrictions of $G(\mathbb{C})$-characters if L embeds in $G(\mathbb{C})$. For instance, on the adjoint module, L must leave invariant a symmetric bilinear form and an alternating trilinear form; these conditions can be expressed in terms of characters. In (Cohen & Wales [1992]) a more detailed relation between the characters that holds for the Lie group but not for "likely" character restrictions for $G_2(3)$, was used to show that $G_2(3)$ cannot be embedded in $E_6(\mathbb{C})$. More specifically, let ψ, χ be the characters of $3 \cdot E_6(\mathbb{C})$ on high weight modules of dimension 27 and 78, respectively. Then $\psi \otimes \bar{\psi}$ contains χ. Assume now that $G_2(3) < E_6(\mathbb{C})$. Then, by character arguments, we see that $L = 3 \cdot G_2(3) < 3 \cdot E_6(\mathbb{C})$, and that there are unique characters ψ_1, χ_1 of L such that $\psi|L = \psi_1$ and $\chi|L = \chi_1$. Now $\psi_1 \otimes \bar{\psi}_1$ does not contain χ_1, a contradiction with $\chi|L$ occurring in $\psi|L \otimes \bar{\psi}|L$.

If such arguments do not help, an explicit model of the group is useful. This model is usually taken to be the smallest dimensional high-weight module of G. □

In conclusion, in establishing the existence of an embedding of a central extension for finite simple groups, we only encounter computational difficulties for the Suzuki group $Sz(8)$ in $E_8(\mathbb{C})$ and for the groups $L = L(2, s)$ with $s = 17, 19, 27, 29, 32, 37, 41, 61$ in the respective cases $X_n = E_6$, E_6, F_4, E_7, E_8, E_7, E_8, E_8. In view of part (i) of the theorem, we have the following result.

4.2. Theorem. *Kostant's conjecture holds.*

To finish off the question marks of Table 1, the following open questions need to be solved.

4.3. Open problems. *Establish that $L(2, 29)$ embeds in $E_7(\mathbb{C})$, and that $L(2, 41)$, $L(2, 32)$, and $Sz(8)$ embed in $E_8(\mathbb{C})$.*

The first of these, the only case left open for E_7, is probably the most straightforward one. The centralizer of the image of the diagonal of $L(2, 29)$ in $E_7(\mathbb{C})$ is a group of type $T_6 A_1$ (that is, a product of a central torus of dimension 6 and $(P)SL(2, \mathbb{C})$). From §6, it will be clear that this slightly complicates the approach to a construction used for cases where the centralizer of a similar image is minimal, i.e., a maximal torus of G.

5. Related embedding problems

Embedding other groups in simple algebraic groups. In (Borovik [1989]; Borovik [1990]) perhaps the most remarkable finite subgroup of any Lie group appeared: it is a finite maximal closed Lie subgroup of $E_8(\mathbb{C})$, whose socle is $Alt_5 \times Alt_6$. It is the only occurrence of a subgroup of a simple complex Lie group $G(\mathbb{C})$ whose normalizer is a finite maximal closed Lie subgroup of $G(\mathbb{C})$ and whose socle is a product of more than one simple group.

Now let L be a finite maximal closed Lie subgroup of a complex simple Lie group $G(\mathbb{C})$ of exceptional type. If L has a nontrivial abelian normal subgroup, then L is known by (Alekseevskii [1974]; Alekseevskii [1975]). If not, then either L has socle $Alt_5 \times Alt_6$ (and G has type E_8), or L has a simple socle, in which case the results of §4 apply. Thus, finite maximal closed Lie subgroups of $G(\mathbb{C})$ are well understood.

Modular representations of finite simple groups. The analog of Theorem 4.1 for algebraic groups over algebraically closed fields of positive characteristic $p > 0$ is more difficult. One extreme is the situation where $(|L|, p) = 1$; by Theorems 3.2 and 3.4, this can be brought back to the Lie group case. At the other extreme, L may be a group of Lie type of the same characteristic. The study in (Seitz [1991]) shows how intricate this situation is.

On the positive side, many constructions as suggested by Table 1 go through due to the results in §3. To indicate that extra embeddings arise, we mention a few, without attempting to be exhaustive. In (Kleidman & Wilson [1993]), the sporadic simple groups embedding in a finite group of exceptional Lie type are determined. Apart from the sporadic simple groups L that can be found by use of Theorems 4.1(i) and 3.2 above, they found M_{22} (in $E_6(4)$), J_1 (in $G_2(11)$), J_3 (in $E_6(4)$), Ru, HS (both in $E_7(5)$), F_{22} (in $E_6(4)$) and Th (in $E_8(3)$). The reader is warned that here, as opposed to [loc. cit.], no exhaustive list is given of the groups of Lie type in which the sporadic groups occur. Earlier, several of the sporadic groups were hypothesised (notably by Steve Smith for Ru and HS) or proven (e.g. by (Thompson [1976]) for Th and by (Janko [1966]) for J_1) to embed in a group of exceptional Lie type. The added value of [loc. cit.] is that it establishes exactly where these groups occur and that the list is complete.

Modular representations of other groups. In the theory of maximal finite subgroups of algebraic groups over fields of positive characteristic, much progress has been made, especially for finite and algebraically closed fields. The cases where L has a nontrivial normal abelian subgroup have been dealt with in (Cohen et al. [1992]). Borovik's remarkable subgroup remains the sole finite maximal closed subgroup whose socle is a product of more than one simple group (cf. (Borovik [1990]; Liebeck & Seitz [1990])). The remaining case, where the socle is simple and there is no abelian normal subgroup, is very hard. See (Seitz [1992]) and references contained therein, for general results in this direction. The determinations of maximal finite subgroups of groups of Lie type G_2, F_4 and E_6 are fairly satisfactory, due to, among others, (Aschbacher [1991]; Kleidman [1988]; Magaard [1990]). Those for E_7 and E_8 are still unfinished. It should be noted that not all finite subgroups are contained in maximal finite subgroups. Many finite subgroups of a torus are examples.

6. Description of hard embeddings

The earliest description of a method for an embedding of an $L(2,s)$ in an exceptional Lie group was perhaps (Meurman [1982]) (although in this case, according to I. of §4, no explicit construction was necessary). A method along the same lines works in principle for most of the hard cases. The starting point for these constructions is a presentation for $L = L(2,s)$ by generators and relations, together with a model for G.

By way of example, consider the group $L(2,s)$, where s is an odd prime. It has a presentation of the form

$$L_1 = \langle u,t,w \mid u^s = t^{(s-1)/2} = 1, \ tut^{-1} = u^{g^2},$$
$$w^2 = 1, \ wtw = t^{-1},$$
$$(uw)^3 = 1, wu^g w = t^a u^i wu^j \rangle$$

for integers g,a,i,j such that

  i. g mod s is a generator of the multiplicative group of Z/s,
 ii. $a \equiv (s-3)/2 \pmod{s}$,
iii. $i \equiv g^{-a} \pmod{s}$ and $j \equiv i^{-1} \pmod{s}$.

In particular, the following map on $\{u,t,w\}$ can be extended to an isomorphism $L_1 \to L$.

$$u \mapsto \pm \begin{pmatrix} 1 & 1 \\ 0 & 1 \end{pmatrix}, \quad t \mapsto \pm \begin{pmatrix} \overline{g} & 0 \\ 0 & \overline{g}^{-1} \end{pmatrix}, \quad w \mapsto \pm \begin{pmatrix} 0 & 1 \\ -1 & 0 \end{pmatrix}.$$

Here $\overline{g}$ stands for g mod s. The subgroup $\langle u,t \rangle$ of L_1 is clearly isomorphic to a Borel subgroup of L, and hence a Frobenius group of order $s(s-1)/2$, and so a Todd-Coxeter enumeration for the presentation with respect to this subgroup of L_1 (yielding that $\langle u,t \rangle$ has index $s+1$) suffices to show that L and L_1 are isomorphic.

Next, a model for $G(\cdot)$ is needed. In almost all cases, $G(F)$, for a suitable field F, is viewed as a subgroup of the linear group $GL(n, F)$ preserving a form or a multiplication on F^n. In (Cohen, Griess & Lisser [1993]) the Lie algebra product μ is used to define G, that is, $G(F)$ is viewed as the subgroup of $GL(n, F)$ of all matrices preserving μ. In (Griess & Ryba [1991]) $G(F)$ is viewed as the subgroup of $GL(n, F)$ of all matrices which leave invariant (under conjugation) the Lie algebra L corresponding to $G(F)$, where L is presented as a linear space of $n \times n$-matrices over F.

In such a setting, a maximal split torus T is fixed, usually the subgroup of all diagonal matrices in $G(F)$, and its normalizer N in $G(F)$ is a monomial group, which can be described explicitly.

It is easy to embed $\langle u, t \rangle$ in $G(F)$ provided F contains s-th roots of unity. Because this group is supersolvable, the theorem of Borel-Serre mentioned in the proof of 4.1(ii) (a variant for algebraic groups is due to (Springer & Steinberg [1970])) yields that, up to conjugacy, we may assume $\langle u, t \rangle$ is contained in N, the normalizer in G of the standard maximal torus T of G. The structure of N then forces the image of u to lie in T. The T-coset of the image of t in $W = N/T$ belongs to a well-studied conjugacy class inside the Weyl group. In most cases, for instance in the Kostant series, the image of t in W is a regular element in the sense of (Springer [1974]). This implies that all elements in the inverse image $Tt \subset N$ are conjugate in $G(F)$.

Now suppose we are given such an embedding of $\langle u, t \rangle$ into N. We shall denote the images of u and t also by u and t. How do we extend it to an embedding of L_1 into G? In order to find an element in G which is the image of $w \in L_1$ under an embedding, we start with $w_0 \in N$ inducing an involution inverting t in W. Due to the good control over N, such an element w_0 is easy to find. The next stage is to look for $w \in w_0 C$, where C is the centralizer in $G(F)$ of t. If t is a Coxeter element, C is a maximal split torus (in general, the dimension of C is at least the Lie rank r of G). Maximal split tori are conjugate. Suppose now that t is a Coxeter element. In order to be able to compute with elements of C, we need to identify this group with an explicit conjugate of T, that is, we need to find $d \in G$ with $dTd^{-1} = C$. For this operation, it is useful to have $(s - 1)/2$-th roots of 1 in F. Thus, an appropriate choice for F would be a field of prime order $1 + ms(s - 1)/2$ for a suitable natural number m. (If $L = L(2, 61)$, we can take $m = 1$ so that $|F| = 1831$.) Here, for the first time, we need an explicit model for G. We can take it to be $G(F) = \operatorname{Aut} g$, where g is the corresponding Lie algebra defined over F.

Computationally, finding d is a hard step. An eigenspace decomposition of F^n with respect to d is needed, but is not enough. In (Cohen, Griess & Lisser [1993]), detailed information regarding the behaviour of the eigenspaces under the multiplication μ was exploited to finish this step. But the result is gratifying in that it enables us to explicitly construct C, so that the embedding problem can be transformed into a set of equations, the unkowns of which are the entries of a matrix representing an element $x \in C = dTd^{-1}$. The number of unknown entries in x is $r = \dim T$ if we regard the diagonal entries of $d^{-1}xd$ as monomials in r independent variables, or n, if we regard that diagonal as $n = \dim G$ linear variables.

The final step consists of solving the equations pertaining to the relation $(uw)^3 = 1$ (often the most complicated relation $wu^g w = t^a u^i wu^j$ is not needed). To this end, rewrite this relation as:

$$uw_0 xu = w_0 xu^{-1} w_0 x$$

for $x \in dTd^{-1}$. By letting these matrices act on vectors $y \in dt$, where $\mathbf{t}$ is the Cartan subalgebra related to T, we get the equations

$$uw_0 xuy = w_0 xu^{-1} w_0 y,$$

which are linear in x. Big systems of linear equations are more easily solved than small systems of polynomial equations. Thus, for the case $s = 61$, the linear equations in $n = 248$ variables were quite manageable, whereas the polynomial equations in $r = 8$ variables were extremely difficult to solve. (Recently, A. Reeves, using the software package Macaulay, managed to solve a set of polynomial equations derived in the course of the work described in (Cohen, Griess & Lisser [1993]); it took her Sparc server little over an hour to find the unique solution.) Seeing to it that $|F|$ is coprime with $|L|$, we can conclude by Theorem 3.4 that L embeds in $G(\mathbf{C})$.

In some cases, the lifting argument was not needed; for instance, in (Cohen & Wales [1993]), the group $L(2, 13)$ could be explicitly embedded in $3 \cdot E_6(\mathbf{C})$.

An entirely different method of construction, based on computer experiments, is to be found in (Kleidman & Ryba [1993]). The method works with a smaller field F, and has a probabilistic portion (for finding an embedding; the resulting existence proof is not probabilistic).

7. Existence by computer

The kind of proof described in the previous section raises the question about construction of an embedding by means of computer. In this section, we discuss some of the issues regarding an existence proof by computer.

It goes without saying that a computer-free proof, if it did not degenerate into a dull stack of computations accounted for on paper, would be much preferable. But given the fact that no such proof of Kostant's conjecture is in sight, we are faced with the question of what is acceptable as a proof when computations are involved that can no longer be checked by a single person using pad and pencil.

For the sake of exposition, it is convenient to revisit the embedding described in the previous section. So, suppose we are given three square matrices u, t and w of size 248 and we wish to verify that they generate a subgroup of the Lie group of type E_8 isomorphic to the finite group $L = L(2, 61)$. Here, by 'being given' a matrix of this size, we mean that there is a simple routine available for generating them, or that they are on file, because the amount of data is simply too large for visual inspection or typing. We need to be able to multiply two such matrices. These computations are useful since it is possible to verify an identity between products of matrices. In particular, checking whether u, t and w satisfy the defining relations for L_1 is feasible.

For the computations involved it is essential that the entries lie in a field of moderate size, such as $\mathbf{Z}/p$ for p a prime less than 10^6. Multiplication of two matrices of this size would otherwise not be practical. This shows the importance of the lifting results: the computer calculations will only explicitly embed L in $G(\mathbf{Z}/p)$; Theorem 3.4 is subsequently used to derive the existence of an embedding in $G(\mathbf{C})$.

In purpose-dedicated software, multiplication of two 248×248-matrices over such prime order fields takes less than a second. In packages like GAP, MAGMA, and LiE which are specially suited for computations with such matrices, it will take several seconds, which is still acceptable.

As a consequence, it is possible for everyone with access to a workstation with one of the abovementioned packages to perform the necessary matrix multiplications in order to be convinced that the defining relations for L_1 are satisfied. So much for the verification that u, t and w generate a subgroup of $GL(248, \mathbf{Z}/p)$ isomorphic to L.

Another part of the verification that L embeds in $G(\mathbf{Z}/p)$ is the check that u, t and w preserve the E_8 Lie algebra product. To this end, the Lie algebra product μ is given as a vector $\mu(x, y)$ of 248 polynomials in the 2×248 variables x, y (representing vectors of $(\mathbf{Z}/p)^{248}$). Then, for $k = 1, \ldots, 248$ and $g = u, t, w$, it is checked whether, for generic vectors x and y, the k-th component of the vectors $\mu(gx, gy)$ and $g\mu(x, y)$ coincide. By reduction of the check to one component at a time, this computation is feasible in a general purpose package (such as Maple or Mathematica).

In general, it can be argued that, provided the source code and the software used is well documented, widely available and implementable, computations that are independently verifiable (with relative ease) can be accepted as parts of a mathematical proof. The argument in defense of acceptance is that, if the intermediate steps documented in the proof suffice for a monastery of mathematically skilled monks to be able to perform the computations within a reasonable time span, the usual proof check is conceivable (albeit blown out of proportion) in times and places where no computers are available.

Although the proof requires relatively little computer effort, finding such a proof can be much more time consuming. Indeed, this has been the case in the computer search for the right matrices yielding the preceding embedding of L in $G(\mathbf{Z}/p)$, especially w. But, once the 'oracle-like' results are establish, the time-consuming constituents of the computer work need not be repeated (with the minor exception that, in (Cohen, Griess & Lisser [1993]), a uniqueness proof (up to conjugacy) of the embedding of $L(2,61)$ in $E_8(\mathbf{C})$ is given that depends on computer computations).

8. Conclusion

Most embedding questions regarding finite simple groups in complex Lie groups of exceptional type have been solved, except for the four persistent problems of §4.3. More detailed questions are still (partially) open, such as minimal splitting fields, the number of conjugacy classes, a description of integral representations, and a geometric interpretation of the existence of such amazingly small groups as maximal closed Lie subgroups of such huge Lie groups.

References

A.V. Alekseevskii [1974], "Finite commutative Jordan subgroups of complex simple Lie groups," *Funct. Anal. and its Appl.* 8, 277–279.

A.V. Alekseevskii [1975], "Maximal finite subgroups of Lie groups," *Funct. Anal. and its Applications* 9, 248–250.

M. Aschbacher [1991], "The 27-dimensional module for E_6, IV," *J. Algebra* 191, 23–39.

A. Borel & J-P. Serre [1953], "Sur certains sous-groupes des groupes de Lie compacts," *Commentarii Math. Helv.* 27, 128–139.

A. Borel & J. de Siebenthal [1949], "Les sous-groupes fermés de rang maximum des groupes de Lie clos," *Commentarii Math. Helv.* 23, 200–221.

A.V. Borovik [1989], "On the structure of finite subgroups of the simple algebraic groups," *Algebra and Logic* 28, 249–279, (in Russian).

A.V. Borovik [1990], "Finite subgroups of simple algebraic groups," *Soviet Math. Dokl.* 40, 570–573, (transl).

R. Brauer [1980], *Collected papers*, MIT Press, Boston, ed. by P. Fong and W.J. Wong.

F. Bruhat & J. Tits [1972], "Groupes réductifs sur un corps local I. Données radicielles valuées," *Publ. Math. IHES* 41, 5–252.

A.M. Cohen & R.L. Griess, Jr. [1987], "On finite simple subgroups of the complex Lie group of type E_8," *AMS Proceedings of Symposia in Pure Math.* 47, 367–405.

A.M. Cohen, R.L. Griess, Jr. & B. Lisser [1993], "The group $L(2, 61)$ embeds in the Lie group of type E_8," *Comm. Algebra* 21, 1889–1907.

A.M. Cohen & A.G. Helminck [1988], "Trilinear alternating forms on a vector space of dimension 7," *Communications in Algebra* 16, 1–25.

A.M. Cohen, M.W. Liebeck, J. Saxl & G.M. Seitz [1992], "The local maximal subgroups of exceptional groups of Lie type, finite and algebraic," *Proc. London Math. Soc.* 64, 21–48.

A.M. Cohen & G.M. Seitz [1987], "The r-rank of groups of exceptional Lie type," *Proceedings of the KNAW* 90, 251–259.

A.M. Cohen & D.B. Wales [1983], "Finite subgroups of $G_2(\mathbb{C})$," *Comm. Algebra* 11, 441–459.

A.M. Cohen & D.B. Wales [1992], "On finite subgroups of $E_6(C)$ and $F_4(C)$," preprint.

A.M. Cohen & D.B. Wales [1993], "Embedding of the group $L(2, 13)$ in the groups of Lie type E_6," *Israel J. Math.* 82, 45–86.

H.S.M. Coxeter [1946], "Integral Cayley numbers," *Duke Math. J.* 13, 561–578.

R.L. Griess, Jr. [1977], "Finite groups as automorphisms of Lie algebras," Vortragsbuch, Oberwolfach.

R.L. Griess, Jr. [1994], "Basic conjugacy theorems for G_2," University of Michigan, report, Ann Arbor.

R.L. Griess, Jr. [1991], "Elementary abelian p-subgroups of algebraic groups," *Geom. Dedicata* 39, 253–305.

R.L. Griess, Jr. & A.J.E. Ryba [1991], "Embeddings of $U_3(8)$, $Sz(8)$ and the Rudvalis group in algebraic groups of type E_7," preprint, U. of Michigan.

M. Isaacs [1976], *Character theory of finite groups*, Pure & Appl. Math. #69, Academic Press, New York.

Z. Janko [1966], "A new finite simple group with Abelian Sylow 2-subgroups, and its characterization," *J. Algebra* 3, 147–186.

P.B. Kleidman [1988], "The maximal subgroups of the Chevalley groups $G_2(q)$ with q odd, of the Ree groups $^2G_2(q)$ and of their automorphism groups," *J. Algebra* 117, 30–71.

P.B. Kleidman & A.J.E. Ryba [1993], "Kostant's conjecture holds for E_7: $L_2(37) < E_7(\mathbb{C})$," *J. Algebra* 161, 535–540.

P.B. Kleidman & R.A. Wilson [1993], "Sporadic simple subgroups of finite exceptional groups of Lie type," *J. Algebra* 157, 316–330.

V. Landazuri & G. M. Seitz [1974], "On the minimal degrees of projective representations of the finite Chevalley groups," *J. Algebra* 32, 418–443.

M.A.A. van Leeuwen, A.M. Cohen & B. Lisser [1992], *LiE, A package for Lie group computations*, CAN, Amsterdam.

M.W. Liebeck & G. Seitz [1990], "Maximal subgroups of exceptional groups of Lie type, finite and algebraic," *Geom. Dedicata* 35, 353–387.

K. Magaard [1990], "The Maximal Subgroups of the Chevalley Groups $F_4(F)$ where F is a Finite or Algebraically Closed Field of Characteristic $\neq 2,3$," PhD Thesis, Caltech.

A. Meurman [1982], "An embedding of $PSL(2,13)$ in $G_2(\mathbb{C})$," in *Lie Algebras and Related Topics*, Lecture Notes in Mathematics #933 , Springer Verlag, Berlin, 157–165.

S.P. Norton [1988], "On the group F_{24}," *Geom. Dedicata* 25, 483–501.

G.M. Seitz [1991], "The maximal subgroups of exceptional algebraic groups," *Memoirs of the Amer. Math. Soc.* 90, 1–197.

G.M. Seitz [1992], "Subgroups of finite and algebraic groups," in *Groups, combinatorics and geometry*, M.W. Liebeck & J. Saxl, eds., London Math. Soc. LNS #165, Cambridge University Press, Cambridge, 316–326.

J-P. Serre [1994], Private Communications.

P. Slodowy [1993], "Two notes on a finiteness problem in the representation theory of finite groups," Universität Hamburg, Hamburger Beiträge zur Mathematik.

T.A. Springer [1974], "Regular elements of finite reflection groups," *Inventiones Math.* 25, 159–198.

T.A. Springer & R. Steinberg [1970], "Conjugacy classes," in *Seminar on Algebraic groups and related finite groups*, Springer Lecture Notes in Math. #131, Springer-Verlag, Berlin, 167–266.

M. Suzuki [1982], *Group Theory I*, Grundlehren der math. Wiss. #247, Springer-Verlag.

D. Testerman [1989], "A construction of certain maximal subgroups of the algebraic groups E_6 and F_4," *J. Algebra* 122, 299–322.

J.G. Thompson [1976], "A simple subgroup of $E_8(3)$," in *Finite groups*, N. Iwahori, ed., Tokyo, 113–116.

J. Tits [1979], "Reductive groups over local fields," in *Automorphic forms, representations and L-functions, Oregon State Univ., Corvallis, Ore.*, A. Borel & W. Casselman, eds., Proc. Symp. Pure Math. #XXXIII, Amer. Math. Soc., Providence, R.I., 29–69.

A. Weil [1964], "Remarks on the cohomology of groups," *Annals of Math.* 80, 149–157.

Flag-transitive extensions of buildings of type G_2 and C_3^*

Hans Cuypers

1. Introduction

In recent work of Weiss, Yoshiara, Meixner and others many sporadic simple groups are characterized as flag-transitive automorphism group of a diagram geometry which is an extension of a building. See for example [1, 11, 13, 15, 16, 17].

In this note we extend Weiss' work [15, 16] on extensions of generalized hexagons (i.e. extensions of buildings of type G_2), see also [1], to extensions of buildings of type G_2 or C_3^*, thereby obtaining geometric characterizations of the sporadic groups HJ, Suz and Co_2. (We use the notation of [15], see also Section 3.)

Theorem 1.1 *Let Γ be a residually connected diagram geometry with diagram*

$$\begin{array}{ccc} & c & \\ \underset{1}{\bullet} & \underset{s}{\rule{1.5cm}{0.4pt}} & \underset{t}{\equiv} \end{array} \quad or \quad \begin{array}{ccc} & c & \\ \underset{1}{\bullet} & \underset{s}{\rule{1cm}{0.4pt}} & \underset{t_2}{\rule{1cm}{0.4pt}} \underset{t_2}{} \end{array}$$

in which the point residue is a classical or dual classical generalized hexagon or a dual polar space with finite parameters s and t, respectively, t_2, where $s \geq 2$. Assume that $Aut(\Gamma)$ is flag-transitive on Γ. If Γ satisfies the following condition:

(∗) there are three pairwise adjacent points not in a circle,

then it is isomorphic to one of the following:

1. the extended generalized hexagon on 24 points related to $GL_2(7)/\langle -I \rangle$;

2. *the extended generalized hexagon on* 64 *points related to* $2^6{:}7{:}6$;

3. *the extended generalized hexagon on* 64 *points related to* $2^6{:}G_2(2)$;

4. *the extended generalized hexagon on* 128 *points related to* $2^7{:}G_2(2)$;

5. *the extended generalized hexagon on* 120 *points related to* $PSp_6(2)$;

6. *the extended dual polar space on* 256 *points related to* $2^8{:}Sp_6(2)$;

7. *the extended dual polar space on* 2300 *points related to* Co_2;

8. *the extended generalized hexagon on* 36 *points related to* $G_2(2)$;

9. *the extended generalized hexagon on* 162 *points related to* $PSU_4(3)$;

10. *the extended generalized hexagon on* 100 *points related to* HJ;

11. *the extended generalized hexagon on* 1782 *points related to* Suz.

For existence of the geometries occurring in the above result, the reader is referred to [1, 3, 15, 16].

Generalized hexagons as well as rank 3 dual polar spaces carry the structure of a near hexagon. This indicates that extensions of buildings of type G_2 and C_3^* can be studied by similar methods.

As already mentioned before, parts of the above theorem also follow from the work of Van Bon, Meixner, Yoshiara, and in particular of Weiss, see [1, 11, 15, 17]. In [15] for example, Weiss studies extensions of the classical generalized hexagons related to the groups $G_2(q)$ and $^3D_4(q)$ satisfying condition $(*)$. He finds the examples related to the groups $2^6{:}G_2(2)$, $2^7{:}G_2(2)$, HJ and Suz. However the example on 120 points related to $PSp_6(2)$ is missing. Weiss informed us that this example should appear in 4.5 of [15], where the coset enumeration of the group $G_{1,1,1}$ leads to a group isomorphic to $PSp_6(2)$ and not to the trivial group, as stated incorrectly in [15].

The proof of the above result is, for the greater part, independent of Weiss' work in which an important rôle is played by coset enumeration on a computer. Here we use geometric and graph theoretical methods to reconstruct the geometry from the local information. Only in the case where the residue of a point is the generalized hexagon of order $(2,1)$ on which a flag-transitive group 7:6 is induced, we use Van Bon's results [1], which do rely on coset enumerations by computer.

Various ideas used in this paper come from [7, 8] where purely geometric characterizations of the extended generalized hexagons and dual polar spaces are given.

2. Preliminaries

In this section we give two preliminary results that will be of use in the next section.

The finite classical or dual classical generalized hexagons are the generalized hexagons associated to the groups $G_2(q)$, $^3D_4(q)$ and $PSL_3(q){:}2$. The dual polar spaces of rank 3 are the dual polar spaces related to the groups $PSp_6(q)$, $O_6^+(q)$, $O_7(q)$, $O_8^-(q)$, $PSU_6(q)$ and $PSU_7(q)$.

Proposition 2.1 *Let Γ be a finite classical or dual classical generalized hexagon or dual polar space of rank 3 with at least 3 points per line, and G a flag- transitive group of automorphisms. Then either G is distance transitive on the associated near hexagon and induces the 3-transitive group $PGL_2(s)$ on each line of the near hexagon or we have*

(i) Γ is a a generalized hexagon of order $(2, 1)$ and $G \simeq 7{:}6$;

(ii) Γ is a generalized hexagon of order $(8, 1)$ and $G \simeq 73{:}18$;

(iii) Γ is the classical generalized hexagon of order $(2, 2)$ and $G \simeq G_2(2)' \simeq PSU_3(3)$.

Proof. This is an easy consequence of [10, 14].

Proposition 2.2 *Let Γ be a flag-transitive one-point-extension of a finite classical generalized quadrangle of order (s, t) with $s, t \geq 2$. Then Γ is isomorphic to the extension of the $PSp_4(2)$ quadrangle related to the group $2^4{:}Sp_4(2)$ or of the $O_6^-(2)$ quadrangle related to $PSp_6(2)$.*

Proof. This can be found at various places, see for example [4, 13].

3. Extensions of near hexagons

In this section we give a proof of Theorem 1.1. Let Γ be a geometry satisfying the hypothesis of that theorem with diagram

and flag-transitive group G of automorphisms. The type 0, 1, 2, respectively, 3 elements of Γ are called the *points*, *edges*, *circles*, respectively, *extended quads* of Γ. Two points on an edge are called *adjacent*.

Let p be a point of Γ. By Proposition 2.1 we find that the action of G_p induced on Γ_p is primitive except in the exceptional cases (i) and (ii) of Proposition 2.1. Van Bon has considered extensions of the generalized hexagon of order $(2,1)$ satisfying $(*)$, in which the stabilizer of a point induces a Frobenius group 7:6 as in $2.1(i)$ on the local hexagon at that point. He finds two such extended hexagons; these are the extended hexagons under (1) and (2) of Theorem 1.1. In [16], Weiss gives an easy argument showing that there exists no extension of the generalized hexagon of order $(8,1)$ in which the point stabilizer induces the Frobenius group 73:18 as in $2.1(ii)$ on the residue of the point. Thus for the rest of this section we can and will assume that the stabilizer G_p of a point p of Γ does not induce a group as in $2.1(i)$ or (ii) on Γ_p. Hence G_p acts primitively on Γ_p, which by standard arguments of [15, 17] implies:

Lemma 3.1 *Two extended quads, circles or edges have the same point shadow if and only if they are the same.*

It follows that we can identify the edges, circles and extended quads with their point shadows. The *point graph* $\mathcal{G}$ of Γ is the graph whose vertices are the points of Γ, and whose edges are the edges of Γ. If p is a point of Γ, then we often identify the edges on p with the points adjacent to p inside $\mathcal{G}$. The set of points adjacent to p is sometimes denoted by $p^\perp$. For each point p the geometry Γ_p carries the structure of a near hexagon with parameters (s, t_2, t), where $t_2 = 0$ if Γ_p is a generalized hexagon and $t = t_2(t_2+1)$ otherwise. Since $\mathcal{G}$ is transitive on the points of Γ, these parameters are independent of p.

The following two lemmas and their proofs are straightforward generalizations of the Lemmas 2.2 and 2.3 of [15].

Lemma 3.2 *Let p be a point of Γ, then G_p acts faithfully on Γ_p.*

Lemma 3.3 $s \leq 4$.

By condition $(*)$ in the hypothesis of the theorem and 2.1, we find that Γ is either a one–point–extension of the local near hexagon or satisfies the following condition for $i = 2$ or 3:

$(*)_i$ $\{x, y, z\}$ is a clique of the point graph not in a circle if and only if the distance between y and z in Γ_x is i;

except possibly when Γ is an extension of the classical generalized hexagon of order $(2,2)$, while the group induced on this hexagon is isomorphic to $PSU_3(3) \simeq G_2(2)'$. The next lemma, however, shows that this exception does not occur. (Compair with the proof of Lemma 4.5 of [15].)

Lemma 3.4 *If Γ is an extension of the classical hexagon of order $(2,2)$, then it is either a one–point–extension or satisfies $(*)_i$ for $i = 2$ or 3.*

Proof. Suppose the contrary. Let p be a point of Γ. The group G_p does not act distance transitively on Γ_p and hence is isomorphic to $PSU_3(3)$. If q is a point of Γ_p, then $G_{p,q}$ is transitive on the points at distance 1, respectively, 2 of q in Γ_p and has two orbits of size 16 on the points at distance 3 from q. Each line of Γ_p containing a point at distance 3 from q meets both orbits of size 16.

Since we assume Γ to be a counterexample to the statement of the lemma, there is a point r of Γ_p at distance 3 from q that is adjacent to q. However, not all points at distance 3 from q are adjacent to q. We have $G_{p,q,r} = G_{p,\{q,r\}}$. Since each line of Γ_p through r contains a point at distance 3 from q not in $r^{G_{p,q}}$, each circle on p and r contains at least one point not adjacent to q. If inside Γ_r the points p and q are at distance 2, then the circle on p and r which in Γ_r is the line on p containing some point at distance 1 from q contains only points adjacent to q, which contradicts the above. Thus in Γ_r the distance between p and q is 3. By the same argument we find that p and r are at distance 3 inside Γ_q. Since there is only one orbit of adjacent point pairs $\{q,r\}$ in Γ_p that are at distance 3, we find that $G_{\{p,q,r\}}$ is transitive on $\{p,q,r\}$, and $G_{\{p,q\},r} = G_{p,q,r}$.

Let C be a circle on p and q. Then in the stabilizer of this circle we find an element g switching p and q, i.e. with $p^g = q$ and $q^g = p$. Now consider $\{p^g, q^g, r^g\} = \{p, q, r^g\}$. The point r^g is at distance 3 form p inside Γ_q. So, by the above, r^g and q are also at distance 3 in Γ_p, and in $G_{p,q}$ there is an element h with $(r^g)^h = r$. But then $gh \in G_{\{p,q\},r} \setminus G_{p,q,r}$, which is empty. A contradiction.

The Propositions 3.5, 3.6 and 3.13 below are concerned with the geometries Γ that are one-point-extensions of the local near hexagon, or satisfy condition $(*)_2$, respectively, $(*)_3$. Together they provide us with a proof of Theorem 1.1.

Proposition 3.5 *If Γ is a one–point–extension, then it is isomorphic to the extension of the classical generalized hexagon of order $(2,2)$ related to the group $2^6{:}G_2(2)$.*

Proof. First assume that Γ is an extension of a dual polar space. Then by Proposition 2.2 we find that $s = 2$, and t_2 is either 2 or 4. So Γ is an extension of the dual polar space related to $PSp_6(2)$ or $PSU_6(2)$. In particular Γ contains $1 + 135$ or $1 + 891$ points. If Γ is locally the $PSp_6(2)$ dual polar space, then it contains $136.63/16$ extended quads. Since this is not an integer, we obtain a contradiction. In case Γ is locally the $PSU_6(2)$ dual polar space, we find that the stabilizer of a point p satisfies $PSU_6(2) \leq G_p \leq PSU_6(2){:}S_3$. So the order of G equals $2^{17+n}.3^{6+m}.5.7.11.223$ where $n, m \in \{0, 1\}$. Consider a Sylow-223-subgroup S of G. The elements of order 5, 7, 9 and 11 can not normalize S, which contradicts Sylow's Theorem. See also [11].

Now suppose Γ is the extension of a generalized hexagon of order (s, t). If $t = 1$, then the number of circles in Γ equals $(1 + (1 + s + s^2)(1 + s))(2(1 + s + s^2))/(s + 2)$. Thus $s \neq 3$. If $s = 2$, then G is a 2-transitive group on 22 points, with point stabilizer G_p isomorphic to $PSL_3(2){:}2$. In particular, $|G| = 2^5.3.7.11$. Consider a Sylow-11-subgroup S of G. The elements of order 3 and 7 can not normalize S, and the number of conjugates of S is thus $2^n.3.7$, for some $n \leq 5$. By Sylow's Theorem this number is 1 mod 11, and thus equal to $2^5.3.7$. Hence G contains $2^6.3.5.7$ elements of order 11. Each element of order 3 in G fixes a unique point of Γ. Thus there are 22.56 elements of order 3 in G. But then $|G| > 2^6.3.5.7 + 22.56 > 2^5.3.7.11$, again a contradiction. If $s = 4$, then the point stabilizer G_p satisfies $PSL_3(4){:}2 \leq G_p \leq Aut(PSL_3(4))$, and the order of G is $2^{8+n}.3^{2+m}.5.7.53$ where $n, m = 0, 1$. If G_p contains a field automorphism σ, then the fixed points of σ are the points of a one–point– extension of the $PSL_3(2)$-hexagon of order $(2, 1)$, which contradicts the above. Thus G_p does not contain such an element. Consider a Sylow-53-subgroup S of G. This group can not be normalized by any element of order 5 or 7. But then again, Sylow's Theorem yields a contradiction.

If Γ is locally the dual classical generalized hexagon of order $(2, 2)$ related to $G_2(2)$ or the generalized hexagon of order $(2, 8)$ related to $^3D_4(2)$, then the stabilizer of a point p contains a subgroup H of order 3, respectively, 7, fixing exactly 9, respectively, 21 points in Γ_p. See [6]. Any subgroup of order 3, respectively, 7 in G_p fixing 9, respectively, 21 points of Γ_p is conjugate to H. The number of conjugates of H in G is therefore equal to $64/10$, respectively, $820/22$ times the number of conjugates in G_p, which is, respectively, 28 and 89856. In both cases this is not an integer, and we obtain a contradiction.

If Γ is locally a generalized hexagon of order $(3, 3)$ related to $G_2(3)$ or of order $(3, 27)$, respectively, $(4, 64)$ related to $^3D_4(3)$, respectively, $^3D_4(4)$, then G_p contains a (root) subgroup R of order s, whose fixed points are p together with the $(s+1)(1+st)$ points at distance at most 1 from some line in Γ_p. Any subgroup of G of order s fixing $1 + (s+1)(1 + st)$ points is conjugate R. But then there are $(1+(s+1)(1+st+s^2t^2))(t+1)(1+st+s^2t^2)/(1+(s+1)(1+st))$ subgroups of G conjugate to R. In all three cases this number is not an integer.

We are left with the case where Γ is locally the classical generalized hexagon of order $(2,2)$. Suppose we are in that situation. Fix two points p and q of Γ. Let r be a point at distance 2 from q in Γ_p. Then $G_{p,q,r}$ fixes a unique other point r' say of Γ_p, and $G_{p,q,r} = G_{p,q,r'}$. The point r is also at distance 2 from q in Γ_p. Let $\mathcal{C}'$ be the set of $\{p,q,r,r'\}^G$. Then $(P, \mathcal{C} \cup \mathcal{C}')$ is a one point extension of the symplectic polar space of $PSp_6(2)$, and by [12, 9], it is the affine symplectic space with automorphism group $2^6{:}Sp_6(2)$. But then G is isomorphic to $2^6{:}G_2(2)'$ or $2^6{:}G_2(2)$ and Γ is the extended generalized hexagon related to $2^6{:}G_2(2)$ on 64 points.

Proposition 3.6 *If Γ satisfies $(*)_2$, then it is isomorphic to the extended generalized hexagon on 128 points related to $2^7{:}G_2(2)$, or on 120 points related to $PSp_6(2)$, or to the extended dual polar spaces on 256 or 2300 points related to $2^8{:}Sp_6(2)$, respectively, Co_2.*

We prove this proposition in a number of steps. Obviously we have:

Lemma 3.7 *If C is a circle and p a point not in C, then p is adjacent to all, two or no point of C.*

Lemma 3.8 $s = 2$.

Proof. If the local near hexagon contains quads, then Γ contains extended quads which are one–point–extensions of a classical generalized quadrangle of order (s, t_2) with $s, t_2 \geq 2$. The stabilizer of such an extended quad induces a flag-transitive automorphism group on this extended quad, and by Proposition 2.2 we find that $s = 2$.

Thus suppose that Γ is locally a generalized hexagon of order (s, t). Let C be a circle on the two points x and y. Consider the set S of points in Γ_x not on C that are collinear with y. The points in $C \cup S$ form a clique of $\mathcal{G}$. If z is a third point of C, then S is contained in Γ_z, and each point of S is at distance 2 from both x and y. Since S is a clique in Γ, the points of S are at mutual distance at most 2 inside Γ_z, and there is a unique point u on C different from x, y and z such that these points in S are all collinear with u inside Γ_z. But, by 2.1 $G_{x,y,z}$ is transitive on the remaining $s - 1$ points in C, which implies that u is the only point of C different from x, y and z. Hence C contains 4 points and $s = 2$.

A *(geometric) hyperplane* of a generalized hexagon is a proper subset of the point set meeting each line of the hexagon in one or all points. A point of a hyperplane is called *deep* if all lines on the point are contained in the hyperplane.

Let p and q be two points at distance 2 in $\mathcal{G}$. By H we denote the complement in Γ_p of the set $p^\perp \cap q^\perp$.

Lemma 3.9   *(i) H is a geometric hyperplane of Γ_p.*

(ii) A point of H is deep or on $t - t_2 - 1$ lines of Γ_p inside H.

Proof. Let r be a point of $p^\perp \cap q^\perp$. Inside Γ_r the distance between p and q is 3. Thus each circle on p and r contains 2 points adjacent to q. But then each line of Γ_p on r meets H in a unique point. This proves (i).

The number of points in H that are at distance 2 from r is the number of points in Γ_r that is at distance 2 from p and 3 from q. This number is $(t+1)(3t-2t_2-2)/(t_2+1)$. Thus, on the average, each point of H collinear with r is collinear with $2(t-t_2-1)$ points of H. The group $G_{p,q,r}$ is transitive on the $t+1$ circles on p and r. So each point of H collinear with r inside Γ_p is collinear with exactly $2(t - t_2 - 1)$ points of H and on $t - t_2 - 1$ lines of H. This proves (ii).

Lemma 3.10   *(i) If Γ is locally the classical generalized hexagon of order $(2,2)$, then it is isomorphic to the extension on 128 points related to $2^7{:}G_2(2)$ or on 120 points related to $PSp_6(2)$.*

(ii) If Γ is locally the $PSp_6(2)$ dual polar space, then it is isomorphic to the extension on 256 points related to $2^8{:}Sp_6(2)$.

Proof. Suppose Γ is as in the hypothesis of (i) or (ii). Then $\mathcal{G}$ is locally the collinearity graph of a polar space related to $PSp_6(2)$, respectively, $PO_8^+(2)$. By [5] we find that $\mathcal{G}$ is the collinearity graph of an affine polar space obtained by removing a hyperplane from either the $PSp_8(2)$ polar space or the $PO_{10}^+(2)$ polar space.

The group G is contained in the automorphism group of this affine polar space. If the removed hyperplane is degenerate, we find that G is a subgroup of $2^7{:}GO_7(2)$ or $2^8{:}GO_8^+(2)$ which leads to the examples related to $2^7{:}G_2(2)$, respectively, $2^8{:}Sp_6(2)$. If the removed hyperplane is nondegenerate, then we find that G is a subgroup of $PGO_8^+(2)$ or $PGO_8^-(2)$ or of $2 \times PSp_8(2)$. In the first case we find the example on 120 points related to $PSp_6(2)$. In the second case, G is a subgroup of index 120 or 2.120 in $PGO_8^-(2)$. By [6] there is no such group. Finally in the last case, the graph $\mathcal{G}$ is antipodal of diameter

3. The group G has index 240 in $2 \times PSp_8(2)$, and induces a 2-transitive subgroup of $PSp_8(2)$ on the 136 pairs of antipodal points. However, it follows from the information in [6] that no such group exists.

Lemma 3.11 *If* Γ *is locally the* $PSU_6(2)$ *dual polar space, then it is isomorphic to the extension on* 2300 *points related to* Co_2.

Proof. Let p and q be points at distance 2 in $\mathcal{G}$. of Γ. Denote by H the complement of $p^\perp \cap q^\perp$ in $p^\perp$ and by D the set of deep points in H.

Fix a line of H. There are 5 quads on this line. The hyperplanes of a quad are subquadrangles of order $(2,2)$ or the 5 lines through a point. If the line contains a deep point, then all the quads on the line are either in H or meet H in the 5 lines on that deep point. But that implies that on any other point of that line there are 1 mod 4 lines inside H. By Lemma 3.9(ii) all the points of such a line are deep. This clearly implies that H can not contain deep points. Double counting of collinear pairs of points yields that $12|H| = 21(891 - |H|)$, from which we deduce that $|H| = 567$. But then $\mathcal{G}$ is of diameter 2 and strongly regular with parameters $(2300, 891, 378, 324)$. The group G acts as a rank 3 permutation group on $\mathcal{G}$.

The group $PSU_6(2)$ contains 3 classes of subgroups of index 1408. These subgroups are isomorphic to $PSU_4(3){:}2$, and they are all conjugate under the automorphism group of $PSU_6(2)$. Each such subgroup has two orbits on the 891 points of the near hexagon, one of length 567 whose points form a hyperplane, and one of length 324, being the set of points in the complement of such a hyperplane.

Let $\mathcal{H}$ be the $PSU_6(2)$-orbit of H. A straightforward computation yields that there are 567 hyperplanes in $\mathcal{H}$ intersecting H in 375 points and 840 intersecting it in 351 points. From the above it is clear that the graph $\mathcal{G}$ is isomorphic to the graph whose vertices are p, the points of Γ_p and the elements of $\mathcal{H}$. The point p is adjacent to all members of Γ_p, two points in Γ_p are adjacent if and only if they have mutual distance at most 2. Two hyperplanes are adjacent if and only if they meet in 375 points, and a point and a hyperplane are adjacent if and only if the point is not in the hyperplane. This proves uniqueness of $\mathcal{G}$, but then uniqueness of Γ follows easily.

Lemma 3.12 *The residue at a point of* Γ *is not isomorphic to the dual classical generalized hexagon of order* $(2,2)$, *or the generalized hexagons of order* $(2,1)$ *respectively,* $(2,8)$.

Proof. Suppose that Γ is locally the generalized hexagon of order $(2,1)$. Let p and q be two points at distance 2 in $\mathcal{G}$, and set H to be the set of points adjacent to p but not to q. The set H is a hyperplane of Γ_p. Clearly there are no lines nor deep points in H. By double counting we find $4|H| = 2(21-|H|)$, and H contains 7 points. Since H contains no deep points, the graph $\mathcal{G}$ has diameter 2, and is strongly regular with $(k, \lambda, \mu) = (21, 12, 14)$. However, such a parameter set is not feasible, see [2]. A contradiction.

Now assume that Γ is locally the hexagon of order $(2, 8)$ related to $^3D_4(2)$. Fix some point p of Γ, and an element g of order 7 in G_p. The fixed points of g in Γ_p form a subhexagon of order $(2, 1)$. Suppose q is a point of Γ_p fixed by g. Then $g \in G_q$ and it fixes p and 12 points at distance at most 2 from p inside Γ_q. The fixed points of g in Γ_q form also a subhexagon of order $(2, 1)$ inside Γ_q. In particular, the points fixed by g form a union of extensions of the generalized hexagon of order $(2, 1)$. However, this contradicts the above.

Finally suppose that Γ is locally the dual classical generalized hexagon of order $(2, 2)$. Let (p, q, r) be a path of length 2 in the graph $\mathcal{G}$. Inside the hexagon Γ_q the distance between p and q is 3. This implies that there is a subgroup S of order 3 of G_q fixing p and r. The group S is in $\mathcal{S}$, the conjugacy class of subgroups of G_p generated by elements of type $3A$ (cf [6]). Each element of $\mathcal{S}$ fixes 9 points of Γ_p forming an ovoid. All points adjacent to both p and r are fixed by some element of $\mathcal{S}$ in $G_{p,r}$. Since the number of points at distance 2 from p is less than 63.32 we find that $G_{p,r}$ has index at least 3 in G_p. By lemma 3.9, there is a point q' collinear with q in Γ_p that is also adjacent to r. This point is not fixed by S, and there are at least 2 members of $\mathcal{S}$ in $G_{p,q}$. These two members generate a group isomorphic to $SL_2(3)$ containing 4 members of $\mathcal{S}$. Together with any other element of $\mathcal{S}$ this group generates G_p'. Thus $G_{p,q}$ contains only the 4 elements of $\mathcal{S}$ in this $SL_2(3)$. The fixed points of these 4 elements are a fixed point r' say, and all the points at distance 3 from that point. Using Lemma 3.9 we find that the set of common neighbours of p and r must be the set of all elements of Γ_p at distance 3 from r'. In particular, there are 63 points at distance 2 from p inside $\mathcal{G}$, and G_p induces the same action on these points as on $p^\perp$. The point r' is the unique point in Γ_p which is at distance 3 from r. Consider $\Gamma_{r'}$. If p' is a point in $\Gamma_{r'}$ at distance 2 from r, then, by the above, all the points at distance 1 or 2 from p' in $\Gamma_{r'}$ are also at distance 2 from r. Since $\Gamma_{r'}$ is connected, we find that all its points are at distance 2 from r. This implies that Γ contains $1 + 63 + 63 + 1$ points. Moreover, each element of $\mathcal{S}$ fixes $1 + 9 + 9 + 1 = 20$ points. Thus the conjugacy class of S in G contains $128.28/20$ elements. A contradiction.

This finishes the proof of Proposition 3.4.

Proposition 3.13 *If Γ satisfies $(*)_3$, then it is isomorphic to one of the* *extended generalized hexagon on* 36, 162, 100, *or* 1782 *points related to* $G_2(2)$, $PSU_4(3)$, HJ, *respectively, Suz.*

We prove this proposition in a series of lemmas. Let p and q be two points at distance 2 in $\mathcal{G}$. Set H to be the set of points in Γ_p that are not in Γ_q.

Lemma 3.14   *(i) Let C be a circle and x a point of Γ not in C. Then $|x^\perp \cap C| = 0$, 2 or $s+1$.*

(ii) Each line of Γ_p meets H in 0, $s-1$ or all points.

(iii) Any point of Γ_p outside of H is on $t_2 + 1$ lines of Γ_p meeting H non-trivially.

Proof.   The proof is similar to the one of 3.9(i), and is left to the reader.

Lemma 3.15 *The group G acts as a rank 3 permutation group on $\mathcal{G}$. The graph Γ is strongly regular with parameters $k = (1+s)((1+t_2)(1+st) + s^2t(t-t_2))/(t_2+1)$, $\lambda = s(t+1) + s^3t(t-t_2)/(t_2+1)$ and $\mu = k - (s^2 - 1)(t^2 + t + t_2 + 1)/(t_2+1)$.*

Proof.   We show that the diameter of $\mathcal{G}$ is 2. Then the first part of the lemma follows immediately.

Let p and q be two points at distance 2 in $\mathcal{G}$, and let r be a point adjacent to q, but at distance 3 from p. Then all points in Γ_q that are collinear with r or at distance 3 from r are contained in the complement of $p^\perp \cap q^\perp$. Fix a point u collinear with both p and q. Then this point is at distance 2 from r inside Γ_q. Since $t_2 < t$, there is a line in Γ_q on u inside $p^\perp \cap q^\perp$, see the above lemma. This line contains a point at distance 1 or 3 from r inside Γ_q, contradicting that $r^\perp \cap p^\perp \cap q^\perp$ is empty.

The valency of the graph $\mathcal{G}$ equals the number of points in the local near hexagon, λ equals the number of points at distance 1 or 3 from a given point in the local near hexagon. So k and λ are as given.

Fix two points p and q at distance 2 in $\mathcal{G}$, and let r be adjacent to both p and q. Inside Γ_r we see that there are $t_2 + 1$ circles on r and p meeting $q^\perp$ in 2 points, while all other circles meet $q^\perp$ in all points but p. Fix a circle on p and r whose only point not in $q^\perp$ is p. This circle is a line l of Γ_p. By the above we see that there are $(t_2 + 1)(s^2 - 1)$ points of Γ_p at distance 1 from l that are not in $q^\perp$. For each point r' on l the set of points at distance 3 from r' and not in $q^\perp$ is the set of points at distance 3 from p and 2 from q inside $\Gamma_{r'}$. There are $(s-1)s(t-t_2)(t+t_2+1)/(t_2+1)$ of such points. This implies that there are $((s+1)/s)(s-1)s(t-t_2)(t+t_2+1)/(t_2+1)$ points at distance 2 from l not in $q^\perp$. Hence $k - \mu = (s^2 - 1)(t^2 + t + t_2 + 1)/(t_2 + 1)$.

Lemma 3.16 Γ *is an extension of a generalized hexagon of order* $(2,1)$, $(2,2)$, $(4,1)$ *or* $(4,4)$.

Proof. The only parameter sets (s, t_2, t) passing the feasibility conditions for the strongly regular graph $\mathcal{G}$ (see [2]) are $(2,0,1)$, $(2,0,2)$, $(4,0,1)$ and $(4,0,4)$.

Now we can finish the proof of Proposition 3.13 in a way similar to the proof of Lemma 3.11. We will only cover the case where Γ is an extension of a generalized hexagon of order $(4,4)$. The other cases are left to the reader, who can find detailed analyses of these cases in [8]. Thus suppose Γ is an extension of the classical hexagon of order $(4,4)$ or its dual. We show that the graph $\mathcal{G}$ is unique up to isomorphism, which clearly implies uniqueness of Γ.

By Lemma 3.15 the graph is strongly regular with parameters $(k, \lambda, \mu) = (819, 1040, 1050)$. Let p be a point of Γ and G_p the stabilizer of p in G. Then G_p is isomorphic to $G_2(4)$ or $G_2(4){:}2$. The group G_p fixes p, is transitive on the points in Γ_p and has a unique orbit of length 416 on the remaining points of $\mathcal{G}$, that are all at distance 2 from p. The group G_p contains a unique conjugacy class of subgroups of index 416, these subgroups are isomorphic to HJ, respectively, $HJ{:}2$. Such a subgroup has two orbits on the points of the classical generalized hexagon of length 525 and 840, and two orbits on the point set of the dual classical hexagon of length 315 and 1050. Fixing an orbit O of size 1050 on the points of the dual classical hexagon, there are 315 of such orbits in $\mathcal{O} = O^{G_p}$ intersecting it in 806 points and 100 intersecting it in 788 points.

Let q be a point at distance 2 from p, then the group $G_{p,q}$ is transitive on the 1050 points in $p^\perp \cap q^\perp$. We conclude that Γ_p is the dual classical hexagon related to $G_2(4)$. Moreover, we can identify the graph $\mathcal{G}$ with the graph whose point set consists of p, the points of Γ_p, and the set of orbits of length 1050 in $\mathcal{O}$. The point p is adjacent to the points of Γ_p, two points of Γ_p are adjacent if and only if their distance is 1 or 3 in Γ_p, a point of Γ_p is adjacent to an orbit of $\mathcal{O}$ if and only if it is in the orbit, and finally two orbits are adjacent if and only if they intersect in 806 points. This shows uniqueness of the graph $\mathcal{G}$ and hence also of Γ.

References

[1] J. van Bon, Some extended generalized hexagons, in *Finite geometry and*

combinatorics, eds. F. De Clerck et al., LMS Lecture Notes Ser., 191, CUP, Cambridge, 1993, pp 395–403.

[2] A.E. Brouwer, A.M. Cohen and A. Neumaier, *Distance-regular graphs*, Ergebnisse der Math., Springer Verlag, 1989.

[3] F. Buekenhout, Diagram geometries for sporadic groups, in *Finite groups – coming of age*, Contemp. Math. **45** (1985), 1–32, AMS, Providence.

[4] P. Cameron, D. Hughes and A. Pasini, Extended generalised quadrangles, *Geom. Dedicata* **35** (1990), 193–228.

[5] A.M. Cohen and E.E. Shult, Affine polar spaces, *Geom. Dedicata* **35** (1990), 43–76.

[6] J. H. Conway et al., *Atlas of finite groups*, Clarendon Press, Oxford, 1985.

[7] H. Cuypers, Extended near hexagons and line systems, in preparation.

[8] H. Cuypers, Extended generalized hexagons and the Suzuki chain, preprint.

[9] H. Cuypers and A. Pasini, Locally polar geometries with affine planes, *Europ. J. Comb.* **13** (1992), 39–57.

[10] D.G. Higman, Flag–transitive collineation groups of finite projective spaces, *Illinois J. Math.* **6** (1962), 432–446.

[11] Th. Meixner, A geometric characterization of the simple group Co_2, to appear in *J. of Algebra*.

[12] A. Pasini, On locally polar spaces whose planes are affine, *Geom. Dedicata* **34** (1990), 35–56.

[13] A. Pasini and S. Yoshiara, Flag–transitive Buekenhout geometries, in *Combinatorics '90*, eds A. Barlotti et al., Elsevier Science Publ., 1992, pp 403–447.

[14] G. Seitz, Flag–transitive subgroups of Chevalley groups, *Ann. of Math.* **97** (1973), 27–56.

[15] R. Weiss, Extended generalized hexagons, *Math. Proc. Camb. Phil. Soc.* **108** (1990), 7–19.

[16] R. Weiss, A geometric characterization of the groups McL and Co_3, *J. London Math. Soc.* **44** (1991), 261–269.

[17] S. Yoshiara, On some extended dual polar spaces I, to appear in *Europ. J. of Comb.*

Hans Cuypers
Department of Mathematics
Eindhoven University of Technology
P.O. Box 513, 5600 MB Eindhoven
The Netherlands

DISCONNECTED LINEAR GROUPS
AND RESTRICTIONS OF REPRESENTATIONS

BEN FORD

1. INTRODUCTION

In [1, 2, 9, 10, 13], Dynkin, Seitz and Testerman classified the maximal closed connected subgroups of the simple algebraic groups over an algebraically closed field K of characteristic $p \geq 0$. The hard part of their analyses, for subgroups of the groups of classical type, concerns an irreducible, closed, connected subgroup G of $SL(V)$ for some K-vector space V. They determine explicitly all possibilities for closed connected overgroups Y of G in $I(V)$ (where $I(V) = SL(V), SO(V)$, or $Sp(V)$ depending upon the form on V preserved by G); the results appear in tables giving the high weights of the modules $V|_G$ and $V|_Y$.

The question of inclusion relations among irreducible subgroups of $SL(V)$, in addition to having implications for the subgroup structures of classical groups, is of interest in its own right. In this paper we present some recent results concerning this question when we allow subgroups that are not connected (the full proofs may be found in [3, 4, 5]). Specifically, in Sections 2 and 3 we discuss the structure and some of the methods of the proof of Theorem 1.

Let G be a non-connected algebraic group with simple identity component X. Let V be an irreducible KG-module with restricted X-high weight(s).

Theorem 1. *Let Y be a simple algebraic group of classical type such that $X < Y < SL(V)$ and $V|_Y$ is irreducible with restricted high weight. Then $Y = SO(V)$, $Y = Sp(V)$, or (X, Y, V) appears in an explicit list, with the embedding $X \hookrightarrow Y$ and the high weights of $V|_Y$ and $V|_X$ identified.*

As is true for connected G, there are surprisingly few examples of such triples.

The problem one faces when approaching this problem is how to limit the embeddings $X \hookrightarrow Y$ which must be considered. In our case, since Y is assumed to be of classical type, it has a natural module W. The action of X on W provides the information we need. The proof of Theorem 1 differs greatly depending on whether $W|_X$ is reducible ([4]) or irreducible ([5]).

Note that if $V|_X$ is irreducible, then we are back in the connected case discussed above, with the extra requirement that X have an outer automorphism

1991 *Mathematics Subject Classification*. 20G05, 20C30.

Supported in part by the National Science Foundation

which acts on V. We can examine the list obtained by Seitz in [9] and determine when this condition is satisfied; this gives several triples (X, Y, V). So we assume below that $V|_X$ is reducible.

In Section 4 we discuss an interesting application of some of the methods developed in this study to a problem in the representation theory of the symmetric groups.

2. WHEN W IS AN IRREDUCIBLE X-MODULE

We assume we have a triple (X, Y, V) as in Theorem 1, with $W|_X$ irreducible.

Assume for this description that X is not of type D_4 (the argument is much the same in that case, but somewhat more complex) and that $G \neq X$ (the case $G = X$ is the "connected" case discussed above). So $G = \mathrm{Aut}(X)$, as except for D_4, simple algebraic groups have at most one non-identity outer automorphism up to conjugation by a group element; and X has type A_m, D_m or E_6. We let t be a representative in G of the non-identity element in G/X. Then let $B_X > T_X$ be t-stable Borel and maximal toral subgroups of X, respectively (see [11, Theorem 7.5]). We let $\{-\beta_1, ..., -\beta_m\}$ denote the simple roots of X relative to B_X and T_X, so we are using the opposite of the standard convention for Borel and parabolic subgroups of X; we use similar conventions for parabolic subgroups of Y. The high weight of $W|_X$ will be denoted δ.

As we are assuming that $V|_X$ is reducible and $X \neq D_4$, we have $V|_X = V_1 \oplus V_2$, with V_1 and V_2 non-isomorphic irreducible X-modules.

The proof in the case $W|_X$ irreducible is based on a particular construction of a parabolic subgroup P_Y of Y containing B_X and satisfying some other nice properties.

The T_X-weights which appear in W have the form $\delta - \sum e_i \beta_i$ for integers $e_i \geq 0$. We will call $\sum e_i$ the *level* of this weight.

Thanks to the results in [12] and [8], which tell us that weights which appear in the Weyl module $W(\delta)$ also appear in W, we know which weights appear at each level. We let

$$W_i = \bigoplus_{\gamma \text{ of level } i} W_\gamma.$$

Then we have a flag

$$W = \bigoplus_{i \geq 0} W_i \geq \bigoplus_{i \geq 1} W_i \geq \cdots \geq 0$$

in W, and we let P_Y be the stabilizer in Y of this flag. Then $B_X \leq P_Y$ and P_Y is a parabolic subgroup of Y.

Now we investigate instead the restriction of V to P_Y and to B_X. The parabolic subgroup P_Y is t-stable, and we find that $[V, U_X] = [V, Q_Y]$ since t interchanges the T_X-high weight spaces in V_1 and V_2 (U_X, Q_Y are the unipotent radicals of B_X, P_Y respectively). Thus

$$V/[V, Q_Y] = V_1/[V_1, U_X] \oplus V_2/[V_2, U_X]$$

has dimension 2. But this implies that one of the simple factors of the derived group of a Levi factor L_Y of P_Y must be of type A_1 (since $V/[V, Q_Y]$ is an irreducible L'_Y-module by 2.1 of [9]). In other words, one of the spaces W_i must have dimension 2. Combinatorial arguments then give strong conditions on the high weight δ of $W|_X$.

Thus we have established a strong connection between V and W, severely limiting the embeddings $X \hookrightarrow Y$ that we must consider.

Having ruled out most embeddings, we deal with those that remain by performing much the same "level" construction as above, except we begin with a parabolic subgroup P_X and define the P_X-level of the weight $\delta - \sum e_i \beta_i$ to be $\sum e_i$ where the sum is taken over those i such that β_i is not in the root system of the Levi factor of P_X. Choosing our parabolic subgroups P_X judiciously (and using some *ad hoc* methods for the few remaining cases), our list of triples (X, Y, V) is arrived at.

The most interesting example that arises here is the following (labels of 0 are omitted):

X	Y	$W\|_X$	$V\|_X$	$V\|_Y$	char(K)
A_3	D_{10}	$2\delta_2$	(diagram)	(diagram)	$p \neq 2, 3, 5, 7$

3. WHEN X ACTS REDUCIBLY ON W

We assume we have a triple (X, Y, V) as in Theorem 1, with $W|_X$ reducible. It is here that the infinite families of examples appear. The hardest part of this analysis is the case $X = D_n < B_n = Y$, and the methods used in that case are somewhat simpler to describe in the context of a different problem to which they are applicable; this is done in the next section. Here we give a brief description of how we get to the point where those methods become applicable.

Let $\{\alpha_1, \ldots, \alpha_n\}$ and $\{\lambda_1, \ldots, \lambda_n\}$ be the sets of simple roots and fundamental dominant weights, respectively, of $Y = B_n$. Let V be an irreducible Y-module with restricted high weight $\lambda = \sum a_i \lambda_i$. Then let D_n be the subgroup of B_n generated by the root subgroups corresponding to long roots, and let $t \in B_n$ be a representative of the Weyl group reflection s_{α_n}.

Each of the sums

$$V(i) = \sum_{\mu = \lambda - b_1 \alpha_1 - \cdots - b_{n-1} \alpha_{n-1} - i \alpha_n} V_\mu$$

is a D_n-submodule of V. So if we hope to have $V|_{D_n\langle t \rangle}$ irreducible, at most two of these can be non-zero. It is not hard to see that it must be exactly two (else $V|_{D_n}$ is irreducible), and that this implies $a_n = 1$.

Given this condition, the question becomes: "When are $V(0)$ and $V(1)$ irreducible as D_n-modules?" It is to this question that the methods outlined below give an answer. The result is the setup below, with the coefficients a_i satisfying the congruences

$$a_i + a_j \equiv i - j \pmod{p}$$

whenever a_i and a_j are non-zero coefficients with only 0's between them and $i < j < n$; and

$$2a_i \equiv -2(n-i) - 1 \pmod{p}$$

when a_i is the last non-zero coefficient before $a_n = 1$. Recall that $V|_X = V_1 \oplus V_2$, with V_1 and V_2 irreducible X-modules, interchanged by the graph automorphism.

X	Y	$W|_X$	$V_1|_X$	$V|_Y$	char(K)

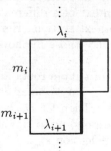

$$D_n \quad B_n \quad \text{usual} \qquad\qquad\qquad\qquad \overset{a_1 \quad a_2 \qquad a_{n-1}\ 1}{\bullet\!-\!\bullet \cdots \bullet\!\Rrightarrow\!\bullet} \qquad p \neq 2$$

4. REPRESENTATIONS OF THE SYMMETRIC GROUPS

In [6], Jantzen and Seitz conjectured that an irreducible representation M of Sym_n is irreducible for $\mathrm{Sym}_{n-1} < \mathrm{Sym}_n$ if and only if the p-regular partition $\lambda = (\lambda_1^{m_1}, \ldots, \lambda_l^{m_l})$ of n corresponding to M satisfies the congruences

$$\lambda_i - \lambda_{i+1} + m_i + m_{i+1} \equiv 0 \pmod{p}.$$

Note that this is just the condition that the hooks pictured here

have length a multiple of p.

This conjecture translates via the Schur functor into a conjecture about restrictions to SL_{n-1} of irreducible SL_n-representations. Jantzen and Seitz prove the "if" direction of this translation, and methods similar to some of those used in the work described above, combined with a theorem of A. Kleshchev ([7]), can be used to prove the "only if" direction.

Let $\{\alpha_1, ..., \alpha_n\}$ be the set of fundamental roots of the root system $\Sigma = \Sigma(\mathrm{SL}_n)$ (with respect to a fixed Borel subgroup B and maximal torus T of SL_n). The fundamental dominant weights of T are denoted $\lambda_1, ..., \lambda_n$. Let $V = V(\lambda)$ be a non-trivial irreducible restricted high weight module for SL_n of high weight $\lambda = \sum a_i \lambda_i$.

The maximal parabolic subgroup of SL_n corresponding to $\alpha_n \in \Pi = \Pi(\mathrm{SL}_n)$ is denoted P, and $P = QL$ is its Levi decomposition with respect to B and T. So $L' \cong \mathrm{SL}_{n-1}$. We let $V^1 = \oplus V_\mu$, with the sum taken over those $\mu = \lambda - \sum b_i \alpha_i$ with $b_{n-1} = 1$. Let v^+ be a non-zero T-high weight vector in V.

To prove the conjecture, it suffices to prove that if V^1 is an irreducible L'-module, then the congruences

$$a_i + a_j + j - i \equiv 0 \pmod{p}$$

hold for all pairs of successive non-zero coefficients (a_i, a_j). We do this working over the Lie algebra of SL_n (which is legitimate as λ is restricted). We use the standard basis $\{e_\alpha, f_\alpha, h_i \,|\, \alpha \in \Sigma^+, \ 1 \le i \le n-1\}$ for the Lie algebra, and we let $V_{i,l}$ be the subspace of $V_{\lambda-(\alpha_i+...\alpha_l)}$ spanned by all elements other than $f_{\alpha_i+\cdots+\alpha_l} v^+$ of the form

$$f_{\alpha_i+\cdots+\alpha_j} f_{\alpha_{j+1}+\cdots+\alpha_{j'}} \cdots f_{\alpha_{j(m)+1}+\cdots+\alpha_l} v^+,$$

with $i < j < \cdots < j^{(m)} < l$. To ease the notational pain, we henceforth write $f_{[i,j]}$ for $f_{\alpha_i+\cdots+\alpha_j}$.

The L'-module V^1 has a filtration

$$
\begin{aligned}
0 &\le \langle f_{[k,n-1]}v^+ \rangle \le \langle f_{[k,n-1]}v^+, f_{[k-1,n-1]}v^+ \rangle \\
&\le \langle f_{[k,n-1]}v^+, f_{[k-1,n-1]}v^+, f_{[k-2,n-1]}v^+ \rangle \\
&\le \cdots \le V^1,
\end{aligned}
$$

where a_k is the "last" non-zero coefficient of λ, i.e. $a_l = 0$ for every $l > k$. Since V^1 is irreducible as an L'-module, this filtration has only one member, i.e. $f_{[i,n-1]}v^+ \in \langle f_{[k,n-1]}v^+ \rangle$ for all $i < k$. In particular, $f_{[i,n-1]}v^+ \in V_{i,n-1}$ for all $i < k$.

Through a series of lemmas, we show that this implies

(1) If $i < k$, then $f_{[i,k]}v^+ \in V_{i,k}$.

(2) Let a_i, a_m be non-zero coefficients with $m > i$; let a_l be the last non-zero coefficient before a_m. Then $f_{[r,l]}v^+ \in V_{r,l}$ for all $i \le r < l$.

(3) With a_i, a_m as above, if $f_{[r,m]}v^+ \in V_{r,m}$ for all $i \le r < m$, then $f_{[i,j]}v^+ \in V_{i,j}$, where a_j is the first non-zero label after a_i.

These facts imply that $f_{[i,j]}v^+ \in V_{i,j}$ for any pair of successive non-zero labels a_i and a_j. Assume we have such a pair. Then the $\lambda - (\alpha_i + \cdots + \alpha_j)$-weight space is spanned by $\{F_l v^+ = f_{[i,l]} f_{[l+1,j]} v^+ \,|\, i \le l < j\} \cup \{f_{[i,j]} v^+\}$. So

$f_{[i,j]}v^+ \in V_{i,j}$ if and only if there is a relation

$$0 = f_{[i,j]}v^+ + \sum_{l=i}^{j-1} b_l F_l v^+.$$

The vector on the right-hand side of this equation is not a high weight vector, so it is 0 if and only if it is killed by e_β for every positive root β.

We apply e_{α_l} to the equation for $i \leq l \leq j$, and obtain a series of equations which reduce to $b_i = \cdots = b_{j-1} = -1/a_j$ and, as we wished, $\frac{1}{a_j}(a_i+1) + \frac{1}{a_j}(j - i - 1) = -1$, or $a_j + a_i = i - j$.

REFERENCES

1. Eugene B. Dynkin, *Maximal subgroups of the classical groups*, Amer. Math. Soc. Transl. Ser. 2 **6** (1957), 245–378.
2. _____, *Semisimple subalgebras of semisimple Lie algebras*, Amer. Math. Soc. Transl. Ser. 2 **6** (1957), 111–244.
3. Ben Ford, *Irreducible restrictions of representations of the symmetric groups*, (to appear).
4. _____, *Overgroups of irreducible linear groups, I*, (to appear).
5. _____, *Overgroups of irreducible linear groups, II*, (to appear).
6. Jens C. Jantzen and Gary M. Seitz, *On the representation theory of the symmetric groups*, Proc. London Math. Soc. (3) **65** (1992), 475–504.
7. Aleksander S. Kleshchev, *On restrictions of irreducible modular representations of semisimple algebraic groups and symmetric groups to some natural subgroups*, Preprint 3(494), Academy of Sciences of Belarus Institute of Mathematics, February 1993.
8. Alexander A. Premet, *Weights of infinitesimally irreducible representations of Chevalley groups over a field of prime characteristic*, Math. USSR-Sb. **61** (1988), 167–183.
9. Gary M. Seitz, *The maximal subgroups of classical algebraic groups*, Mem. Amer. Math. Soc. **67** (1987), no. 365, 1–286.
10. _____, *Maximal subgroups of exceptional algebraic groups*, Mem. Amer. Math. Soc. **90** (1991), no. 441, 1–197.
11. Robert Steinberg, *Endomorphisms of linear algebraic groups*, Mem. Amer. Math. Soc. (1968), no. 80.
12. Irene D. Suprunenko, *The invariance of the weight system of irreducible representations of algebraic groups and Lie algebras of type A_l with restricted highest weights, under reduction modulo p*, Vestsĭ Akad. Navuk BSSR Ser. Fĭz.-Mat. Navuk **2** (1983), 18–22 (Russian).
13. Donna M. Testerman, *Irreducible subgroups of exceptional algebraic groups*, Mem. Amer. Math. Soc. **75** (1989), no. 390, 1–190.

PRODUCTS OF CONJUGACY CLASSES IN ALGEBRAIC GROUPS AND GENERATORS OF DENSE SUBGROUPS

Nikolai L.Gordeev

Abstract. The survey of results contained in [**G1**] is given here. We consider two connected problems in algebraic groups. The first is the decomposition of groups into products of conjugacy classes. The second is the problem on the minimal numbers of generators of a dense subgroup from a given conjugacy class. We show, also, that both problems are connected with some questions arising from the theory of actions of algebraic groups on algebraic varieties.

1. Multiclasses of Algebraic Groups

Let G be a group and $C_1, \ldots, C_k$ be conjugacy classes of G. The product

$$M_k = C_1 C_2 \ldots C_k = \{\, g_1 g_2 \ldots g_k \mid g_i \in C_i \,\}$$

is called a k-class (if $k > 1$ we suppose $C_i \not\subset Z(G)$). Any k-class we also call a multiclass of G. When $C_1 = C_2 = \ldots = C_k = C$ we will also write C^k instead of M_k. There are some questions in group theory connected with multiclasses (see, for instance, [**AH**]). One of them is when $G = M_k$ for some M_k.

Our interest in multiclasses is inspired by the following problem connected with actions of algebraic groups on algebraic varieties. Let an algebraic group G acts regularly on an algebraic variety X. Sometimes we need to know or, at any rate, to estimate the number corank g = codimension of the subvariety of fixed points of g, where $g \in G$ (sometimes we need to know corank g for a given element g and sometimes to estimate coranks for some set of elements). First of all, we can see that coranks are the same for the elements of a conjugacy class. Let C be a conjugacy class of the group G and $g \in C$. Suppose that $G = C^k$ for some k. Then any element of G can be represented as a product $g_1 g_2 \ldots g_k$ where all elements $g_1, \ldots, g_k$ belong to C and, hence, have the same corank. In cases when the action is "good" (see 5. Examples of c-actions) we can apply the inequality for the codimension of the intersection of subvarieties to the intersection of the subvarieties of fixed points of elements g_i. Thus, corank $g' \leq k$ corank g for every $g' \in G$. Suppose we know the corank for some element $g_0 \in G$. Then corank $g \geq$ (corank g_0)/k. Thus, if we can decompose the group G into the product C^k we can estimate (under some conditions) coranks of elements from the class C through the number k and some known corank g_0. To calculate or even to estimate the smallest number k (if it exists) such that $G = C^k$ is rather difficult. It is more easy to calculate this number "up to Zariski closure", that is, to calculate the number l such that

$G = \overline{C^l}$ where $\overline{C^l}$ is the closure of C^l with respect to Zariski topology. One can see that if $G = \overline{C^l}$ then $C^{2l} = G$ (it is a simple fact following from the elementary theory of algebraic groups). Therefore, $k \leq 2l$, and, if we know l, we can take the inequality corank $g \geq (\text{corank } g_0)/2l$ instead of corank $g \geq (\text{corank } g_0)/k$. In fact, we can use the inequality corank $g \geq (\text{corank } g_0)/l$ if we consider more special actions (say, actions on linear or projective spaces). There is a different approach to the same problem which is also connected with a products of conjugacy classes. Let we know that r elements $g_1, \ldots, g_r \in C$ generate a gense subgroup (with respect to Zariski topology) in $\langle C \rangle$. Suppose also $\langle C \rangle = G$. Since the action G on X is regular, $X^{\langle g_1, \ldots, g_r \rangle} = X^G$. If we can apply the inequality for the codimensions of intersections of $\{X^{g_i}\}$, then we can obtain the inequality corank $g \geq$ codim X^G/r. On the other hand, the smallest number r, such that there are r elements from the class C generating a dense subgroup of G, could be estimated (for the semisimple groups) through the smallest number k such that $\overline{C^k} = G$ (see 4, Theorem 3). Thus, the generating of dense subgroup by elements from the given class C (where $\langle C \rangle = G$), the decomposition G into C^k (up to closure) and the estimate of coranks of elements of C are connected. In fact, using this connections we can enlarge the area of applying these constructions (see 5, Examples of ac-actions and Theorem 5).

Thus, the problems of estimating of coranks and the problem of generating of dense subgroups by elements from the same conjugacy classes lead us to the problem of the representation $G = \overline{C^k}$. This, in its turn, rises the interest to the behaviour of affine varieties like $\overline{C^k}$. The natural step here is to consider the more general type of affine varieties, namely, $\overline{M_k}$ (e.g. the closure of a multiclass of G). Except the analogue of the traditional question when $G = \overline{M_k}$ (see 2, Theorem 1), there are natural questions on the structure of $\overline{M_k}$, the law of multiplication, etc. The description of the semigroup generated by the semisimple classes is $SL_n(K)$ (see 3, Theorem 2) shows, for instance, that we have here rather rich and slim structure.

The use of the closures of multiclasses has already given the possibility to carry out some results on linear actions to some non-linear cases (see 5, Theorems 6 and 7). We hope that the investigation of multiclasses in algebraic groups which is the self-interesting problem give us some more possibilities to study non-linear actions.

Here we will consider algebraic groups over an algebraically closed field K of characteristic zero.

2. COVERING NUMBERS OF SEMISIMPLE ALGEBRAIC GROUPS

The smallest integer k such that $G = C^k$ for every conjugacy class C which generates G is called the covering number of G ([**AH**]) and denoted by $cn(G)$. The smallest integer k such that $G = M_k$ for every k-class which consists of classes $C_1, \ldots, C_k$ such that $\langle C_1 \rangle = \ldots = \langle C_k \rangle = G$ is called the extended covering number and denoted by $ecn(G)$ ([**AH**]). In the case when G is an algebraic group we can define "topological" covering and extended covering numbers in the following way

$$\overline{cn}(G) = \min\{k \mid \overline{C^k} = G \text{ for any class } C \text{ such that } \langle C \rangle = G\}$$

$$\overline{ecn}(G) = \min\{k \mid \overline{C_1 C_2 \ldots C_k} = G$$
$$\text{for all classes } C_1, \ldots, C_k \text{ such that } \langle C_1 \rangle = \ldots = \langle C_k \rangle = G\}.$$

It is easy to see that

$$cn(G) \leq ecn(G), \overline{cn}(G) \leq \overline{ecn}(G), cn(G) \leq 2\overline{cn}(G), ecn(G) \leq 2\overline{ecn}(G).$$

Moreover, for a perfect algebraic group G (that is, G is connected and $G = [G,G]$) it is easy to show $\overline{ccn}(G) \leq \dim G$.

THEOREM 1. *Let G be a semisimple algebraic group over a field K $(\overline{K} = K$, char $K = 0)$, let r be the maximum of the ranks of simple components of G. Then*

$$\overline{cn}(G) \leq 2r, \overline{ecn}(G) \leq 2r + 1.$$

Remark. If G does not contain simple components of types B_n, F_4, G_2 and if $C_1, \ldots, C_k$ are semisimple conjugary classes such that $\langle C_1 \rangle = \ldots = \langle C_k \rangle = G$ then $\overline{C_1 C_2 \ldots C_k} = G$ if $k \geq r+1$.

The proof of this theorem is based on the consideration of the action of G on G by conjugation. Here we have the algebraic factor G/G and the factor morphism $\pi: G \rightarrow G/G$ (see [**Kr**], Ch. II). Moreover, there is a natural isomorphism between G/G and T/W where T is a maximal torus and W is the Weyl group of G ([**SprSt**], §3). Thus, the dimension of G/G is equal to the rank of G. Further, it is easy to see that $\overline{M_k} = G$ if $\overline{\pi(M_k)} = G/G$. Now the proof can be obtained by the estimating of the growth of $\dim \pi(M_k)$ with respect to k.

Conjecture. $cn(G/Z(G)) \leq 2r$, $ecn(G/Z(G)) \leq 2r + 1$ (that is, the estimates of theorem 1 are true for the covering and extended covering numbers). Moreover, we suppose that the assumption char$K=0$ could be omitted.

It would be interesting to obtain the estimates for $cn(G)$, $ecn(G)$ in the case of Chevalley groups, and for "topological" covering numbers in the case of Lie groups over C and R (with respect to the usual topology).

3. THE EXAMPLE: $SL_n(K)$

Here we consider the multiplication of semisimple conjugacy classes in $SL_n(K)$. The description is given in the language of Young diagrams.

Let $p, q \leq n$; $\lambda_p = (p_1, \ldots, p_k)$, $\gamma_q = (q_1, \ldots, q_e)$ be partitions where $p_1 \geq p_2 \geq \cdots \geq p_k$, $p = p_1 + p_2 + \cdots + p_k$, $q_1 \geq q_2 \geq \cdots \geq q_e$, $q = q_1 + q_2 + \cdots + q_e$. Let

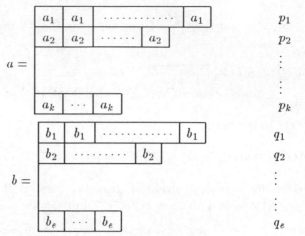

be the corresponding Young diagrams with $a_i, b_j \in K$, $a_i, b_j \neq 0$, $a_i \neq a_s$, $b_j \neq b_r$ if $i \neq s$, $j \neq r$.

Suppose $p_1 \geq q_1$. In all cases, except $k = e = 2$, $p = q = n$, we define

(the table for $a \cdot b = c$ with rows labelled $s_1 = p_1 + q_1 - n$, $s_2 = p_1 + q_2 - n$, ..., $s_t = p_1 + q_t - n$, entries $a_1 b_1$, $a_1 b_2$, ..., $a_1 b_t$)

if $p_1 + q_1 - n > 0$ (c is a right corner of b "corrected" by a_1; here $s_t > 0$, $p_1 + q_{t+1} - n \leq 0$), and

$$a \cdot b = (0)$$

if $p_1 + q_1 - n \leq 0$. In the case $k = e = 2$, $p = q = n$, we define

$$a \cdot b = \langle c, \delta \rangle$$

where c is obtained by the same rule as above and $\delta = a_1 a_2 b_1 b_2$. Let

$$M[a] = \{g \in SL_n(K) \mid \mathrm{rank}(g - a_i 1) \leq n - p_i\},$$

$$M[(0)] = SL_n(K).$$

If $n - p = 21$, we define

$$M[\langle a, \delta \rangle] = M[a] \cap \{g \in SL_n(K) \mid \det(g - x1)$$

$$= (-1)^n \prod_{i=1}^{k} (x - a_i)^{p_i} \prod_{j=1}^{l} (x^2 + \varepsilon_j x + \delta) \quad \text{for arbitrary } \varepsilon_j \in K\}.$$

It is easy to see that $M[a]$, $M[\langle a, b \rangle]$ is closed, connected and invariant under conjugations. (If $p = n$, $M[a]$ is a conjugacy class of $SL_n(K)$.)

THEOREM 2. *Let M_k be a product of semisimple conjugacy classes in $SL_n(K)$. Then $\overline{M}_k$ coincides with some $M[a]$ or $M[\langle a, \delta \rangle]$. Moreover,*

$$M[a]M[b] = M[a \cdot b],$$

$$M[\langle a, \delta \rangle]M[b] = M[a \cdot b],$$
$$M[\langle a, \delta \rangle]M[\langle b, \varepsilon \rangle] = M[a \cdot b].$$

The first step of the proof here is to show that the closure of the product of any two semisimple conjugacy classes coincides with $SL_n(K)$ if the sum of the maximal multiplicities of the eigenvalues of both classes less or equal to n and if one of them has more than 2 different eigenvalues. This can be proved by induction with respect to n. The next step is the consideration of a product of two elements in "general position" from two semisimple conjugacy classes. Then we prove our statement, comparing the results of both steps and using the well known theorem of Blichfeldt on the groups generated by matricies with two eigenvalues ([**Bl**]).

4. MULTICLASSES OF SEMISIMPLE ALGEBRAIC GROUPS AND DENSE SUBGROUPS

Let G be a semisimple algebraic group, and C be a conjugacy class of $G \setminus Z(G)$. We define

$$\mathrm{gen}\, C = \min \{ n \mid \text{there exist elements } g_1, \dots, g_n \in C \text{ such that the group}$$
$$\langle g_1, \dots, g_n \rangle \text{ is dense in } \langle C \rangle \},$$

$$k(C) = \min \{ m \mid \overline{C^m} = \langle C \rangle \}.$$

THEOREM 3. $k(C) \le \mathrm{gen}\, C \le k(C) + 1$.

The first inequality can be proved in the following way. We consider the morphism

$$\Theta : \underbrace{G \times G \times \cdots \times G}_{k \text{ factors}} \to G$$

(where $k = k(C)$) defined by the formula

$$\Theta((x_1, \dots, x_k)) = \left(\prod_{i=1}^{k} x_i h_i x_i^{-1} \right) \left(\prod_{j=0}^{k-1} h_{k-j}^{-1} \right)$$

where $h_1, \dots, h_k \in C$ are elements in "general position". Here we have $\mathrm{Im}\, \Theta = C^k h_k^{-1} h_{k-1}^{-1} \dots h_1^{-1}$. Thus $\dim \mathrm{Im}\, \Theta = \dim C^k$. Therefore it is enough to prove $\dim \operatorname{Im} \Theta = \dim G$. It can be done by the consideration of the differential $d\Theta$ (in the point $(1, 1, \ldots, 1)$). The second inequality follows from two facts. The first is the existence of an element $g \in C^k$ such that g^2 is semisimple and regular if $\overline{C^k} = G$. The second is $\langle g, h \rangle = G$ for some $h \in C$.

Remark. In fact, we have more general results on multiclasses of different classes and their connections with dense subgroups.

We can also obtain some estimates for the minimal number of generators of dense subgroups from a given conjugacy class in the case of perfect groups.

Let H be a perfect algebraic group (that is H is connected and $[H, H] = H$), $G = H/R_u(H)$ where $R_u(H)$ be the unipotent radical of G, $V = R_u(H)/[R_u(H), R_u(H)]$. The action (by conjugation) of H on $R_u(H)$ defines the structure of G-module on V. Let V_i be the set of irreducible components of G-module V. Denote by $\langle V_i, V \rangle$ the multiplicity of V_i in V. We put

$$n_i = \min\{n \in Z \mid n \geq \frac{\langle V_i, V \rangle}{\dim V}\},$$

$$n(H) = \max\{n_i\}.$$

The proof of the following theorem can be done by the reduction to the case $[R_u(H), R_u(H)] = 1$ and by the consideration of G-module V.

THEOREM 4. *Let $Q \subset H$ be a conjugacy class of H and C be the image of Q in G. Suppose $\langle C \rangle = G$. Then*

$$\operatorname{gen} Q \leq (\operatorname{gen} C)(n(H) + 1).$$

5. APPLICATIONS

Let G be an algebraic group which acts regularly and faithfully on an algebraic variety X (we suppose that X is irreducible and separated). The number

$$\min(G, X) = \min \{\operatorname{corank} g \mid g \in G, g \notin Z(G)\}$$

where

$$\operatorname{corank} g = \dim X - \dim X^g,$$

$$X^g = \{x \in X \mid g(x) = x\}$$

is one of the characteristics of the action. In particular, this number can play an important role in invariant theory (see [**G2**]). Here we use the results mentioned above to estimate $\min(G, X)$ in some cases. Let $Y, Z \subset X$. We define

$$Y \propto Z = \{g \in G \mid Y \cap g(Z) \neq \emptyset\}.$$

We say that the action of G on X is *almost centred* or *ac*-action if

$$\overline{(X^H)^{\max}} \propto (X^F)^{\max} = G$$

for all subgroups $H, F \leq G$ such that

$$\text{codim } X^H + \text{codim } X^F < \text{codim } X^G$$

(here $U^{\max}$ is the union of irreducible components of U which have the maximal dimension; if $U = \emptyset$ we put dim $U = 0$, codim $U = $ dim X). We say that the action G on X is *centred* or *c-action* if

$$\dim ((X^H)^{\max} \cap (X^F)^{\max}) \geq \dim X^H + \dim X^F - \dim X$$

for all subgroups $H, F \leq G$. Obviously, every centred action is almost centred.

EXAMPLES OF c-ACTIONS.

(1) $X = A_k^n$, *(affine space) and the action of G is analytically equivalent to a linear action.*
(2) X *is an algebraic group and* $G \leq \text{Aut } X$.
(3) $X = P(V)$ *where V is a G-module.*

EXAMPLES OF ac-ACTIONS WHICH ARE NOT c-ACTIONS.

(1) $X = A_k^n$, $G = \text{Aff } A_k^n$.
(2) *Let $G = SL_2(C)$, $Y = SL_2(C)/N$ be the homogeneous space corresponding to the subgroup $N = N_G(T)$, $X = Y \times C$. Let G act on X by the formula $g(y \times c) = gy \times c$. Then the action is ac but not c-action.*

In fact, ac and c-actions are "near" to affine and linear actions correspondently with respect to the behaviour of intersections of subvarieties X^H where $H \leq G$.

Suppose that the action of G on X be a c-action. Let C be a non-central conjugacy class of G, $g \in C$, $\langle C \rangle = H \leq G$. It is easy to see that

$$\text{corank } g \geq \text{codim } X^H / \text{gen } C.$$

If, in addition, $X^H = X^G$ for every $H \triangleleft G$, $H \not\subset Z(G)$, then

(*) $$\text{corank } g \geq \text{codim } X^G / \text{gen } C.$$

The resembling inequality can be obtained for ac-actions of semisimple groups on smooth varieties. Namely, using the inequality for the codimension of an intersection on a smooth variety and Theorem 3 one can prove.

THEOREM 5. *Let the action of a semisimple group G on a smooth algebraic variety X be an ac-action. Let $\overline{C^k} = G$ for some conjugacy class C of G and let $g \in C$. Then*

$$\text{corank } g \geq \text{codim } X^G/(k+1)$$

(if $X^G = \varnothing$, we put $\dim X^G = 0$).

Remark. We can take the smallest k such that $\overline{C^k} = G$. Then $k \leq \text{gen } C \leq (k+1)$ according to theorem 3. Thus, the inequality of theorem 5 is not much worse than (*).

Comparing theorem 5 and 1 we obtain

THEOREM 6. *Let the conditions of theorem 5 hold. Suppose $X^H = X^G$ for every non-central normal subgroup $H \lhd G$. Then*

$$\min(G, X) \geq \text{codim } X^G/(2r+1)$$

where r is the maximum of the ranks of the simple components of G.

Using the inequality of Theorem 6 we can obtain

THEOREM 7. *Let the conditions of Theorem 6 hold. Suppose that the stabilizer of general position in X exists and not trivial. Then*

$$\text{codim } X^G \leq (\dim G - \text{rank } G)(2r+1).$$

Theorems 6 and 7 are the generalisations of the results of E.Andreev and V.Popov which were obtained in the case when X is a G-module V or $P(V)$. ([**AP**]).

REFERENCES

AP E.Andreev and V.Popov, *The stationary subgroups of points in general position in a representation space of a semisimple Lie group*, J.Functional Anal.i Prilozhen. **5(4)** (1971), 1-8.
   English transl. in, Functional Anal. Appl. **5** (1971).
AH Z.Arad and M.Herzog, Eds, *Products of conjugacy classes in groups*, Lecture Notes in Mathematics (1985, no.1112), New York, Springer Verlag.
Bl H.F.Blichfeldt, "Finite collineation groups," Chicago, Univ. Chicago Press, 1917.
G1 N.Gordeev, *Products of conjugacy classes in algebraic groups 1,2 (to appear)*.
G2 N.Gordeev, *Corank of elements of linear groups and the complexity of algebras of invariants*, Algebra and Analysis **2(2)** (1990), 39-64.
   English transl. in, Leningrad Math.J. **2(2)** (1991), 245-267.
Kr H.Kraft, "Geometrishe Methoden in der Invariantentheorie," Brauschweig / Wiesbaden, 1985.
SprSt T.A.Springer and R.Steinberg, *Conjugacy classes*, Lect. Notes Math. **131** (1970), 167-267.

Department of Mathematics,
Russian State Pedagogical University,
Sankt-Petersburg, Moijka-48, 191-186
Russia

Monodromy Groups of Polynomials

Robert M. Guralnick[1] and Jan Saxl[2]

1. Introduction

Let F be a field of characteristic p and let $f(x)$ be a polynomial in $F[x]$ of degree n. Assume that f is not a polynomial in x^p; then $F(x)/F(f(x))$ is a finite separable extension. Let K be the Galois closure of $F(x)$ over $F(f(x))$. We are interested in the Galois group of $K/F(f(x))$ – it turns out there are severe restrictions on the possible groups.

We first refine the problem. We say $f(x)$ is indecomposable (over F) if f is not the composition of two nonlinear polynomials. It is an easy exercise (using Lüroth's Theorem) to show that this is equivalent to saying that $F(x)$ is a minimal field extension of $F(f(x))$. Let G be the Galois group (or arithmetic monodromy group) of $K/F(f(x))$. We can pass to the algebraic closure $\bar{F}$ of F and consider the extensions over $\bar{F}$. Let $\bar{G}$ be the corresponding Galois group (geometric monodromy group) . Note that $\bar{G}$ is normal in G with quotient isomorphic to the Galois group of F'/F, where F' is the algebraic closure of F in K. Let G_x denote the subgroup of G fixing x (and similarly for $\bar{G}_x$). Then G (and $\bar{G}$) act transitively on Ω, the set of n conjugates of x in K.

We can view f as a map from $\mathbb{P}^1$ to $\mathbb{P}^1$. The fact that it is a polynomial just asserts that $f^{-1}(\infty) = \infty$.

Let H be the (geometric) decomposition group in $\bar{G}$ for a prime of K over ∞. Since we are now in the case of algebraically closed residue fields, inertia groups and decomposition groups coincide. In the characteristic zero situation, H is cyclic. We are mainly interested in the case where K has characteristic $p > 0$; then $H/O_p(H)$ is cyclic, where $O_p(H)$ is the maximal normal p-subgroup of H (see [S, IV.2 Corollaries 2 and 4]).

The fact that $f^{-1}(\infty) = \infty$ implies that H acts transitively on the conjugates of x ([S, II.3 Theorem 1 and Corollary 4]). Thus, we have the following group theoretic situation:

1. G is a primitive subgroup of S_n.
2. H is a transitive subgroup of $\bar{G}$ with $H/O_p(H)$ cyclic.

In this article we classify all the possibilities of G and Ω. The problem remains to determine when this group theoretic data corresponds to a polynomial. We have not yet used the fact that the field $F(x)$ has genus zero. This is not so useful in positive characteristic, but is very powerful in

[1] partially supported by NSF
[2] both authors thank the Italian government for its support of the Como conference

characterisitic zero. Indeed, for F algebraically closed of characteristic zero, Feit [Fe] found essentially all possibilities (see also [M2]). The point is that in characteristic zero the Riemann-Hurwitz formula and Riemann's existence theorem allow one to convert the existence of such an f completely to a group theoretic problem. Moreover, in characteristic zero, H is cyclic. This implies that either G is doubly transitive or n is prime and G is metabelian. In either case, there is a quite manageable collection of groups to consider.

Our main result is the following:

Theorem A. *Let p be a prime. Let G be a primitive subgroup of S_n containing a transitive subgroup H with $H/O_p(H)$ cyclic. Then one of the following holds:*

(i) $n = p^a$;

(ii) $F^*(G)$ *is simple and is given in Theorem 3.1;*

(iii) $F^*(G) = L \times L$, $n = 4p^{2a}$, p *is odd and L is given in Theorem 4.2;*

(iv) $F^*(G) \cong P\Omega_4^+(p^a) \cong L_2(p^a) \times L_2(p^a)$, $n = p^a(p^{2a} - 1)/(2, p - 1)$ *for some a with $p^a \geq 4$;*

(v) $n = r$ *is prime and $G'' = 1$, or $n = 4$ and $G = S_4$.*

All cases above do arise group theoretically – we do not know which groups correspond to polynomials. Only a few are known to occur. It is well known that S_n and A_n do occur. Also, the affine groups which are subgroups of $PGL_2(F)$ do occur (see §4). Abhyankar [A1, A2, A3] has produced many polynomials with simple groups as monodromy groups. The theorem shows that most simple groups cannot be obtained as Galois groups of the Galois closure of $F(x)/F(f(x))$ with F finite or algebraically closed. In particular, exceptional Chevalley groups cannot occur. This is in contrast to a recent result of Raynaud [R] that any group generated by its Sylow p-sugbroups can be realized as the monodromy group of an unramified covering of the affine line (this was first conjectured by Abhyankar; Harbater [H] has extended the result to other affine curves).

The primitive groups of degree p^a have been classified – we discuss them in §4. These include the primitive affine permutation groups. In particular, we do obtain some results about when affine groups do occur.

The geometric form of the theorem can be stated as follows. Since we never use the polynomial hypothesis, we state this as:

Theorem B. *Let F be an algebraically closed or finite field of characteristic $p > 0$. Let X and Y be smooth, projective, nonsingular curves defined over F. Let $\varphi : X \to Y$ be a branched covering of degree n defined over F. Let G be the arithmetic monodromy group of the cover and $\bar{G}$ the geometric monodromy group. Assume that the covering is minimal (i.e. does not factor through an intermediate curve over F) and that for some F-point $y_0 \in Y$, $|\varphi^{-1}(y_0)| = 1$. Then one of the following occurs.*

(i) $n = p^a$;

(ii) $F^*(G)$ is simple and is given in Theorem 3.1;

(iii) $F^*(G) = L \times L$, $n = 4p^{2a}$, p is odd, and L is given in Theorem 4.2;

(iv) $F^*(G) \cong P\Omega_4^+(p^a) \cong L_2(p^a) \times L_2(p^a)$, $n = p^a(p^{2a} - 1)/(2, p - 1)$ for some a with $p^a \geq 4$;

(v) $n = r$ is prime and $G = 1$, or $n = 4$ and $G = S_4$.

We mention two related problems. A polynomial $f(x)$ is called exceptional over F if every irreducible factor $\varphi(x, y) := (f(x) - f(y))/(x - y)$ in $F[x, y]$ factors further over $\bar{F}$. If F is a finite field, these polynomials turn out to be permutation polynomials (i.e. are bijections as functions on F). Moreover, any permutation polynomial of a sufficiently large field (compared to the degree) is exceptional. The group theoretic interpretation of this property is that G_x stabilizes no nontrivial orbit of $\bar{G}_x$. See [FGS] for further details. In [FGS, 13.6 and 14.1], the following result was proved:

Theorem. [FGS] *Let p be a prime and f an indecomposable exceptional polynomial of degree n over the finite field $\mathbb{F}_q$ of characterisitic p. Let G be the arithmetic monodromy of f. Then one of the following holds:*

(a) *G is an affine group and $n = p^a$;*

(b) *G is metabelian and n is prime with $(n, q - 1) = 1$;*

(c) *$p \leq 3$, $F^*(G) = L_2(p^a)$ with $a > 1$ odd and $n = p^a(p^a - 1)/2$.*

In particular, the conjecture of Carlitz was proved – an exceptional polynomial over a field of odd characteristic has odd degree. Indeed, if $F = \mathbb{F}_q$ has characteristic > 3, then an exceptional polynomial has degree relatively prime to $q - 1$ (this stronger version was conjectured by Wan and is still open in characteristic 2 and 3 – all that remains to verify the conjecture is to consider polynomials associated with the groups in (c) above). It is not difficult to obtain this result as a Corollary to Theorem A (using the methods developed in [FGS]).

Cohen [Co] and Wan [W] also obtained certain results about exceptional polynomials. In particular, [Co] also used group theoretic tools to study the problem. Fried [F1, F2] had indicated that group theoretic ideas would prove useful in this and related problems. Müller [M1] showed that in fact the first group in (c) above for $p = 2$ does correspond to an exceptional polynomial of degree 28. Recently Cohen and Matthews [CM] have produced exceptional polynomials for each group in that family in characteristic 2.

In fact, [FGS] does not use the genus zero hypothesis (and neither do we) and so in fact the same result was proved for 'exceptional' covers where $f^{-1}(\infty) = \infty$. Let X and Y be smooth projective curves defined over $\mathbb{F}_q$. A cover $\varphi : X \to Y$ is called exceptional if $\{(u, v) \in X \times X | \varphi(u) = \varphi(v), u \neq v\}$ has no absolutely irreducible components defined over $\mathbb{F}_q$. If Y has genus zero, then this implies that for infinitely many t, X has exactly one $\mathbb{F}_{q^t}$ point over each $\mathbb{F}_{q^i}$ point of Y. It would be interesting to classify the exceptional covers without the condition at ∞. We hope to address this issue in the future. Fried [F3] has obtained results about constructing exceptional covers.

The second problem involves indecomposability. Let F be a finite field. If $f(x) \in F[x]$ is indecomposable over F, can it decompose over $\bar{F}$? It was shown in [FGS, 4.1] that this can only happen if the degree of f is divisible by the characteristic. Such examples were given in [FGS, §11] with affine groups (the polynomials have degree p^a for any $a \geq 2$). Müller (see [FGS, 11.5]) produced an example of such a polynomial of degree 21 (over $\mathbb{F}_7$). The main result here is:

Theorem C. *Let F be a finite field of characteristic p. If $f \in F[x]$ has degree n and is indecomposable over F but decomposable over $\bar{F}$, then one of the following holds:*
  (i) $n = 21$ and $p = 7$;
 (ii) $n = 55$ and $p = 11$;
(iii) $n = p^a$ with $a \geq 2$; or
(iv) $n = 4p^{2a}$ with $a \geq 1$.

As we have remarked above, there are such examples in (i) and (iii). We do not know whether the corresponding polynomial in case (ii) exists. One can produce examples in (iii) using affine groups. It may also be possible to produce examples where the monodromy group is nonsolvable (eg., where the geometric monodromy group is a direct product of alternating or symmetric groups of prime power degree). The possible examples in (iv) arise from the groups in Theorem 4.2.

Note that if $p = 2$, then any indecomposable polynomial which decomposes over the algebraic closure must have degree a power of 2. Similarly, if p does not divide 77, then any such polynomial has degree $p^a, a > 2$ or degree $4p^{2a}$.

While there is some overlap with the group theoretic methods used in [FGS], we should point out the differences. In [FGS], we used whichever property (factorization or exceptionality) was more convenient to eliminate the possible groups. Here we focus on the factorization property. Secondly, in [FGS], we appealed to the factorization results in [LPS] to eliminate many of the almost simple groups. Since we are interested in very special factorizations (but which are not in general maximal), in many cases we give direct arguments to eliminate the groups.

The article is organized as follows. Section 2 contains some preliminary results – in particular, a version of the Aschbacher - O'Nan - Scott Theorem on the structure of primitive groups. Sections 3-6 consider the various cases described in the AOS theorem (almost simple groups, affine groups, product

structure preserving groups and diagonal groups). The final section shows
how the main theorems follow.

We thank Mike Fried for many very helpful discussions.

2. Preliminary Results

We shall use the Aschbacher-O'Nan-Scott Theorem on the structure of
primitive permutation groups in the form below (see [AS] or [LPS2]). If
G acts on a set Ω and $w \in \Omega$, let G_w denote the stabilizer of w. Recall
that a component of a finite group is a quasisimple subnormal subgroup.
The Fitting subgroup of G is the maximal normal nilpotent subgroup of
G. We set $E(G)$ to be the normal subgroup generated by the components
of G (this is a central product). Then, $F^*(G) := E(G)F(G)$ is called the
generalized Fitting subgroup of G. It has the property that it contains its
own centralizer. In particular, $F^*(G) = L$ nonabelian simple is equivalent to
saying that $L \triangleleft G \leq \mathrm{Aut}(L)$ (G is said to be *almost simple*).

Proposition 2.1. *Let G be a primitive group of permutations on a set Ω
of order n. Let $E = F^*(G)$. Then one of the following holds:*
(a) *G preserves an affine structure on Ω, $n = p^a$, $E = O_p(G)$ is elementary
    abelian, $G - EG_w$ with $E_w = 1$;*
(b) *E is simple and nonabelian;*
(c) *G preserves a product structure on Ω. In particular, $G \leq S_\ell \wr S_t$, with
    $n = \ell^t, t > 1, \ell > 4$. Also, $E = L \times \ldots \times L$ with L nonabelian simple.
    Either $n = |L|^m, m > 1$ or G has exactly t components. In the latter
    case we may assume that $E_w = U \times \ldots \times U$, the product of t copies of
    a sugroup U of index ℓ in L;*
(d) *G is of diagonal type – i.e. G has $t \geq 2$ components each isomorphic
    to the simple nonabelian group L, $E_w \cong L$ is a full diagonal subgroup of
    E and $n = |L|^{t-1}$. Moreover, either $t = 2$ or G acts primitively on the t
    components.*

We will need some results about exponents of Sylow subgroups of $\mathrm{Aut}(L)$
for L a nonabelian simple group.

If G is a finite group, let $e(G)$ denote its exponent and $e_p(G) = e(P)$ for
some Sylow p-subgroup P of G. If m is a positive integer, let m_p denote the
largest power of p dividing m.

Lemma 2.2. *Let L be a finite nonabelian simple group which is alternating
or sporadic. Then the set of primes r such that $e_r(\mathrm{Aut}(L)) < |L|_r$ and
$2e_r(\mathrm{Out}(L)) < |L|_r$ has cardinality at least two unless $L = A_5, A_6$ or J_1.*

Proof. First consider the case that $L = A_n$ with $n \geq 7$. Then $e_r(S_n) <$
$(n!/2)_r$ for $r = 2, 3$ (this is just the simple observation that the Sylow 3-

subgroup of A_n is not cyclic and a Sylow 2-subgroup has exponent at most $n_2 < (n!/2)_2$.

If L is sporadic, it follows (cf [AT]) that except in the case J_1, a Sylow 3-subgroup of L is not cyclic and so (since $|\mathrm{Aut}(L) : L| \le 2$), $r = 3$ satisfies the condition. If $L = \mathrm{Aut}(L)$, then 2 also satisifies the condition since a Sylow 2-subgroup cannot be cyclic. In the remaining cases, one verifies (cf [AT]) that 2 also satisfies the condition.

Lemma 2.3. *Assume that $F^*(G) = L$, a simple Chevalley group in characteristic r.*
(a) $e_r(G) < |L|_r$ *unless* $L = L_2(4)$, $L = L_2(r)$, $L = {}^2G_2(3)'$ *or* $L_3(2)$.
(b) $e_2(\mathrm{Out}(L)) < |L|_2$.
(c) $2|\mathrm{Out}(L)|_r < |L|_r$ *unless* $L = L_2(4)$.

Proof. If $L = L_2(r^a)$, $|L|_r = r^a$ and $e_r(\mathrm{Aut}(L)) \le ra_r$ and $|\mathrm{Out}(L)| = a_r$. Thus the results hold in this case.

In general, if L is defined over $\mathbb{F}_{r^a}$, $|\mathrm{Out}(L)|_r \le a_r\delta$, where δ is the order of the r-subgroup of the group of graph automorphisms of L. It is straightforward to verify that (a)-(c) hold (compare this with the more difficult [FGS, 12.8]).

Lemma 2.4. *Assume that $F^*(G) = L$ a simple Chevalley group in characteristic r. One of the following holds:*
(a) *There exists a prime s dividing $|L|$ such that s does not divide the order of the normalizer of any parabolic subgroup of G;*
(b) $L = L_6(2)$ *or* $L = L_2(r)$ *with r a Mersenne prime.*

Proof. This follows easily from the existence of primitive divisors.

3. Almost Simple Groups

We consider almost simple groups in this section. If L is a simple Chevalley group, we let P_i denote a maximal parabolic subgroup of L corresponding to the ith node of the Dynkin diagram (in particular, if $L = L_m$, then P_1 is the stabilizer of a point in projective space and P_{m-1} is the stabilizer of a hyperplane). If L is a classical group preserving a form (i.e. L is an orthogonal, sympletic or unitary group), then N_1 denotes the stabilizer of a nonsingular 1-space (or hyperplane). This is unambiguous except in the case $L = \Omega_{2m+1}$, where there are two orbits of nonsingular 1-spaces. We denote the two types of subgroups by N_1^ϵ where $\epsilon = \pm$ depending upon the type of the orthogonal complement to the 1-space.

Theorem 3.1. *Let $L \lhd G \le \mathrm{Aut}(L)$ with L a non-abelian simple group. Suppose that G factorizes as $G = XY$, where X is a maximal subgroup of G of index n, $L \not\le X$, and for some prime p, the factor group $Y/O_p(Y)$ is a* cyclic group of order prime to p. Then (up to an isomorphism), one of the following:

(A) $L = A_c$ and one of:

    i) $n = c$, Y is a transitive subgroup of S_c;

    ii) $n = \binom{c}{2}$, $c = p^a$, Y is a 2-homogeneous subgroup of S_c contained in $AGL_a(p)$;

(B) L is classical, and one of:

    i) $X = N_1$ in G with L one of U_{2m}, $P\Omega_{2m}^+$ or Ω_{2m+1} (with $X = N_1^-$ in the last case);

    ii) $X = P_1$ (or P_{m-1}), $L = L_m$;

    iii) one of the exceptions in Table B;

(C) L is sporadic, in actions given in Table C.

Table B: Exceptional factorizations of groups of Lie type.

G	X	Y	n
$L_3(2).2$	$P_{1,2}.2$	7.6	21
$L_4(2)$	$GL_2(4).2$	$2^3.7$	56
$L_5(2)$	P_2	31.5	155
$U_3(3)$	$L_3(2)$	$3^{1+2}.8$	36
$U_3(5)$	A_7	$5^{1+2}.8$	50
$U_4(2)$	P_2	$[3^4]$	27
$U_4(3).2$	$L_3(4).2$	$3^4.2$	162
$Sp_6(2)$	S_8	$3^{1+2}.8$	36
$\Omega_7(3)$	$Sp_6(2)$	$3^{3+3}.13$	$3,159$
$\Omega_8^+(2)$	A_9	$2^6.15$	960
$P\Omega_8^+(3)$	$\Omega_8^+(2)$	$3^{6+3}.13$	$28,431$
$L_2(23)$	S_4	23.11	253
$L_2(11)$	A_5	11	11
$L_2(19)$	A_5	19.3	57
$L_2(29)$	A_5	29.7	203
$L_2(59)$	A_5	59.29	$1,711$
$L_2(11).2$	S_4	11.5	55
$L_2(16).4$	$(A_5 \times 2).2$	17.8	68

Table C: Factorizations of sporadic groups.

G	X	Y	n
M_{11}	M_{10}	11	11
M_{11}	$L_2(11)$	$3^2.8$	12
M_{11}	$M_9.2$	11.5	55
M_{12}	M_{11}	$3^2.8$	12
M_{12}	M_{11}	$2^2 \times 3$	12
$M_{22}.2$	$M_{21}.2$	11.2	22
M_{23}	M_{22}	23	23
M_{23}	$M_{21}.2$	23.11	253
M_{23}	$2^4.A_7$	23.11	253
M_{24}	M_{23}	$2^6.3$	24
$J_2.2$	$U_3(3).2$	$5^2.4$	100
$HS.2$	$M_{22}.2$	5×5.4	100
$He.2$	$Sp_4(4).4$	$7^{1+2}.6$	2,058
$Suz.2$	$G_2(4).2$	$3^5.22$	1,782

Remarks.

1. There are further examples, which are obtained from the factorizations given in the Theorem by applying an exceptional automorphism between simple groups, or by applying an outer automorphism. These include:

   a) the important family $L_2(q)$ acting on the set of cosets of the dihedral subgroups of order $(2, q)(q + 1)$ in [FGS] is listed here as $\Omega_3(q)$ on N_1^-;

   b) the family $Sp_{2m}(q)$ on the cosets of $O_{2m}^-(q)$ (q even) is given as $\Omega_{2m+1}(q)$ on N_1^-;

   c) the various examples in $L = P\Omega_8^+(q)$ of $X = O_7(q)$ are all given as N_1;

   d) the "sporadic" examples of A_c with $c = 6$ and $c = 8$ are given in (B).

2. All the cases listed in the Theorem lead to relevant factorizations, but in some cases an outer automorphisms needs to be present in G, either to facilitate the factorizations, or to make X maximal. These instances are not always indicated in the statement of the Theorem but are visible in the proof. In Tables B and C, the group G is taken to be minimal so that the claim holds; it may be possible to replace G by one of its overgroups in $\text{Aut}(L)$. The subgroups Y are not in general unique.

3. In most cases described, the prime p is uniquely determined. However, in cases A(i), B(ii) and the cases $L_2(11), M_{11}$ and M_{23}, the prime is arbitrary (since Y can be taken cyclic – this may force G to contain outer automorphisms). In B(i), p is the characteristic of the group.

Proof.

A) The case where L is the alternating group A_c.

Let Σ be the set of c points on which L acts naturally.

Consider first the case where n is $\binom{c}{k}$, X is the stabilizer of some k-subset of Σ and Y is k-homogeneous on Σ, with $k \leq c/2$. If $k > 1$ then Y is primitive on Σ, so $c = p^a$ for some a and $Y \leq AGL_a(p)$. Also, if $c \neq 8$, we have $k \leq 2$, since $AGL_a(p)$ contains no cycle of order $(p^a - 1)(p^a - 2)/6$ (or $(p^a - 1)(p^a - 2)/2$ if $p = 3$). And if $c = 8$ and $k = 3$, then $n = 56$, leading to an example in the Theorem (given there as $L_4(2)$ in (B)).

Assume then that n is not $\binom{c}{k}$. By [LPS, Theorem D], either A_{c-k} is normal in Y for some $k \leq c/2$, or c one of 6, 8 and 10. Since $Y/O_p(Y)$ is cyclic, it remains to consider $c \leq 10$, $c \neq 9$. If $c = 5$, since X is maximal, $n = 6$ (which leads to the example $L_2(5)$ on the cosets of P_1). If $c = 7$, X would be a 3-transitive subgroup of A_7, which is impossible. Remark 2 of [LPS, p.9] deals with the cases $c = 8$ (note that the example with $n = 15$ is $L_4(2)$ on the set of cosets of P_1) and $c = 10$. Finally, let $c = 6$. From [At], n is one of 6,10,15,36 and 45. The third and fifth possibilities are out - there are no suitable subgroups Y in G. The other three appear as examples in the Theorem, the second as $L_2(9)$ on P_1 and the fourth as $\Omega_3(9)$ on N_1^- in (B).

B) Here L is a Chevalley group of characteristic r. All factorizations of exceptional groups are known [HLS]. None satisfy the conclusion of the theorem. Thus we may assume that L is a classical simple group of dimension d, of characteristic r, with $q = r^a$. In view of A) we may assume that L is not alternating. Much of the analysis here is similar to that in [LPS, Chapters 3 - 6] but is easier. We shall make use of primitive prime divisors. As in [LPS] (see in particular 2.4 and 2.5 there), we shall write q_i for any prime divisor of $r^{ai} - 1$ not dividing any $r^j - 1$ with $j < ai$. We also write q_i^* for the product of all such primitive prime divisors of $r^{ai} - 1$ (counting multiplicities). Further, we write q_i^{**} for the product of all prime divisors of $q^i - 1$ not dividing any $q^j - 1$ for $j < i$. Finally, we shall reserve the symbol e for the largest i for which q_i divides $|L|$.

We proceed in a series of steps. For various arithmetic reasons it will be convenient to put

$$\mathcal{L} = \{L_2(q), L_3(4), U_4(2), Sp_6(2), L_6(2), \Omega_8^+(2)\}$$

and consider those L in $\mathcal{L}$ separately for much of the proof.

Step 1. If $d \geq 12$ then $|G : Y| \geq q^{2d+4}$; if moreover G is unitary then $|G : Y| \geq q^{4d+8}$.

This is immediate from the structure of Y. (In fact, the smallest such index is at least as large in magnitude as the index of the Borel subgroup.)

Step 2. If $d \geq 12$ then X is a geometric subgroup of G (that is, X is in one of the Aschbacher families C_i, with $1 \leq i \leq 8$ – cf. [LPS, 2.2.5]).

Since $G = XY$, we have $|X| \geq |G : Y|$. We apply Liebeck's bounds (cf. [LPS, 2.2.9]). If X is not geometric, we deduce quickly that $F^*(X)$ is the alternating group A_c with c either $d+1$ or $d+2$. Now r divides $|X|$ to power less than $d+2$. Consideration of the r-exponent of G (and structure of Y) forces $p = r$. Hence Y lies in (the normalizer of) a parabolic subgroup. It follows that q_e^{**} must divide $|X|$. On the other hand, $e \geq d - 2$ and $|X|$ divides $(d+2)!$. By a result Hering (cf. [LPS, 2.4D(i)]) we have $q = 2$ and d is 12 or 18. These are clearly impossible.

Step 3. Case where q_e^* divides Y, with $L \notin \mathcal{L}$.

Since q_e can never divide the order of a parabolic subgroups, we see that $p \neq r$ and Y is not in (the normalizer of) a parabolic subgroup of L.

We claim that $Y \leq N(T)$, where T is a torus of order divisible by q_e, so that $|Y \cap L|$ divides $|T|.c$, where c is in fact in most cases d, except that it is $d-1$ for L either $U_d(q)$ with d even or $\Omega_{2m+1}(q)$ with q odd, and it is $d-2$ for $L = P\Omega_{2m}^+(q)$. Assume first that $Y \cap L$ is irreducible. If p is a primitive prime divisor of $q^e - 1$, then the normalizer of any p-subgroup (and hence of Y) is well known to be (contained in) the normalizer of a suitable cyclic torus as claimed. Suppose then that p is not a q_e. If $O_p(Y)$ is reducible, we see that since q_e divides $|Y|$, we have Y imprimitive with d components of dimension 1, and $Y \cap L < (q-1)^d d$. Hence $q_e^* = d$, an odd prime. Also, $L = \Omega_d(q)$ (cf. [LPS, 2.5]) and $e = d - 1$. Hering's result [LPS, 2.4D] now implies that $d = 7$ and q is 3 or 5. These are clearly impossible (note that X must be parabolic now). Now assume that $O_p(Y)$ is irreducible, so d is a power of p. Using the fact that some q_e-elements normalize $O_p(Y)$, we get Y to be a subgroup of a group in Aschbacher's class C_6; moreover, $p = 2$, $q_e^* = e + 1$. We see as before that either $r = 3$ and e is 4 or 6, or $r = 5$ and $e = 6$. It follows (cf. [LPS, Table C_6]) that $Y \cap L < 2^6 A_8$ with $L = P\Omega_8^+(r)$ (note that $L \neq PSp_4(3)$ here). Then $Y \cap L = 2^6.7$, which is impossible. Finally, if $Y \cap L$ is reducible, the presence of q_e^* forces L to be one of U_{2m}, Ω_{2m+1} or $P\Omega_{2m}^+$ and $Y \cap L$ in the stabilizer of an orthogonal partition $W_1 \perp W_2$ with W_1 of codimension 1 (possibly 2 for $L = P\Omega_{2m}^+$). We apply the above analysis to the intersection of Y with the classical group on W_1, and the claim follows also in this case.

Thus $Y \leq N_G(T)$, with T a torus of order divisible by q_e. We claim that X is (the normalizer of) a parabolic subgroup as in the Theorem – in fact $L = L_d(q)$, and X is P_1 (or P_{d-1}), with one small exception. The strong upper bound on the order of Y gives a strong lower bound on the order of X, and this is used to determine X by applying the bounds of Cooperstein and Liebeck (cf. [KL, 5.2.2] and [L, Section 5]). The only cases that remain for further consideration have L equal to one of

$L_3(5), L_5(2), U_3(q), U_4(3), U_4(5), U_4(8), Sp_4(4), \Omega_8^-(2), Sp_8(2)$.

The first of these is ruled out in [LPS, 5.1.2(a)]. The second leads immediately to the well-known exceptional example with $X = P_2$ (or P_3) in Table B. The groups $U_3(q)$ are dealt with in [LPS, p.98] and there are no examples (note that while $\text{Aut}(U_3(8))$ has an exact factorization involving the normalizer of our torus, we must have $p = 19$ and $Y/O_p(Y)$ elementary abelian 3^2). Next, $U_4(3)$ is impossible by [At, p.52]. If L is $U_4(5)$ or $U_4(8)$ then $X = P_2$ of index 7.108 and 19.243, respectively; we see that Y equals (respectively has index at most 2 in) the normalizer of the torus in $\text{Aut}(L)$, and we deduce that $Y/O_p(Y)$ cannot be cyclic. Finally, the last three cases are eliminated in the following way: from [At], the only possibility in each of these is the exact factorization $G = X.(17.8)$, where X has index 136 in G; but in each of these cases, X contains representatives of all the involution classes in G.

Step 4. Case X geometric, with $L \notin \mathcal{L}$ and q_e dividing $|X|$.

Here $X \in \mathcal{C}_i$, with $i \in \{1, 2, 3, 6, 8\}$, and further restrictions apply (cf. [LPS, 2.5]).

If $X \in \mathcal{C}_1$ then one of

$X = N_1$ in G of type O_{2m}^+ or U_{2m},

$X = N_1^-$ in G of type O_{2m+1}, or

$X = N_2^-$ in G of type O_{2m}^+.

The possibilities in the first two lines give examples in the Theorem (see Step 7 below). In the last case, the index of X in G is $q^{2m-2}(q^m - 1)(q^{m-1} - 1)/2(q + 1)$. Since G has no elements of order either $q^{2m-2}/(2, q)$ or $(q^m - 1)(q^{m-1} - 1)/(q + 1)(2, q - 1)$, this is impossible.

If $X \in \mathcal{C}_2$, an imprimitive group, then $L = \Omega_{2m+1}(q)$ with q odd, $2m + 1 \geq 7$ an odd prime and X is of type $O_1 \wr S_{2m+1}$ (cf. [LPS, 2.5]). Since $|Y| \geq |G : X|$, we have $2m + 1 = 7$ and q is 5 or 3. These are impossible: in the former case, there are no elements of order either 5^8 or 31.13 in G; in the latter, the index is $3^7.13$ - but 13 does not normalize any 3-subgroup of order $\geq 3^7$ (cf [At, p.109]).

If $X \in \mathcal{C}_3$, an extension field subgroup, we use the argument of the last case in $\mathcal{C}_1$ to get $d \leq 4$. The case $d = 3$ is ruled out as in Step 3 by reference to [LPS, 5.1.2 and 5.1.10] (the exact factorization of $\text{Aut}(U_3(8))$ fails again to be an example, since $N(P)/O_2(N(P))$ is not cyclic). So let $d = 4$ and L be linear or symplectic. In the former the index is $q^4(q^3 - 1)(q - 1)/2$. Now G has no element of order $q^4/(q, 2)$ (since $q > 2$), so $Y \leq P_1$ and Y contains an element of order $(q^3 - 1)(q - 1)/(q - 1, 2)$. It follows that $q = 3$ - but even here the 13 element normalizes no 3-subgroup of order $\geq 3^4$. In the latter, we get the examples given as N_1^- in Ω_5. The other related possibility, the $\mathcal{C}_3'$ subgroups $L \cap X = (q^2 + 1).4$ in $L = Sp_4(q)$ with q even, is easily ruled out as $q > 2$.

If $X \in \mathcal{C}_6$, so X is of extraspecial type, then by [LPS, 2.5] d is a power of 2, q is odd and either $q_e = d+1$ a prime, G linear or symplectic, or $q_e = d-1$ a prime, G orthogonal of type $+$. In the former case we get G symplectic with $d = 4$ and $q = 3$, a case not considered here. In the latter case $d = 8$ and $q = 3$ – however $O_8^+(3)$ has no elements of order 3^{10} or 65.

If finally $X \in \mathcal{C}_8$, a classical subgroup, the above argument concerning orders of elements in G forces either $L \cap X = PSp_4(q)$ with $L = L_4(q)$, or $L \cap X$ of type O_{2m}^- with $L = Sp_{2m}(q)$, q even. These are examples in the Theorem, given there as N_1 in $P\Omega_6(q)$ and N_1^- in $\Omega_{2m+1}(q)$, respectively.

Step 5. Case $L \notin \mathcal{L}$, $X \in \mathcal{S}$, q_e divides $|X|$.

Here $d \leq 11$ by Step 2. Much is known about the modular representations of almost simple groups of such small dimensions. Indeed, Kleidman's unpublished manuscript [K1] contains a list of maximal subgroups of classical groups with $d \leq 11$. This is what we used in [LPS]; since [K1] has not been published, we shall not use it here explicitly. Much of what we need is in fact well known – cf. [KL, 5.3 and 5.4], the modular atlas [MAt] of Parker et al, [Ko1], [Ko2]. We remark that a forthcoming manuscript [PPS] will contain much more information concerning subgroups X in this Step (cf. also [MSW]).

We consider separately the cases where $F^*(X)$ is of Lie type in characteristic r, alternating, sporadic and of Lie type in different characteristic.

i) $F^*(X)$ of Lie type in characteristic r. Using [KL, 5.4] it is quite easy to list all examples with q_e dividing $|X|$ here:

type of X	type of L	restrictions	m_i
$^2B_2(q)$	$B_2(q)$	$q = 2^{2a+1}$	$q^2, (q^2-1)(q+1)$
$G_2^\epsilon(q)$	$O_7(q)$	$\epsilon = +$ unless $q = 3^{2a+1}$	$q^3, (q^4-1)/(q-1,2)$
$U_3(q)$	$O_7(q)$	characteristic 3	$q^6, q_4 \cdot q_3$
$U_3(q)$	$O_8^+(q)$	$q \equiv 2(3)$	$q^9, q_4 \cdot q_3$
$O_7(q)$	$O_8^+(q)$		

(We remark that the irreducible representation of $L_2(q^3)$ in dimension 8 gives $SL_2(q^3)$ as a subgroup of $Sp_8(q)$ but not of $O_8^+(q)$.) The last line yields examples: a triality automorphism moves these X to N_1 in $O_8^+(q)$ given in the list. The other possibilities are easily ruled out: the last column of the table gives two coprime divisors m_i of $|G : X|$ such that G has elements neither of order m_1 nor m_2.

ii) $F^*(X)$ is alternating, say A_c. Since $d \leq 11$, we have $c \leq 13$; moreover, $c \leq 10$ unless the module is the nontrivial irreducible constituent of the natural permutation module. If $q_e = 13$ then $d \geq 12$ – not so. If $q_e = 11$ then $c \geq 11$, so $d \geq 9$, whence $c = 10$, so d is 10 or 11; however there is no element in G of order divisible by q^{10} or by $q_8 \cdot q_5$.

If $q_e = 7$ then $c \geq 7$; hence $e = 6$. If $L = U_3(q)$, we get the only possibility $A_7 < U_3(5)$; this gives an example ([At, p.34]). If $L = U_4(q)$ then $c \leq 7$; in all the other cases $d \geq 6$. It follows that Y is in (the normalizer of) a parabolic subgroup. Hence $q_6^* = 7$, and q is 3 or 5. The case $L = U_4(q)$ is now ruled out in [LPS, p.112]. In the other cases there is no element of order $q_4.q_3$ in G.

If $q_e = 5$, as $d > 2$, we have $e = 4$, so $d = 4$ and L is $L_4(q)$ or $PSp_4(q)$. Since $q > 2$ and $q_4 = 5$, we have q odd. Also, $q > 3$ if $L = PSp_4(q)$. The above argument now forces $q_4^* = 5$, $q = 3$, so $L = L_4(3)$. From [At, p.69], $c = 6$ and the index of X in G is $3^4.104$. However, G has no elements of order 104.

iii) $F^*(X)$ is a sporadic simple group. Since $d \leq 11$, $F^*(X)$ is a Mathieu group or one of the first three Janko groups. The only possibility for J_3 is $J_3 < U_9(2)$; however, G has no elements of order 2^{28} or $2_{14}.2_{10}$. Also, J_1 can only arise in $O_7(11)$ - but $11_6^* = 37$. Next, $J_2 < PSp_6(q)$ for suitable q (eg. $q \neq 3$, and here $q > 2$). As before, $q_6^* = 7$, so $q = 5$; but G has no elements of order 31.13.

If $F^*(X)$ is M_{23} or M_{24} then $L = L_{11}(2)$; but here $2_{11}^* = 23.89$, so Y is not in a parabolic, which is impossible. Now $M_{22} < U_6(2)$ has index $2^8.3^4$ and G has no elements of order 2^8 or 3^4; the only other possibility for M_{22} has $d = 10$ and is easily ruled out. If $F^*(X) = M_{11}$ then either $M_{11} < L_5(3)$ or $d \geq 10$. In the former, G has no elements of orders either 3^8 or 11.13; the latter is clearly impossible. Finally, if $F^*(X) = M_{12}$ then $q_e = 11$; since $d \geq 6$, have $e = 10$ and $d \geq 10$; this again is clearly impossible.

iv) $F^*(X)$ of Lie type in characteristic different from r (but not an alternating group). From the Landazuri - Seitz bounds, the relevant $F^*(X)$ are (cf. [Ko2]):

$L_2(q)$ with $q \in \{7, 8, 11, 13, 17, 19, 23\}$,

$L_3(3), L_3(4), U_3(3), U_4(2), U_4(3), Sz(8), Sp_6(2)$ and $\Omega_8^+(2)$.

Assume first that $q_e = 5$; then $e = 4$, so $d = 4$, and the only candidates for $F^*(X)$ are $L_3(4)$ in characteristic 3 and $U_4(2)$ in odd characteristic. The former leads to the embedding $L_3(4) < U_4(3)$ and the factorization $U_4(3).2 = (L_3(4).2)(3^4.2)$ in Table B: for, by [LPS, p. 113], $L_3(4) \cap 3^4.A_6 = A_6$, so the normal 3-subgroup of $3^4.A_6$ has 2 orbits of size 81 on $L : X$, and these are interchanged by an outer involution in the normalizer. In the latter, we may take the characteristic to be ≥ 5; as usual, $q_4^* = 5$, so Hering's result gives a contradiction.

If $q_e = 11$, we have $F^*(X) = L_2(s)$ with $s = 11$ or $s = 23$. The usual argument reduces to $L_2(11) < U_5(2)$ – but there is no factorization here [LPS, p.99]. If $q_e = 13$ then $F^*(X)$ is $L_2(13)$ or $Sz(8)$, and $e = 6$. In the former, [LPS, 2.4D(ii)] gives $q = 4$ – however, $L_2(13) < U_6(2)$ here. In the latter, get $Sz(8) < O_8^+(5)$ – but $13 = 5_2$. If $q_e > 13$ then $F^*(X) = L_2(q_e)$ of dimension $\geq (q_e - 1)/2$, which is easily ruled out.

Finally suppose that $q_e = 7$. If $e = 3$ then $d = 3$, so we have the possibility $L_2(7) < L_3(q)$, where q is not a power of 2 or 7. However, [LPS, 2.4D(ii)] leads to $q \in \{2,4\}$, a contradiction. So $e = 6$. If $L = U_d(q)$ with d either 3 or 4 then either $F^*(X) = L_3(4) < U_4(3)$ or $F^*(X) = L_2(7)$ in odd characteristic. We discussed the former case already. In the latter case, [LPS, 2.4D(i)] forces q to be 3 or 5. The embedding $L_2(7) < U_3(3)$ yields an example. Consider $L = U_3(5)$ next: the maximality of X forces $G = U_3(5).2$ ([At, p.34]), with $n = 750$; however, the parabolic subgroup order is not divisible by 3 unless 3 divides $|G/L|$. Next, X is not maximal if $L = U_4(3)$ [At, p.52]; and if $L = U_4(5)$, there is no element in G of order 117. Thus $6 \leq d \leq 8$. Again, Y is contained in a parabolic, so q is 3 or 5. The latter is easily ruled out: since G has no elements of order 31.13, we have 13 dividing $|X|$ - so $F^*(X)$ is $L_2(13)$ or $Sz(8)$; however, G also has no elements of order 31.3^3.

Thus L is one of $L_6(3)$, $PSp_6(3)$, $\Omega_7(3)$ and $P\Omega_8^+(3)$. It is now easy (using the above arguments) to dispose of the remaining possibilities $X = L_2(s)$ (with s one of 7, 8 and 13) – so $F^*(X)$ is one of $L_3(4), Sp_6(2)$ and $\Omega_8^+(2)$, leading to the possibilities

$L_3(4) < L_6(3)$,

$Sp_6(2) < \Omega_7(3)$ of index $3^5.13$, and

$\Omega_8^+(2) < P\Omega_8^+(3)$ of index $3^7.13$.

The first of these is not maximal – we have $F^*(X) < U_4(3) = P\Omega_6^-(3) < L_6(3)$. The other two yield examples in Table B: If $L = P\Omega_8^+(3)$ then $L = X.P_4$. Write $U = O_3(P_4)$ and let K be a Levi complement. We have $P_4 \cap X = (3 \times PSp_4(3)).2$, and $U \cap X = 3$; it follows that U has 117 orbits (of size 3^5), and these are permuted transitively by K and hence by a subgroup $3^3.13$ therein (cf. (c) in Step 7). Similarly, if $L = \Omega_7(3)$ then $L = X.P_3$ with $P_3 = 3^{3+3}L_3(3)$. Writing $U = O_3(P_3) = 3^{3+3}$ and $U_1 = U'$, we have U_1 consisting of elements 3A, so that $U_1 \cap X = 1$. Now $X \cap P_3 = 3_+^{1+2}.2S_4$. Writing $Q = O_3(X \cap P_3)$, have $Q \cap U$ equal either $Z(Q)$ or Q. The latter is impossible, as here $Q < U$ and $U_1 \cap Q = 1$ would force $U = QU_1$ and hence $U/U_1 \simeq Q$, which is not so. In the former case, U has 13 orbits on $G : X$, and these are permuted transitively by $L_3(3)$ and hence by a 13-cycle therein.

Step 6. The remaining cases.

a) First assume that $L = P\Omega_8^+(q)$ with $q > 2$, G involves a triality, $X \notin \mathcal{S}$ and q_6 divides $|X|$. By [K1], $X \cap L$ is either $2^{3+6}L_3(2)$ or $(\frac{1}{d}GU_3(q) \times \frac{1}{d}(q + 1)).2^d$. Neither is possible: G contains no elements of order q^9 or divisible by the odd part of $(q^2 + 1)(q^3 - 1)$.

b) Next we remark that the assertions concerning the linear groups of dimension 2 follow immediately from [LPS, 5.1.1].

c) If $L = L_3(4)$, we use the information in [At, p.23]. If 7 divides $|Y|$ then $|Y|$ divides 42; it follows that X is P_1 (or P_2), an example in the Theorem. So assume that 7 divides $|X|$. If $X \cap L = L_2(7)$ then the maximality of X implies that there are no outer automorphisms in G of order 3. But $|G : X| = 120$, whereas there are no elements in G of order 15, 24 or 40. Hence $|X \cap L| = 21$, $|G : X| = 2^6.15$. It follows that Y is in the normalizer of a parabolic. However, an element of order 15 cannot normalize a subgroup of order $\geq 2^6$.

d) If $L = U_4(2)$, there are maximal subgroups in G of index 27, 36, 40, 40 and 45 (see [At, p.26]). The first three are examples in the Theorem (the first in Table B, the second is N_1^- in $\Omega_5(3)$, the third is N_1 in $U_4(2)$). In the fourth case, the permutation character is $1 + 15_b + 24_a$. Now $Y = 2^4.5$, where the normal subgroup $U = O_2(Y)$ contains 5 elements of type 2A and 10 of type 2b. A simple character calculation shows that U has 10 orbits on $G : X$, so $2^4.5$ is intransitive. The last possibility is also out: no 5-subgroup normalizes a 3-subgroup in G and vice versa.

e) If $L = Sp_6(2)$, all possibilities for X in [At, p.46] other than $X = O_6^\epsilon(2)$ yield to the usual considerations concerning Y. The remaining two possibilities give examples in the Theorem: for $\epsilon = -$ this is the usual example (cf. Step 7), whereas for $X = O_6^+(2)$ the factorization $Sp_6(2) = S_8(3^{1+2}.8)$ is inherited from the factorizations $Sp_6(2) = S_8(U_3(3).2)$ and $U_3(3) = L_3(2)(3^{1+2}.8)$.

f) If $L = L_6(2)$, it is quite easy to show that there are no maximal subgroups in $\mathcal{S}$. Hence $X \in \mathcal{C}$, and we see as usual that either X is parabolic, or Y is in the normalizer of a parabolic in G. The former leads to the usual example P_1. The latter is ruled out as usual, using the fact that G has no elements of order 105 and 217.

g) Finally let $L = \Omega_8^+(2)$. If 7 divides $|Y|$, we see from [At, p.85] that $|Y|$ divides 42 or $2^9.7$; there is no suitable X. So 7 divides $|X|$. If X is N_1 (or one of its triality images), we get examples as usual. If X normalizes a parabolic, we see that 135 divides $|G : X|$. However, there is no element in G of order 27, and no 5-element normalizes a 3-subgroup of order at least 27. Finally, let $X = A_9$ in L, of index 960. We claim that there is a subgroup $Y = 2^6.15$ which gives an example. We consider $P_1 = 2^6.A_8$ and write $U = O_2(P_1)$. Then U consists of involutions in classes 2A and 2B. We choose X so that the permutation character of G is $1 + 84_b + 175_a + 700_c$, so that $L = XP_1$. Now U is semiregular on $L : X$, so has 15 orbits of size 64. It follows that a Levi complement in P_1, and hence also a 15-cycle therein, acts transitively on the set of these orbits of U, and the assertion follows.

Step 7. General examples.

a) Case $L = Sp_{2m}(q)$, $X = SO_{2m}^-(q)$, of index $\frac{1}{2}q^m(q^m - 1)$, with q even. It is quite easy to see that $L = XZ$, with $Z = Sp_2(q^m)$ and $X \cap Z = SO_2^-(q^m)$

(cf. [LPS, p.48]). Hence the Borel subgroup $Y = q^m.(q^m - 1)$ of Z is transitive on $L : X$.

b) Case $L = \Omega_{2m+1}(q)$, $X = N_1^-$, q is odd and the index is $\frac{1}{2}q^m(q^m - 1)$. We have $L = XP_m$ [LPS, p.57]. Also, $P_m = U.K$, where $U = O_p(P_m)$ has order $q^{\frac{1}{2}m(m+1)}$ and K, a Levi complement, is the subgroup of index 2 in $GL_m(q)$ (denoted by $\frac{1}{2}GL_m(q)$).

Taking the usual basis $\{e_1, \ldots, e_m, d, f_1, \ldots, f_m\}$ and P_m to be the stabilizer of the subspace

$< e_1, \ldots, e_m >$, we see that U consists of matrices $\begin{pmatrix} I & 0 & 0 \\ b & 1 & 0 \\ A & -b^t & I \end{pmatrix}$ subject

to $A + A^t + b^t.b = 0$, and K consists of the matrices $\begin{pmatrix} A & 0 & 0 \\ 0 & 1 & 0 \\ 0 & 0 & A^{-t} \end{pmatrix}$ with

$det(A)$ a square. Taking $v = e_1 + \lambda f_1$ (for a suitable λ) so that $X = L_{<v>}$, we see that $U_{<v>}$ consists of those matrices in U with $b_1 = 0$ and $a_{1i} = 0$ for all i. It follows that $|U_{<v>}| = q^{\frac{1}{2}m(m-1)}$ and the orbits of U on $L : X$ have size q^m, and there are $(q^m - 1)/2$ of them. Any vector of minus type projects nontrivially either on $< e_1, \ldots, e_m >$ or on $< f_1, \ldots, f_m >$. If m is even then K contains an element x of order $q^m - 1$. No power of x other than $\pm I$ can fix a 1-space of minus type. It follows that $U. < x >$ is transitive. If m is odd and q is 3 mod 4, taking $x \in K$ of order $\frac{1}{2}(q^m - 1)$ works, since $-I$ is not in K. More care is needed though for m odd when q is 1 mod 4: we need to replace L by $SO_{2m+1}(q)$; then $P_m = U.GL_m(q)$, and taking $x \in GL_m(q)$ of order $q^m - 1$ works.

c) Case $L = P\Omega_{2m}^+(q)$, $X = N_1$, of index $q^{m-1}(q^m - 1)/(2, q - 1)$. The argument here is similar to (but easier than) that above.

d) Case $L = U_{2m}(q)$, $X = N_1$ of index $q^{2m-1}(q^{2m} - 1)/(q + 1)$. We have $L = XP_m$ [LPS, p.53]. Also, taking the usual unitary basis and P_m, we have $P_m = U.K$, with U of order q^{m^2}, consisting of the matrices $\begin{pmatrix} I & 0 \\ A & I \end{pmatrix}$ with

$A + \bar{A}^t = 0$ and K consisting of matrices $\begin{pmatrix} A & 0 \\ 0 & \bar{A}^{-t} \end{pmatrix}$ with $A \in GL_m(q^2)$ having $det(A)^{q-1} = 1$. Taking $X = L_{<v>}$ with $v = e_1 + \lambda f_1$ for a suitable $\lambda \in F_{q^2}$ we see that $U_{<v>}$ consists of those matrices $\begin{pmatrix} I & 0 \\ A & I \end{pmatrix}$ in U with $a_{1i} = 0 = a_{i1}$ for all i. Hence all the orbits of U on $L : X$ have size q^{2m-1}, and there are $(q^{2m} - 1)/(q + 1)$ of them. The argument proceeds as before, though we have to enlarge X in some cases – certainly $G = PGU_{2m}(q)$ always works, taking x a Singer cycle in K as before.

C) The case where L is a sporadic simple group.

By [LPS, Chapter 6], we only need to consider 12 of the sporadic groups.

Let $L = M_{11}$. If 11 divides $|Y|$ then $|Y|$ is 11 or 55, giving two of the examples in the Theorem. So assume that 11 divides $|X|$; as X is maximal, we get $n = 12$. We claim that $M_{11} = L_2(11)(3^2.8)$: the 3^2 normalizer in M_{11} contains the Sylow 2-subgroup, and $L_2(11)$ contains no elements of order 4. On the other hand, Y cannot be a 2-local, since the 3-elements are not semiregular.

Let $L = M_{12}$. Then 11 does not divide $|Y|$, since otherwise n divides 110. By [At], since its order is divisible by 11, $L \cap X$ must be M_{11} or $L_2(11)$, and n is 12 or 144, respectively. The latter is impossible, since $Aut(M_{12})$ contains no elements of order 9 or 16. In the former, the factorization $M_{12} = M_{11}(3^2.8)$ is inherited from the corresponding factorization of M_{11} mentioned above (since $M_{12} = M_{11}M_{11}$ and the two M_{11} intersect in a $L_2(11)$). And, $M_{12} = M_{11}(2^2 \times 3)$: The centralizer of a 3B-element is $A_4 \times 3$ (the normalizer is a 4×3 array group). The four-subgroup is 2A-pure, so we have a $2^2 \times 3$ acting regularly on either class of subgroups M_{11}.

Let $L = M_{22}$. From [At] we get 11 to divide $|Y|$, so n divides 110, and in fact n divides 55 unless G contains outer automorphisms. Thus $G = M_{22}.2$ and $n = 22$. In fact, $M_{22}.2 = (M_{21}.2)D_{22}$ is an exact factorization.

If $L = M_{23}$, we see that 23 must divide $|Y|$, so n is 23 or 253, as claimed in the Theorem. Let $L = M_{24}$. If 23 divides $|Y|$ then n divides 253, which is not possible [At]. So 23 divides $|X|$, and in fact $X = M_{23}$, $n = 24$. Now the trio subgroup $2^6.(L_3(2) \times S_3)$ is transitive in the natural action of M_{24} and the base group has 3 orbits of size 8, permuted transitively by the S_3. It follows that $M_{24} = M_{23}(2^6.3)$.

Let $L = J_2$. By [LPS, 6.6], $n = 100$; also, $p = 5$, so $Y \le N(5^2)$. The argument in [LPS, 6.6] shows that $G = J_2.2$ and $J_2.2 = (U_3(3).2)(5^2.4)$.

If $L = HS$, then again $n = 100$ and $p = 5$. It is not hard to see that L has no relevant factorizations; however, we have $HS.2 = (M_{22}.2)(5 \times 5.4)$: For, $S_5 \times 5.4$ is transitive on $G : X$, so the 5^2 is semiregular there; and, by [At], the 2-elements in the A_5 are 2B, whence the elements of order 4 in the S_5 are 4F and hence semiregular.

Let $L = He$. Using 17, we get $n = 2058$, so $p = 7$ and $Y \cap L \le 7^{1+2}(S_3 \times 3)$. By [LPS, 6.8], we have $G = He.2 = (Sp_4(4).4)(7^{1+2}.6)$.

If $L = Suz$, then $n = 1782$, so $p = 3$ and $Y \cap L \le 3^5 M_{11}$. Hence $G = Suz.2$, and in fact $Suz.2 = (G_2(4).2)(3^5.22)$: by [LPS, 6.10], $3^5 M_{11}$ is transitive on $G : X$, and the base group 3^5 has 22 orbits of size 81 there; and the element of order 22 in $3^5(M_{11} \times 2)$ is semiregular there by [At], so the assertion follows.

Finally the possibilities where L is one of Ru, F_{22} and Co_1 are easily ruled out.

4. Product Actions

Let $\Omega = \Delta^t$ be a set of order $n = \ell^t$ with $|\Delta| = \ell > 1$ and $t > 1$. Let G be a primitive group of permutations of Ω preserving the product structure on Ω. So G is a subgroup of $S_\ell \wr S_t$. Let $E = F^*(G)$.

The following result is [FGS, Lemma 13.5]

Lemma 4.1. *Let G be a primitive group acting on Ω. Assume that H is a transitive subgroup of G and $H/O_p(H)$ is cyclic. Then one of the following holds:*

(a) $\ell = p^a$; *or*

(b) $\ell = 2p^a$, p *odd and* $t = 2$.

Theorem 4.2. *Let G be a primitive group acting on Ω. Assume that H is a transitive subgroup of G and $H/O_p(H)$ is cyclic. If $t = 2$ and $\ell = 2p^a$ with p odd, then $E = L \times L$, where L is one of*

(i) A_ℓ *and p is the odd prime dividing ℓ;*

(ii) $L_2(q)$ *with $\ell = q + 1 = 2p^a$;*

(iii) A_5 *with $\ell = 10$ and $p = 5$;*

(iv) $U_3(5)$ *with $\ell = 50$ and $p = 5$;*

(v) M_{22} *with $\ell = 22$ and $p = 11$; or*

(vi) $U_4(3)$ *with $\ell = 162$ and $p = 3$.*

Proof. The first statement follows from the Aschbacher-O'Nan-Scott Theorem (see 2.1). Moreover, it also implies that if E_w is the stabilizer of some $w \in \Omega$, then we may assume that $E_w = U \times U$ with $|L : U| = 2p^a$. Let A be the subgroup of G acting coordinatewise on Δ. So $E \leq A \leq S_\ell \times S_\ell$. Let π_i be the projection of $S_\ell \times S_\ell$ onto the ith factor. Set $X_i = \pi_i(A)$ and $Y_i = \pi_i(H \cap A)$.

We claim that Y_i is transitive on Δ. Assume that Y_i has v_i orbits on Δ. It follows that H has at least $v_1 v_2 / d$ orbits on Ω, where $d = |H : H \cap A|$. Since H is transitive, this implies that $v_1 v_2 \leq d \leq 2$. If $d = 1$, then $v_1 = v_2 = 1$. If $d = 2$, then Y_1 and Y_2 are conjugate in G, whence $v_1 = v_2 < 2$. This proves the claim.

The primitivity of G implies the primitivity on Δ of $N_i = N_{X_i}(U)$. Thus, $N_i = UY_i$ and $Y_i/O_p(Y_i)$ is a homomorphic image of $H/O_p(H)$, whence is cyclic. Since $F^*(N_i) = L$, we can apply Theorem 3.1. Thus, L occurs in the conclusion. It is straightforward to compute which groups in 3.1 satisfy the additional restriction that Δ has cardinality $2p^a$ for some odd prime p.

Remarks: (1) All groups in the conclusion of Theorem 4.2 yield examples. Let M be a group with $F^*(M) = L$ and U a subgroup of index $2p^a$ given in 3.1 so that $M = UY$ with $Y = \langle y, X \rangle$, $X = O_p(Y)$ and y a 2-element. Set $G = M \wr S_2$, $G_w = U \wr S_2$. Let $h \in G$ be the involution interchanging the two components. Set $g = (y, 1)h$ and $H = (X \times X)\langle g \rangle$. Then $G = G_w H$ and $H/O_p(H)$ is cyclic.

(2) In 4.2(ii), there are restrictions on q. For example, it is easy to see that $q = r^{2^s}$ for r an odd prime.

We now turn to the case that $\ell = p^a$. The existence of H is guaranteed in this case (eg., take H to be a Sylow p-subgroup). An immediate consequence of [Gu] (see also [Ka]) is:

Theorem 4.3. If $\ell = p^a$, then $F^*(G) = L \times \ldots \times L$ where L is one of the following:
(a) A_ℓ;
(b) $L_d(q)$ with $\ell = (q^d - 1)/(q - 1)$;
(c) M_{11} with $\ell = 11$;
(d) M_{23} with $\ell = 23$;
(e) $U_4(2)$ with $\ell = 27$;
(f) $L_2(11)$ with $\ell = 11$.

We also take this opportunity to record the following result. This is an immediate consequence of [AS] and the previous result (see also [Ka]). Let $AGL_d(q)$ denote the group of affine transformations of a d dimensional vector space over $\mathbb{F}_q$.

Theorem 4.4. Let $n = p^a \geq 5$. Let $G = A_n$ or S_n. Let M be a maximal subgroup of G not containing A_n. Then one of the following holds:
(a) M is the stabilizer of a subset of size $k < n/2$;
(b) M is the stabilizer of a partition of $\{1, \ldots, n\}$ into subsets of size p^b for $b = 1, \ldots, a - 1$;
(c) $M = (S_{p^b} \wr S_{a/b}) \cap G$;
(d) $F^*(M)$ is simple and is one of the groups listed in 4.3(b)-(e);
(e) $M = AGL_a(p) \cap G$.

We note that $L_2(11) < M_{11}$ and so does not give rise to a class of maximal subgroups. The groups in 4.3(b)-(e) each give rise to a unique conjugacy class of maximal subgroups in S_n (even when two different representations occur, the point stabilizers are conjugate in the automorphism group). These classes may split into two different classes in A_n (e.g., this happen with M_{11}).

5. Diagonal Actions

In this section, we assume that G acts primitively on Ω and is of diagonal type. Thus, G has $t \geq 2$ components isomorphic to L and if $n = \Omega$, $n = |L|^{t-1}$. Moreover, G embeds in $\mathrm{Aut}(L) \wr S_t$.

Theorem 5.1. *Assuming that G contains a transitive subgroup H, with $H/O_p(H)$ cyclic. Then $F^*(G) = L_2(p^a) \times L_2(p^a)$ with $p^a \geq 4$. Moreover, H normalizes each component of G.*

Proof. It follows from [FGS, 12.5] that $t = 2$.

By Proposition 2.1, we may assume that $E = F^*(G) = L \times L$, and E_w is the diagonal subgroup of E. Thus, G embeds in $\text{Aut}(L) \wr S_2$. Let $x \in \text{Aut}(L) \wr S_2$ be the involution interchanging the copies of L and fixing E_w. Let $D \cong \text{Aut}(L)$ the diagonal subgroup of $\text{Aut}(L) \times \text{Aut}(L)$. Note that $M = \langle E, D, x \rangle$ contains G and acts on Ω. Also, note that since H is transitive, it has order divisible by $|L|$.

Observe that $e_r(G) \leq e_r(\text{Aut}(L))$ for r odd. Set $f = e_2(\text{Out}(L))$. Let $y \in M$ is a 2-element. If $y \in DE < \text{Aut}(L) \times \text{Aut}(L)$, then y has order dividing $e_2(\text{Aut}(L))$. If $y \notin DE$, then $y^{2f} = (a,b) \in L \times L$ with a and b conjugate in L – thus, (a,b) is conjugate to an element of E_w. Hence y^{2f} has a fixed point.

Let r be a prime dividing $|L|$ with $r \neq p$. Let $R = \langle y \rangle$ be a Sylow r-subgroup of H. Note that the transitivity of H implies that $H/O_p(H)$ is a cyclic group acting transitively on the orbits of $O_p(H)$. It follows that R/R_u must have order a multiple of $|L|_r$. In particular, the previous paragraph implies that $e_r(\text{Aut}(L)) \geq |L|_r$ or $r = 2$ and $2e_2(\text{Out}(L)) \geq |L|_2$.

It now follows from Lemma 2.2 that L is not alternating of degree at least 7, and that L is not a sporadic simple group other than J_1. Now consider $L = J_1$. Since $L = \text{Aut}(L)$ and $e_2(L) < |L|_2$, $r = 2$. Moreover, H must contain an element z of order $|L|/|L|_2$. Thus $z \in E$. Hence there exist elements $z_i \in L$ of order d_i such that $d_1 d_2 = |L|/|L|_2$. Since the subgroups of L of order 11 and 19 are self centralizing, this cannot happen.

Since $A_5 \cong L_2(4) \cong L_2(5)$ and $A_6 \cong L_2(9)$, we consider these groups as Chevalley groups. We also consider the smallest Ree group as $L_2(8)$ (and so not as a characteristic three group) and $L_3(2)$ as $L_2(7)$ (and so not as a characteristic two group).

First consider $L = L_2(4) = L_2(5)$. This group is allowed in the conclusion with $p = 2$ or 5. It is straightforward to check that there is no transitive subgroup with $H/O_3(H)$ cyclic (there is no element of order 20 in a 3-local subgroup).

So now assume that L is a simple Chevalley group in characteristic r with L not one of $L_2(4)$, $L_2(5)$, or the smallest Ree group. We claim that we may assume that $p = r$. It follows from Lemma 2.3 that $p = r$ except possibly in the cases $L = L_2(r)$. We need only consider $L_2(r)$ with $r \geq 7$. Assume $p \neq r$. Let u be an element u of order r in H. Of course, $u \in E$. Since $H = O_p(H)C_H(u)$ and the Sylow r-subgroup of L is self centralizing, it follows that u is contained in a component of G. Since, u normalizes no

p-subgroup of $\text{Aut}(L)$, H centralizes u Note that $C_G(u) = \langle u \rangle \times K$, where K is a component of G. The largest r'-subgroup of K has order $r + 1$. Thus $|H| \leq r(r + 1) < |L|$ and H can not be transitive. (note that we have eliminated $L_3(2)$ with $p \neq 7$). So we may assume that $p = r$.

Let π_i denote the ith projection from E onto L. We claim that $r \,\|\, |H \cap E|$. If not, $|L|_r \leq |H|_r \leq 2|\text{Out}(L)|_r$. It follows by Lemma 2.3 that $L = L_2(4)$. In this case, it is easy to check that the claim is valid by considering the centralizer of an involution. So assume that $\pi_1(H)$ has order divisible by r.

If $L = L_2(r^a)$, then r cannot divide $\pi_2(H \cap E)$ (or $H \leq N_G(B)$, where B is the product of Borel subgroups of each component of G; this subgroup is however not transitive). This shows that H must normalize each component of G (since otherwise, the two projections would be conjugate).

From now on, we assume that L is not $L_2(r^a)$ and aim for a contradiction (this also eliminates $L_3(2) = L_2(7)$ and the smallest Ree group). Note first that $L \neq L_6(2)$; here $e_3(G) = 9 < |L|_3$, contrary to $p = r$. In every other case, it follows from Lemma 2.4 that there exists a prime s dividing $|L|$ such that s does not divide the order of the normalizer of any parabolic subgroup of L. It follows that s divides $|\pi_2(H \cap E)|$ but s does not divide $|\pi_1(H \cap E)|$. Thus, as above, H normalizes each component.

Since we may assume that $G = HE$, we may also assume that each component of G is normal. Thus $G \leq HE = AE$, for some subgroup A of $D \cong \text{Aut}(L)$. We may assume that A contains the diagonal subgroup $J = D \cap E$ of E (this determines A uniquely – in fact, $A = G_w$). Since H is transitive, it follows that $G = HA$.

We consider $G \leq B \times B$ with $L \leq B \leq \text{Aut}(L)$. Then A is the diagonal subgroup of $C \times C$ for some $C \leq B$. In particular, $A \cong C$. Let H_i be the projection of H into the ith copy of C.

We claim that $C = H_1 H_2$. First, note that if $x \in L$, then $(x, 1) \in AH$ implies that $(x, 1) = (w, w)(h_1, h_2)$ where $(w, w) \in A$ and $h_i \in H_i$. Thus $x = h_1^{-1} h_2 \in H_1 H_2$. If u is an arbitrary element of C, then $(u, v) \in G$ for some $v \in C$. Since $G = HE$, $(u, v) = (x, y)(h_1, h_2)$ with $(x, y) \in E$ and $h_i \in H_i$. Thus $u = (ux^{-1})x \in H_1 L \subseteq H_1 H_2$, as desired. This proves the claim.

Since $O_r(H_1 \cap L) \neq 1$, it follows that $H_1 \leq Q$, the normalizer of a parabolic subgroup of L. By conjugating, we may assume that the Sylow r-subgroup of H_2 is contained in Q. Thus, $C = QH_3$, where H_3 is a Hall r'-subgroup of H_2. In particular, H_3 is cyclic. Thus, by a classical result of Schur (see [Wi, §25]), C is 2-transitive on the cosets of a maximal subgroup containing Q. Moreover, L must occur in Theorem 3.1. This forces $L = L_d(q)$.

Choose a prime s as follows. Let $s = 3$ if $q = 2$. Otherwise, let s be the largest prime dividing $q - 1$. Since $d > 2$ and also $(d, q) \neq (3, 2)$, a straightforward computation shows that $e_s(G) < |L|_s$. This final contradiction completes the proof.

Remarks: (1) Let $J = PGL_2(r^a)$. Let B be the upper triangular matrices in J (so $B/O_r(B)$ is cyclic of order $q-1$). Let T be a maximal nonsplit torus of J (so T is cyclic of order $q+1$). Let $H = (B \times T) \cap DE \leq J \times J$ and $G = H(L \times L) \leq DE$ (in the notation of the proof). Then G acts transitively on the cosets of G_w, the normalizer of the diagonal subgroup of $L \times L$, H is transitive and $H/O_r(H)$ is cyclic.

(2) Note that $L_2(q) \times L_2(q) = P\Omega_4^+(q)$. Thus, this family of examples is actually in one of the families given in Theorem 3.1 (but the group happens not to be simple).

6. Affine Groups

Let G be a primitive group of affine transformations in characteristic r of degree r^a. So $E = O_r(G)$ is elementary abelian and $G = EG_0$, where G_0 is the stabilizer of the origin and acts irreducibly on E. Let H be a transitive subgroup with $H/O_p(H)$ cyclic. If $p = r$, then we can take H to be any p-subgroup containing E (other choices may be possible). If $p \neq r$, then the Sylow p-subgroup of H has fixed points. Since the set of fixed points is invariant under H, it follows that $O_p(H) = 1$, i.e. H is cyclic. It follows easily that $a = 1$, or $r^a = 4$. If $a = 1$, then $G'' = 1$; if $r^a = 4$ then $G = S_4$.

For the rest of this section, we assume that $p = r$. Set $q = p^a = |E|$. Although the group theoretic conditions are satisfied, we do not, in general, know when a group corresponds to a polynomial. However, there are two cases where we can say more. Consider the case where G_0 is cyclic of order d. Then G embeds into the upper triangular matrices of $PGL_2(q)$ and so acts as a group of automorhpisms of $\mathbb{F}_q[y]$. Indeed, we see that the fixed ring of G will be $\mathbb{F}_q[h(y)]$ where $h(y) = (y^q - y)(y^q - \alpha y) \cdots (y^q - \alpha^{d-1} y)$, where α is a primitive dth root of 1 in $\mathbb{F}_q$. The fixed ring of G_0 is $\mathbb{F}_q[y^d]$. Thus $h(x) = f(x^d)$ for some polynomial f. It follows that G is the (geometric) monodromy group of the extension $\mathbb{F}_q(x)/\mathbb{F}_q(f(x))$.

Note that $h(y)$ is invariant under any automorphism of $\mathbb{F}_q$ and so has coefficients in $\mathbb{F}_p$. Thus, the same is true for $f(x)$. In this construction, we did not need to assume that G_0 acted irreducibly. Then the field extension $\mathbb{F}_p(x)/\mathbb{F}_p(f(x))$ has (arithmetic) monodromy group $G.a$. In particular, this group does act primitively while G may not – this leads to examples of indecomposable polynomials which become decomposable over a finite extension.

Now assume that G is the monodromy group of $F(x)/F(f(x))$, where F is algebraically closed of characteristic p. We are still assuming that G is an affine primitive group. We view $f : X \to Y$ where $X = Y = \mathbb{P}^1$ as a branched covering. Let $B \subset Y$ be the set of branch points. So $B = \{y_0, y_1, \ldots, y_r\}$ with $y_0 = \infty$. We follow the treatment in [FGS].

We recall the Riemann-Hurwitz formula in this context:

$$2(p^a - 1) = \sum_{x \in X} \operatorname{ord}(D_x), \qquad (*)$$

where D_x is the (local) different of the cover. It is convenient to break up this sum. Let $y \in Y$ and set $\lambda(y) = \sum_{x \in f^{-1}(y)} \operatorname{ord}(D_x)$. Then $\lambda(y) = 0$ if $y \notin B$. If the cover is tamely ramified at y_i, then the contribution to the righthand side from the points over y_i will be equal to $p^a - \operatorname{orb}(g_i)$, where $\operatorname{orb}(g_i)$ is the number of orbits of g_i and g_i is a generator of a decomposition group for some point over y_i. If the cover is wildly ramified, then it is more complicated to compute the righthand side, but we shall use the following straightforward result (this was pointed out to us by Fried).

Let $y \in B$. Let D be a decomposition group at some point over y. Define $\operatorname{orb}'(D)$ to be the number of orbits of D whose length is not divisible by p. Then

$$\lambda(y) \geq n - \operatorname{orb}'(D). \qquad (**)$$

In particular, since the decomposition group at y_0 is transitive, it follows that $\lambda(y_0) \geq p^a$.

With these results in hand, it is a simple matter to show (using the fact that this is a group of affine transformations – see [GT, §2] or [N]):

Lemma 6.1. *Assume $y \in B$ is tamely ramified. Let $g \in G$ be a generator for the decomposition group. Assume g has order d.*
(a) $\lambda(y) \geq (1/2)p^{a-1}(p-1)$ *for p odd.*
(b) *If $p = 2$, $\lambda(y) \geq 2^{a-1}$.*
(c) *If p is odd and $d > 2$, then $\lambda(y) \geq (3/5)p^a$.*

Lemma 6.2. *Assume $y \in B$ is wildly ramified with D a decomposition group.*
(a) *If every orbit of D has cardinality divisible by p, then $\lambda(y) \geq p^a$.*
(b) $\lambda(y) \geq p^a - p^{a-1} \geq (1/2)p^a$.

It now follows from $(*)$ and Lemmas 6.1 and 6.2 that:

Theorem 6.3. *Assume that G is the monodromy group of $F(x)/F(f(x))$, where $f(x)$ is an indecomposable polynomial in $F[x]$ of degree p^a with F algebraically closed of characteristic p. Assume that G acts on the conjugates of x as a group of characteristic p affine transformations. We consider $f :$ $X \to Y$ where $X = Y = \mathbb{P}^1$ as a branched covering. Let $B \subset Y$ be the set of branch points. So $B = \{y_0, y_1, \ldots, y_r\}$ with $y_0 = \infty$. Then $|B| \leq 3$. If $|B| = 3$, then p is odd, y_0 is the only wildly ramified point and the decomposition groups at the tamely ramified points have order 2.*

If we now consider the case where G/E is a p'-group, we can show:

Theorem 6.4. *Let G be the monodromy group of $F(x)/F(f(x))$ with F algebraically closed of characteristic p and f of degree p^a. Assume that the permutation representation of G on the conjugates of x is an affine group in characteristic p. If $G/O_p(G)$ is a p'-group, then $G/O_p(G)$ is cyclic or dihedral. Moreover, if $p = 2$, then $G/O_p(G)$ is cyclic.*

Proof. Note that any p-element in G is in $O_p(G)$ and so is fixed point free in the permutation action. So if y_i is wildly ramified, $\lambda(y_i) \geq p^a$. Since y_0 is wildly ramified, this implies it is the only wildly ramified branch point. Let L be the Galois closure of $F(x)/F(f(x))$. Let M be the fixed field of $O_p(G)$. Then M is a Galois extension of $F(f(x))$ (since $O_p(G)$ is normal in G) with Galois group G_0. The hypothesis implies that this extension is tamely ramified.

First assume that there are three branch points. Then p is odd. Hence, $G/O_p(G) = \langle g_0, g_1, g_2 \rangle$ where g_i is the generator of the decomposition group. Moreover, by choosing the g_i appropriately, we may also assume that the product is trivial. It follows by Theorem 6.3 that $d_1 = d_2 = 2$. Thus, $G/O_p(G)$ is dihedral.

If there are only two branch points, then the argument of the previous paragraph shows that $G/O_p(G)$ is cyclic. This completes the proof.

We now consider what one can say about exceptional indecomposable polynomials over $\mathbb{F}_q$ with affine (arithmetic) monodromy group G. Let $\bar{G}$ denote the geometric monodromy group. Assume that f has degree p^2 with $p > 3$. If p divides the order of $|G : O_p(G)|$, then G contains $SL_2(p)$. It follows that $\bar{G}$ does as well. Then $\bar{G}$ is 2-transitive and f cannot be exceptional (see [FGS, §3]). Otherwise, we can apply 6.4 to conclude that G_0 is cyclic or dihedral.

Thus, any exceptional polynomial of degree p^2 with affine monodromy group has a solvable monodromy group.

7. Proofs of the Theorems

Note that Theorem A follows from Propostion 2.1, Theorem 3.1, Theorem 4.2 and Theorem 5.1. Theorem B is essentially a translation of Theorem A to the geometric setting.

We now consider Theorem C. Let $f \in \mathbb{F}_q[x]$ be indecomposable. Assume that f decomposes over $\bar{F}$. This is equivalent to saying that the arithmetic monodromy group G is primitive but the geometric monodromy $\bar{G}$ is not.

We can apply our main result. Assume that the degree of f is not a power of p. Since $G/\bar{G}$ is cyclic, it follows that $F^*(G) = F^*(\bar{G})$. In the diagonal case, it follows that $F^*(\bar{G})$ already acts primitively, a contradiction.

The case where f has degree $4p^{2a}$ is allowed in the conclusion. So it suffices to assume that $F^*(G)$ is simple. One checks that in all cases in Theorem 3.1, except for those allowed in the theorem, $F^*(G)$ already is primitive. Thus, $\bar{G}$ is primitive and so f is indecomposable over $\bar{F}$.

References

[A1] S. S. Abhyankar, Galois theory on the line in nonzero characteristic, Bull. Amer. Math. Soc. 27 (1992), 68-133.

[A2] S. S. Abhyankar, Fundamental group of the affine line in positive characteristic, preprint.

[A3] S. S. Abhyankar, Mathieu group coverings and linear group coverings, preprint.

[AS] M. Aschbacher and L. Scott, Maximal subgroups of finite groups, J. Algebra 92 (1985), 44-80.

[At] J.H. Conway, R.T. Curtis, S.P. Norton, R.A. Parker and R.A. Wilson, Atlas of finite groups, O.U.P. 1985.

[Co] S. D. Cohen, Permutation polynomials and primitive permutation groups, Arch. Math. 57 (1991), 417-423.

[CM] S. D. Cohen and R. W. Matthews, A class of exceptional polynomials, preprint.

[Fe] W. Feit, On symmetric balanced incomplete block designs with doubly transitive automorphism groups, J. of Comb. Theory Series A 14 (1973), 221-247.

[F1] M. D. Fried, On a conjecture of Schur, Mich. Math. J. 17 (1970), 41-55.

[F2] M. D. Fried, Exposition on an Arithmetic-Group Theoretic Connection via Riemann's Existence Theorem, Proceedings of Symposia in Pure Math # 37: Santa Cruz Conference on Finite Groups, Amer. Math. Soc., Providence, 1980, 571-601.

[F3] M. D. Fried, General exceptional covers, preprint.

[FGS] M.D. Fried, R. Guralnick and J. Saxl, Schur covers and Carlitz's conjecture, Israel J.Math. 82 (1993), 157 - 225.

[G] R. Guralnick, Subgroups of prime power index in a simple group, J. Algebra 81 (1983), 304-311.

[GT] R. Guralnick and J. Thompson, Finite groups of genus zero, J. Algebra 131 (1990), 303-341.

[H] D. Harbater, Abhyankar's conjecture on Galois groups over curves, Inventiones Math. 117 (1994), 1-25.

[HLS] C. Hering, M.W. Liebeck and J. Saxl, The factorizations of the finite exceptional groups of Lie type, J. Algebra 106 (1987), 517 - 527.

[Ka] W. Kantor, Some consequences of the classification of finite simple groups, pp. 159-173 in Finite Groups—Coming of Age (ed. J. McKay), Contemporary Math. 45, AMS 1985.

[K1] P.B. Kleidman, The maximal subgroups of the finite 8-dimensional orthogonal $P\Omega_8^+(q)$ and of their automorphism groups, J. Algebra 110 (1987), 173 - 242.

[K2] P.B. Kleidman, The low-dimensional finite simple classical groups and their subgroups, unpublished.

[KL] P.B. Kleidman and M.W. Liebeck, The subgroup structure of the finite classical groups, C.U.P., Cambridge, 1990.

[Ko1] A.S. Kondrat'ev, Modular representations of degree ≤ 27 of the finite quasisimple groups of alternating and sporadic type (in Russian), preprint Sverdlovsk 1992.

[Ko2] A.S. Kondrat'ev, Modular representations of small degrees of the finite groups of Lie type (in Russian), preprint Sverdlovsk 1989.

[L] M.W. Liebeck, On the orders of maximal subgroups of the finite classical groups, Proc. London Math. Soc. 50 (1985), 426 - 446.

[LPS] M.W. Liebeck, C.E. Praeger and J. Saxl, The maximal factorizations of the finite simple groups and their automorphism groups, Memoirs Amer. Math. Soc. 432(1990).

[LPS2] M.W. Liebeck, C.E. Praeger and J. Saxl, On the O'Nan-Scott reduction theorem for finite primitive permutation groups, J. Australian Math. Soc. A 44 (1988), 389-396.

[MAt] K. Lux, R. A. Parker, R. A. Wilson et al, Modular Atlas, to appear.

[MSW] G. Malle, J. Saxl and T. Weigel, Generation of classical groups, Geom. Ded. 49 (1994), 85-116

[M1] P. Müller, New Examples of Exceptional Polynomials, preprint.

[M2] P. Müller, Primitive Monodromy Groups of Polynomials, preprint.

[N] M. Neubauer, On monodromy groups of fixed genus, J. Algebra 153 (1992), 215-261.

[PPS] T. Penttila, C.E. Praeger and J. Saxl, Linear groups with orders divisible by certain large primes, in preparation.

[R] M. Raynaud, Revêtements de la droite affine en caractéristique $p > 0$ et conjecture d'Abhyankar, Inventiones Math. 116 (1994), 85-116

[S] J.-P. Serre, Local Fields, Springer-Verlag, New York, 1979.

[W] D. Wan, Permutation polynomials and resolution of singularities over finite fields, Proc. Amer. Math. Soc. 110 (1990), 303-309.

[Wi] H. Wielandt, Finite Permutation Groups, Academic Press, New York, 1964.

Robert M. Guralnick
Department of Mathematics
University of Southern California
Los Angeles, CA 90089-1113 USA
e-mail guralnic@mtha.usc.edu

Jan Saxl
Department of Pure Mathematics
Cambridge University
Cambridge CB2 1SB, England
e-mail saxl@pmms.cam.ac.uk

Subgroups of Exceptional Algebraic Groups

Martin W. Liebeck and Gary M. Seitz

February 17, 1994

Throughout this paper, let G be a simple algebraic group of exceptional type over an algebraically closed field K of characteristic p. The maximal closed connected subgroups of these groups were determined in [Se], subject to some mild restrictions on the characteristic p. In this article we describe some results from [LS2] concerning arbitrary closed connected reductive subgroups of G, again assuming mild characteristic restrictions (in particular, $p = 0$ or $p > 7$ covers all the restrictions).

Before giving detailed statements, we give a general description of the results. Theorems 5 and 6 below determine the embeddings of arbitrary closed connected semisimple subgroups in G: if X is such a subgroup, then X is embedded in an explicit way in a "subsystem subgroup" of G - that is, a semisimple subgroup which is normalized by a maximal torus of G. Subsystem subgroups are constructed naturally from subsystems of the root system of G; this therefore determines the embedding of X in G. As a consequence, when $p = 0$ there are only finitely many conjugacy classes of such subgroups X, whereas there are infinitely many when $p > 0$.

The proofs are based on Theorem 1, which states that if the reductive subgroup X lies in a parabolic subgroup $P = QL$ of G, with unipotent radical Q and Levi subgroup L, then some conjugate of X lies in L. This result can also be used to prove that $C_G(X)$ is always reductive (Theorem 2).

The other results concern the actions of simple closed connected subgroups X of G on the Lie algebra $L(G)$ of G. Following [Dy], a "labelled diagram" is associated with each such subgroup; this is a labelling of the Dynkin diagram of G with certain non-negative integers which are determined by the action of a suitably chosen 1-dimensional torus of X on $L(G)$. Theorems 7 and 8 reveal the extent to which X is determined up to conjugacy by its labelled diagram. Finally, Theorems 3 and 4 give information about centralizers and composition factors.

We now state the results. In order to specify our assumptions on the characteristic p, we define, for certain pairs (X, G), an integer $N(X, G)$, as given in the following table.

$X =$	$G = E_8$	E_7	E_6	F_4	G_2
A_1	7	7	5	3	3
A_2	5	5	3	3	
B_2	5	3	3	2	
G_2	7	7	3	2	
A_3	2	2	2		
B_3	2	2	2	2	
C_3	3	2	2	2	
B_4, C_4, D_4	2	2	2		

For example, $N(A_2, E_7) = 5$, and so on. (This is the table of [Se, Theorem 1], with a few additional entries.) If (X, G) is not in the table, set $N(X, G) = 1$. And if X is a non-abelian closed connected reductive subgroup of G, and $X' = X_1...X_t$, a commuting product of simple groups X_i, then we define

$$N(X, G) = \max(N(X_i, G) : 1 \le i \le t)$$

In particular, if $p > 7$ then $p > N(X, G)$ for all X, G.

Theorem 1 *Let X be a closed connected reductive subgroup of G, and assume that either $p = 0$ or $p > N(X, G)$. Suppose that X lies in a parabolic subgroup $P = QL$ of G, with unipotent radical Q and Levi subgroup L. Then all closed complements to Q in the semidirect product QX are Q-conjugate. In particular, X is Q-conjugate to a subgroup of L.*

The proof of Theorem 1 is based on the fact that Q has a filtration by particular high weight modules for L. Choosing P to be minimal (subject to containing X), we find the embedding of X in L modulo Q, so we can restrict each of these modules to X. Then if V is a composition factor of such a restriction, we show that with a few exceptions, all closed complements to V in the semidirect product VX are V-conjugate, and the desired conclusion follows from this. The proof is similar in spirit to those of [LS1, Theorems 2.1 and 6.1], but the representation theory involved is much more complicated, and a great deal of calculation is required.

The following result is an immediate consequence of Theorem 1 (compare [LS1, Theorem 8.1]).

Corollary *Let X be a closed connected reductive subgroup of G, and assume that $p = 0$ or $p > N(X, G)$. Suppose that X normalizes a closed unipotent* subgroup U of G. Then all closed complements to U in the semidirect product UX are G-conjugate.

Taken together with [Se, Theorem 1], Theorem 1 leads to a description of all closed semisimple subgroups of G (see Theorems 5 and 6 below). The next result is also a consequence of Theorem 1.

Theorem 2 Let X be a closed connected reductive subgroup of G, and assume that $p = 0$ or $p > N(X, G)$. Then

(i) $C_G(X)$ is reductive;

(ii) if $p > 0$ then $O_p(C_G(X)) = 1$;

(iii) if X is semisimple, then the rank of $C_G(X)$ is equal to the maximal co-rank among subsystem subgroups of G containing X.

The determination of all closed simple subgroups of G leads to the next result.

Theorem 3 Let X be a simple closed connected subgroup of G with $\operatorname{rank}(X) \geq 2$, and assume that either $p = 0$ or p is a good prime for G and $p > N(X, G)$. Then

$$C_{L(G)}(X) = L(C_G(X)).$$

Remark The analogues of Theorems 2 and 3 for classical groups are not in general true. For example, if $G = SL(V)$ and X is a subgroup of G such that V is indecomposable for X with composition series $0 < V_1 < V_2 < V$, where $V_1 \cong V/V_2$, then $C_G(X)$ is not reductive. And if $G = SL_n$ with $p = n$, then $C_{L(G)}(G) \neq L(C_G(G))$.

The next theorem and its corollary concern the connection between Aut G-conjugacy and linear equivalence on $L(G)$ for subgroups of G.

Theorem 4 Let X_1 and X_2 be closed connected simple subgroups of G of the same type, and assume that $p = 0$ or $p > N(X_1, G)$. Suppose that X_1 and X_2 have the same composition factors on $L(G)$ (counting multiplicities). Then either X_1 is conjugate to X_2 in Aut G, or $G = E_8$ and $X_1 \cong X_2 \cong A_2$, with both X_1 and X_2 lying in subsystem groups $D_4 D_4$ and projecting irreducibly in each factor.

Corollary Let X_1, X_2 be closed connected simple subgroups of G, and assume that $p = 0$ or $p > N(X_1, G)$. If X_1 and X_2 are conjugate in $GL(L(G))$, then either they are also conjugate in Aut G, or $G = E_8$ and $X_1 \cong X_2 \cong A_2$ lying in subsystem groups $D_4 D_4$.

Remarks 1. The cases with $G = E_8$ and $X_1 \cong X_2 \cong A_2$ lying in $D_4 D_4$ really are exceptions to the conclusion of Theorem 4 and the corollary. Indeed, E_8 has two conjugacy classes of such subgroups A_2 which have the same composition factors on all E_8-modules.

2. A result like Theorem 4 is not in general true for subgroups of classical groups. For example, let $p > 0, q = p^a > 1$ and let $X_1 \cong X_2 \cong SL(W)$, where W is a vector space of dimension $m \geq 3$ over the algebraically closed field K of characteristic p. Embed X_1 and X_2 in $D = SL_{m^2}(K)$ via the modules $W \otimes W^{(q)}$ and $W^* \otimes W^{(q)}$ respectively, (where $W^{(q)}$ is a Frobenius q-power twist of W). Then X_1 and X_2 are not Aut D-conjugate, but have the same composition factors on $L(D)$.

We come now to the theorems which describe the embeddings of arbitrary semisimple closed connected subgroups of G. In order to state the results, we need to make one further definition.

Definition Let $Y = Y_1 \ldots Y_k$ be a commuting product of simple algebraic groups Y_i, and let X be a closed semisimple subgroup of Y. For each i, let $\hat{Y}_i$ be the simply connected cover of Y_i. If A is a subgroup of Y, write $\bar{A}$ for the group $AZ(Y)/Z(Y)$, and for $i = 1, \ldots, k$ let $\pi_i : \bar{X} \to \bar{Y}_i$ be the ith projection map. We call the connected preimage of $\bar{X}\pi_i$ in Y_i the *projection* of X in Y_i. We say that X is *essentially embedded* in Y if the following hold for all i:

(i) if Y_i is of classical type, with natural module V_i (taken to be the natural $2n$-dimensional symplectic module if $(Y_i, p) = (B_n, 2)$), then either the projection of X in Y_i lifts to a subgroup of $\hat{Y}_i$ which is irreducible on V_i, or $Y_i = D_n$ and the projection of X in Y_i lies in a natural subgroup $B_r B_{n-r-1}$ for some $r \geq 0$, irreducible in each factor with inequivalent representations;

(ii) if Y_i is of exceptional type, then the projection of X in Y_i is either Y_i or a maximal connected subgroup of Y_i not containing a maximal torus (and hence is given by [Se, Theorem 1]).

Theorem 5 *Let X be a closed connected semisimple subgroup of G, such that every simple factor of X has rank at least 2, and assume that $p = 0$ or $p > N(X, G)$. Choose a subsystem subgroup Y of G, minimal subject to containing X (possibly $Y = G$ of course). Then either*

(i) X is essentially embedded in Y, or

(ii) $Y = G = E_6$, $p = 7$, $X = G_2$ and $X < F_4 < G$ with X maximal in F_4.

The conjugacy classes and centralizers in G of the simple subgroups X of rank at least 2 are explicitly listed in [LS2, Section 8]; also given there are the minimal subsystem subgroups containing X and the restrictions $L(G) \downarrow X$.

The next result is the analogue of Theorem 5 for subgroups with a factor A_1.

Theorem 6 *Let X be a closed connected semisimple subgroup of G with a factor A_1, and assume that $p = 0$ or $p > N(A_1, G)$. Then one of the following holds:*

(i) there is a subsystem subgroup Y of G containing X, such that Y is a product of classical groups and X is essentially embedded in Y, or

(ii) there is a subgroup $Y_0 = F_4, E_6, E_7$ or E_8 of G, and a semisimple subgroup Y_1 of $C_G(Y_0)$, such that either

(a) $X = Y_0 Y_1$, or

(b) X is essentially embedded in ZY_1, where Z is a maximal connected subgroup of Y_0 not containing a maximal torus;

(iii) $G = G_2$, $X = A_1$ and X is maximal in G.

Remark In (ii) of the theorem, the possibilities for $Y_0 C_G(Y_0)$ are given by Theorem 5, and those for Z by [Se, Theorem 1]. They are:

Y_0	possibilities for Z	$C_G(Y_0)$ $(G = E_8, E_7, E_6, F_4)$
F_4	$A_1, A_1 G_2, G_2$	$G_2, A_1, 1, 1(\text{resp.})$
E_6	$A_2, G_2, A_2 G_2, C_4, F_4$	$A_2, T_1, 1, -$
E_7	$A_1, A_2, A_1 A_1, A_1 G_2, A_1 F_4, G_2 C_3$	$A_1, 1, -, -$
E_8	$A_1, B_2, A_1 A_2, G_2 F_4$	$1, -, -, -$

The final two results concern labelled diagrams for simple closed connected subgroups X of G. These are defined as follows. Let T_0 be a maximal torus of X, and let $\Sigma(X)$ be the corresponding root system of X. Choose a fundamental system $\Pi(X)$ in $\Sigma(X)$, and denote by $\Sigma(X)^+$ the set of positive roots relative to $\Pi(X)$. If $\alpha \in \Sigma(X)^+$, and U_α is the corresponding root subgroup of X, then $\langle U_{\pm\alpha} \rangle$ is an image of SL_2, and we write $h_\alpha(c)$ for the image of the matrix $\text{diag}(c, c^{-1})$ $(c \in K^*)$. For $c \in K^*$, define $T(c) = \prod h_\alpha(c)$, where the product is taken over all $\alpha \in \Sigma(X)^+$. Then

$$T_X = \langle T(c) : c \in K^* \rangle$$

is a 1-dimensional torus of X.

Now let T be a maximal torus of G containing T_0, and let $\Sigma = \Sigma(G)$ be the root system of G relative to T. For $\beta \in \Sigma$, let e_β be a weight vector for T corresponding to β. Then there exist integers l_β such that for all $\beta \in \Sigma, c \in K^*$,

$$T(c)e_\beta = c^{l_\beta} e_\beta.$$

We can choose a fundamental system $\Pi = \Pi(G)$ such that $l_\alpha \geq 0$ for all $\alpha \in \Pi$, and define the *labelled diagram* of X to be the Dynkin diagram of G, with each node $\alpha \in \Pi$ labelled by the non-negative integer l_α. It can be shown that the labelled diagram of X is uniquely determined up to graph automorphisms of G.

Theorem 7 *Assume that $p = 0$ or $p > N(A_1, G)$. Then any subgroup A_1 of G is determined up to conjugacy in Aut G by its labelled diagram.*

It is a fact that subgroups A_1 with the same composition factors on $L(G)$ have the same labelled diagram (see [LS2, Theorem 7.3]). Thus Theorem 7 is closely related to Theorem 4 in the A_1 case.

Theorem 8 *Let X_1, X_2 be closed connected simple subgroups of G of the same type, and of rank at least 2. Assume that X_1 and X_2 have the same labelled diagram, and that $p = 0$ or $p > N(X_1, G)$. Then with two exceptions, there is a subsystem subgroup Y of G such that*

(i) Y contains X_1 and a conjugate X_2^g of X_2,

(ii) Y is minimal subject to containing X_1, and

(iii) X_1 and X_2^g are conjugate in Aut Y.

In the exceptions, $G = E_7$ (resp. E_8), $X_1 \cong X_2 \cong A_2$ and X_1, X_2 lie in subsystem subgroups A_5, A_5' (resp. $A_2 A_5, D_4 D_4$).

In the last sentence of Theorem 8, A_5 and A_5' are representatives of the two conjugacy classes of subsystem subgroups of type A_5 in E_7.

References

[**Dy**] E.B. Dynkin, "Semisimple subalgebras of semisimple Lie algebras", Amer. Math. Soc. Translations 6 (1957), 111-244.

[**LS1**] M.W. Liebeck and G.M. Seitz, "Subgroups generated by root elements in groups of Lie type", Annals of Math., to appear.

[**LS2**] M.W. Liebeck and G.M. Seitz, "Reductive subgroups of exceptional algebraic groups", to appear.

[**Se**] G.M. Seitz, "Maximal subgroups of exceptional algebraic groups", Memoirs Amer. Math. Soc., No. 441 (1991).

The geometry of traces in Ree Octagons

H. Van Maldeghem *

May 31, 1994

Abstract

In this paper, we prove some geometric properties of traces of perfect Ree octagons. It is shown, for instance, that a derived geometry can be defined and that it is isomorphic to the generalized quadrangle $T_3(\mathcal{O})$ of Tits- type, where $\mathcal{O}$ is a Suzuki-Tits ovoid.

1 Introduction

Generalized polygons were introduced by TITS [6] and have since then been studied by several authors. The main examples arise from groups of Lie type or their twisted analogues. A great deal of research concerning polygons is devoted to characterizing these Lie polygons in a geometric fashion. One of the most beautiful and complete results in this direction is RONAN's [4] characterization of all Moufang hexagons by ideal lines. Also various classes of "classical" generalized quadrangles, mainly finite ones, are characterized by the same idea of looking at intersections of traces. No such characterization of the Moufang octagons is known. From a geometric point of view

---

*Research Associate at the National Fund for Scientific Research (Belgium)

however, it is already an interesting question to ask what kind of properties traces have in the Moufang octagons. We will answer this question in the present paper for a large subclass of Moufang octagons, thus establishing the geometric foundation necessary to reconstruct the ambient metasymplectic space for these geometries, which should eventually lead to a geometric characterization of all such octagons. We will briefly sketch at the end of the paper how to do this.

2   Notation

We will assume that the reader is familiar with the definition of a generalized polygon, in particular a generalized octagon. Also, we will use some common building terminology such as *opposite elements* for 2 points or 2 lines at maximal distance; the projection of an element x onto a non-opposite element y for the unique element incident with y closest to x (see e.g. TITS [8]). Also, for any point x, we will denote by $x^{\perp}$ the set of points collinear to x.

One important class of thick generalized octagons arises from the Ree group of type $^2F_4$, see TITS [10], and we call the members of this class the *Ree octagons*. For every field K of characteristic 2 and every endomorphism σ in K whose square is the Frobenius endomorphism, there exists such a Ree octagon, which we will denote by $O_R(K,\sigma)$. In the infinite case, we also have some other examples arising from some 'free' constructions, see TITS [9].

It is known that the Ree octagon $O_R(K,\sigma)$ can be viewed as the set of absolute points and lines of a certain polarity in the metasymplectic space over K, see e.g. SARLI [5]. In the present paper, we want to clear the way for reconstructing this metasymplectic space entirely in terms of the geometry of the Ree octagon. This will establish the foundation for a geometric characterization of these octagons, which will be done elsewhere. However, in order not to drown in notation and technicalities, we restrict ourselves to the perfect case, i. e. the case where σ is an automorphism. The non-perfect case is – geometrically – very much more complicated and so, in this paper, we do not want to spend twice as much space for objects only half as important (as a figure of speech).

3 The Ree octagon $O_R(K, \sigma)$.

The following description of $O_R(K, \sigma)$, K and σ as in the previous section, is due to JOSWIG & VAN MALDEGHEM [2].

Let $K_\sigma^{(2)}$ be the group on the set of all pairs $(k_0, k_1) \in K \times K$ with operation law $(k_0, k_1) \oplus (l_0, l_1) = (k_0 + l_0, k_1 + l_1 + l_0 k_0^\sigma)$. For $k = (k_0, k_1)$, set $tr(k) = k_0^{\sigma+1} + k_1$ (the *trace* of k) and set $N(k) = k_0^{\sigma+2} + k_0 k_1 + k_1^\sigma$ (the *norm* of k). Define a multiplication $a \otimes k = a \otimes (k_0, k_1) = (ak_0, a^{\sigma+1} k_1)$. Also write $(k_0, k_1)^\sigma$ for (k_0^σ, k_1^σ). Then the points of $O_R(K, \sigma)$ are the elements of $\{(\infty)\} \cup K \cup K_\sigma^{(2)} \times K \cup \ldots \cup K_\sigma^{(2)} \times K \times K_\sigma^{(2)} \times K \times K_\sigma^{(2)} \times K \cup K \times K_\sigma^{(2)} \times K \times K_\sigma^{(2)} \times K \times K_\sigma^{(2)} \times K$ (and these are all denoted by round parantheses); the lines of $O_R(K, \sigma)$ are the elements of $\{[\infty]\} \cup K_\sigma^{(2)} \cup K \times K_\sigma^{(2)} \cup \ldots \cup K \times K_\sigma^{(2)} \times K \times K_\sigma^{(2)} \times K \times K_\sigma^{(2)} \cup K_\sigma^{(2)} \times K \times K_\sigma^{(2)} \times K \times K_\sigma^{(2)} \times K \times K_\sigma^{(2)}$ (and denoted by square brackets); incidence is given by the sequence

$$(a, l, a', l', a'', l'', a''') \ I \ [a, l, a', l', a'', l''] \ I \ (a, l, a', l', a'') \ I \ \ldots$$

$$\ldots (a) \ I \ [\infty] \ I \ (\infty) \ I \ [k] \ I \ (k, b) \ I \ \ldots$$

$$\ldots [k, b, k', b', k''] \ I \ (k, b, k', b', k'', b'') \ I \ [k, b, k', b', k'', b'', k'''],$$

and the rule : $(a, l, a', l', a'', l'', a''')$ is incident with $[k, b, k', b', k'', b'', k''']$ if and only if the following six equations hold:

$$(k_0''', k_1''') = (l_0, l_1) \oplus a \otimes (k_0, k_1) \oplus (0, al_0' + a^\sigma l_0'') \tag{I1}$$

$$\begin{aligned} b'' = \ & a' + a^{\sigma+1} N(k) + k_0(al_0' + a^\sigma l_0'' + tr(l)) \\ & + a^\sigma(a''' + l_0 k_1) + al_0''^\sigma + l_0 l_0' \end{aligned} \tag{I2}$$

$$\begin{aligned} (k_0'', k_1'') = \ & a^\sigma \otimes (k_1, tr(k)N(k)) \oplus k_0 \otimes (l_0, l_1)^\sigma \\ & \oplus (0, tr(k)N(l) + a^{\sigma+1} l_0 N(k)^\sigma \\ & + tr(k)(aa' + a^\sigma l_0 l_0'' + a^{\sigma+1} a''') \\ & + tr(l)(k_1^\sigma a + a''') + k_1^\sigma a^{\sigma+1} l_0'' + k_0^{\sigma+1} a^2 l_0''^\sigma \\ & + k_0(a' + al_0''^\sigma + k_1 a^\sigma l_0 + a^\sigma a''')^\sigma \\ & + k_0^\sigma l_0(a' + al_0''^\sigma + k_1 a^\sigma l_0 + a^\sigma a''') \\ & + a(l_1'' + a'''^\sigma l_0 + a''' l_0') \\ & + l_0''(a' + a^\sigma a''') + a'' l_0 + l_0 l_0' l_0'') \\ & \oplus (l_0', l_1') \end{aligned} \tag{I3}$$

$$\begin{aligned} b' = \ & a'' + a^{\sigma+1} N(k)^\sigma + a(k_0 l_0'' + l_0 k_1 + a''')^\sigma \\ & + tr(k)(l_1 + a^\sigma l_0'') \\ & + k_0^\sigma(a' + a^\sigma a''') + l_0' l_0'' + l_0^\sigma a''' \end{aligned} \tag{I4}$$

$$\begin{aligned} (k_0', k_1') = \ & (l_0'', l_1'') \oplus a \otimes (tr(k), k_0 N(k)^\sigma) \oplus l_0 \otimes (k_0, k_1)^\sigma \\ & \oplus (0, N(k)(a^\sigma l_0'' + l_1) \\ & + k_0(a'' + l_0' l_0'' + aa''''^\sigma + l_0^\sigma a''') \\ & + k_1(k_1 l_0 a^\sigma + a' + al_0''^\sigma + a^\sigma a''') \\ & + k_0 k_1^\sigma a l_0^\sigma + a''''^\sigma l_0 + a''' l_0' \end{aligned} \tag{I5}$$

$$b = a''' + aN(k) + l_0 k_1 + k_0 l_0'' \tag{I6}$$

where $a, a', a'', a''', b, b', b'' \in K$ and $k, k', k'', k''', l, l', l'' \in K_\sigma^{(2)}$ and $k = (k_0, k_1)$, etc. These elements are called *the coordinates*.

This description is also valid in the non-perfect case. From now on however, we will assume that K is perfect. This includes every finite field $GF(2^{2e+1})$. In the latter case, the corresponding generalized octagon is denoted by $O_R(2^{2e+1})$.

Note that every Ree octagon has a lot of automorphisms (it has the *Moufang property* and it is characterized in this way, see TITS [10]). In particular, the twisted Chevalley group $^2F_4(K, \sigma)$ is an automorphism group of $O_R(K, \sigma)$. It acts transitively on the set of opposite pairs of points, and also dually, on the set of pairs of opposite lines. The stabilizer of a pair of opposite points acts on the set of lines incident with either one of these points as a doubly transitive automorphism group of a Suzuki-Tits ovoid (see next section) and the stabilizer of a pair of opposite lines acts on the set of points of either one of these lines as $PGL_2(K)$. We will use these properties in order to choose certain arbitrary elements in a suitable way "without loss of generality".

4 Geometric properties of $O_R(K, \sigma)$

4.1 Properties of the Suzuki-Tits ovoids

Let $W(K)$ be the symplectic generalized quadrangle over K, i. e. the generalized quadrangle arising from a symplectic polarity τ in $PG(3, K)$. Let ρ be a polarity in $W(K)$, then it is known that the set of absolute points (resp. lines) of $W(K)$ (i. e. the points (resp. lines) incident with their image under ρ) forms an ovoid $\mathcal{O}_{ST}$ (resp. spread $\mathcal{S}_L$) in $W(K)$, called the *Suzuki-Tits ovoid* (resp. *Lüneburg spread*), see TITS [7]. Let π be a plane of $PG(3, K)$. It is easily seen that the intersection of π with $\mathcal{O}_{ST}$ is exactly the set of points of $\mathcal{O}_{ST}$ collinear in $W(K)$ to the point π^τ. We call such a set a *circle*. Now, every three points of $\mathcal{O}_{ST}$ determine a unique plane, and hence a unique circle. So we obtain an inversive plane. But the Lüneburg spread puts an extra structure on this inversive plane, indeed, given a circle C lying in the plane π, the point π^τ is incident with a unique element M of $\mathcal{S}_L$. And M is incident with a unique point x of $\mathcal{O}_{ST}$, which belongs to C. Hence every circle C contains a special element x which we will call the *corner* of the circle and denote by ∂C. We list now some immediate properties.

LEMMA 4.1 *Let $\mathcal{O}_{ST}$ be a Suzuki-Tits ovoid and let $x \in \mathcal{O}_{ST}$. Let $\mathcal{C}_\S$ be the set of circles C with $\partial C = x$. Then the $C \setminus \{x\}$ partitions $\mathcal{O}_{ST} \setminus \{\S\}$.*

In other words, any circle is uniquely determined by its corner and a second point, and, conversely, every pair of points (x, y) in $\mathcal{O}_{ST}$ defines a unique circle C such that $y \in C$ and $\partial C = x$.

LEMMA 4.2 *Let $\mathcal{O}_{ST}$ be a Suzuki-Tits ovoid and let $x \in \mathcal{O}_{ST}$ be the corner of a circle C. Let $\mathcal{D}_C$ be the set of circles C' with $\partial C = y \in C \setminus \{x\}$ and $x \in C'$. Then the $C' \setminus \{x\}$ partitions $\mathcal{O}_{ST} \setminus \{\S\}$.*

PROOFS. Every circle is determined by a point u in $W(K)$, $u \notin \mathcal{O}_{ST}$. For the first lemma, let u vary along the line x^ρ of $\mathcal{S}_{\mathcal{L}}$; for the second lemma, let u vary along the line y^ρ, where y is the point of $W(K)$ defining C.

REMARK. Lemmas 4.1 and 4.2 allow one to reconstruct $W(K)$ in a more axiomatized setting.

Following TITS [7], we can describe $\mathcal{O}_{ST}$ by the set $K_\sigma^{(2)} \cup \{\infty\}$. Using, e.g., the coordinates in HANSSENS & VAN MALDEGHEM [1], one calculates that the circle containing ∞, $(0, 0)$ and (k_0, k_1), $k_0, k_1 \in K$, contains, besides ∞, all points (tk_0, tk_1), $t \in K$. The circle containing $(0, 0)$ with corner ∞ contains, besides ∞, the points $(0, k_1)$, $k_1 \in K$. By the action of the Suzuki group, one obtains the other circles. But we will need no explicit description of them.

Now let L_1 and L_2 be two elements of the Lüneburg spread $\mathcal{S}_{\mathcal{L}}$ and let $x_i \ I \ L_i$, $i = 1, 2$, be the corresponding points of the Suzuki-Tits ovoid $\mathcal{O}_{ST}$. The set of lines in the quadrangle $W(K)$ meeting all lines which meet both L_1 and L_2 is a *regulus* $\mathcal{R}$ and each element of $\mathcal{R}$ is incident with one point of $\mathcal{O}_{ST}$. The set T of these points will be called a *transversal* with extremeties x_1 and x_2. It is completely determined by x_1 and x_2. Now consider the set $\{x_1, x_2\}^{\perp\perp}$ of points colinear to all points collinear to both x_1 and x_2 (this is the *span* of x_1 and x_2, see PAYNE & THAS [3],p.2) and let $y \in \{x_1, x_2\}^{\perp\perp} \setminus \{x_1, x_2\}$. The circle C defined by y (via intersection with y^τ) has as corner a point incident with y^ρ; but $y^\rho \in \mathcal{R}$, hence the corner of C lies in T. Now note that the plane y^τ contains all points of the hyperbolic line H consisting of all points collinear to both x_1 and x_2 and H does not meet the ovoid. Hence the set of circles defined by the elements of $\{x_1, x_2\}^{\perp\perp} \setminus \{x_1, x_2\}$ partitions $\mathcal{O}_{ST} \setminus \{x_1, x_2\}$ and the set of corners of the circles is the transversal $\mathcal{T}$. We logically call this partition the *transversal partition* with extremities x_1 and x_2.

Following HANSSENS & VAN MALDEGHEM [1], we can take for the symplectic polarity ρ the bilinear form

$$x_0 y_1 + x_1 y_0 + x_2 y_3 + x_3 y_2.$$

We choose $x_1 = (1, 0, 0, 0)$ and $x_2 = (0, 1, 0, 0)$. The line H is determined by $(0, 0, 1, 0)$ and $(0, 0, 0, 1)$. The Suzuki-Tits ovoid can be chosen to contain the points (see [1],5.6)

$$\{(N(k), 1, k_1, k_2) | k = (k_0, k_1) \in K_\sigma^{(2)}\} \cup \{(1, 0, 0, 0)\}.$$

It is then clear that a circle of the transversal partition described above contains the points

$$\{(N(k), 1, k_1, k_0) | N(k) = \text{Constant}\}.$$

Since every line of $PG(3, K)$ which is not a line of $W(K)$ meets $\mathcal{O}_{ST}$ in either two or zero points, the following lemma is readily verified.

LEMMA 4.3 *Let x be a point of the Suzuki-Tits ovoid $\mathcal{O}_{ST}$ and let C be a circle of $\mathcal{O}_{ST}$ not containing x. Then there exists a unique transversal partition containing C and having x as one of its extremities.*

A Suzuki-Tits ovoid with the additional structure of the inversive plane, corners for all circles and transversal partitions for each pair of points will be called a *Suzuki-Tits inversive plane*, or briefly, an *STi-plane*.

4.2   Properties of $O_R(K, \sigma)$

Let $O_R(K, \sigma)$ be the *perfect* Ree Octagon described in section 3 (i.e. K is a perfect field). The lines through the point (∞) are parametrized by the set

$K_\sigma^{(2)} \cup \{\infty\}$. By the preceding paragraph, we can give this set the structure of an STi-plane $\mathcal{P}_{(\infty)}$ (in an algebraic fashion). We will now reconstruct $\mathcal{P}_{(\infty)}$ geometrically. Note that, by transitivity, all points p define an STi-plane $\mathcal{P}_{\surd}$.

Let p be a point of $O_R(K, \sigma)$ opposite (∞). Clearly p has seven coordinates. The *trace* of p, denoted by $(\infty)^p$, (with respect to (∞)) is the set of projections of p onto the lines incident with (∞). Let o be the point with coordinates $(0,0,0,0,0,0,0)$ and suppose p has coordinates $(a, l, a', l', a'', l'', a''')$. The set of lines incident with (∞) and with one of the points of the intersection $(\infty)^o \cap (\infty)^p$ will be called the *support* of that intersection and, by (I6), it consists, besides possibly $[\infty]$, of all lines $[k]$, $k \in K_\sigma^{(2)}$, such that

$$f(k) =: a''' + aN(k) + l_0 k_1 + k_0 l_0'' = 0. \qquad (*)$$

If $a = 0$, then $[\infty]$ belongs to the support. So assume $a \neq 0$. Put $L = (L_0, L_1) = (\frac{l_0}{a}, \frac{l_0''}{a})$, then we have $f(k \oplus L) = aN(k) + f(L)$. We deduce from this that $k = (L_0, L_1 + \left(\frac{f(L)}{a}\right)^{\sigma^{-1}})$ is a solution of $(*)$. Hence we have shown that every two traces meet in at least one point. Without loss of generality we can take this point to be (0), i. e. $p = (0, l, a', l', a'', l'', a''')$. For any further points in the intersection, the equation $(*)$ reduces to

$$a''' + l_0 k_1 + l_0'' k_0 = 0. \qquad (**)$$

If $l_0 = l_0'' = a''' = 0$, then the two traces coincide; if $l_0 = l_0'' = 0$ and $a''' \neq 0$, then the two traces meet only in (0); if $(l_0, l_0'') \neq (0,0)$, then clearly, the set of lines through (∞) incident with a point of $(\infty)^o \cap (\infty)^p$ is a circle C in $\mathcal{P}_{(\infty)}$ and by the transitivity of the stabilizer of (∞), all circles of $\mathcal{P}_{(\infty)}$ arise in this way. So without loss of generality, let us consider the circle $C = \{[\infty]\} \cup \{[(0, k_1)] | k_1 \in K\}$. The set of points p such that the support of $(\infty)^p \cap (\infty)^o$ contains C is, by (I6), equal to

$$\{(0, (0, l_1), a', l', a'', l'', 0) | l_1, a', a'' \in K, l', l'' \in K_\sigma^{(2)}\}.$$

The projection of the point $q = (0, (0, l_1), a', l', a'', l'', 0)$ onto the line $[0]$ is the line $[0, 0, l'']$, hence every line through $(0, 0)$ (except for $[0]$ of course) arises in this way. This remains true for all points $((0, k_1), 0)$, by transitivity. However, the projection of q onto $[\infty]$ is the line $[0, (0, l_1)]$, and here, not all lines arise. In fact, only lines of a circle in $\mathcal{P}_{(t)}$ with corner $[\infty]$ arise. This characterizes the corner of C in a geometric fashion. It also follows that the corner, defined in this geometric way, is independent of the chosen intersection of traces with C as support. We will call the intersection of two traces *trivial* if it contains only one element, or if the two traces are equal. Let X be a non-trivial intersection of two traces. We shall refer to the set of lines M through $x \in X$ such that M is the projection of a point p whose trace contains X *the gate set with respect to X through x*. We say such a gate set is trivial if it contains all lines through x except its support. Then we can summarize the above results as follows:

LEMMA 4.4 *Let x be any point of a perfect Ree octagon. Consider traces with respect to x. Then two traces always meet. If two traces X and Y meet non- trivially, then there exists a unique point u in $X \cap Y$ such that the gate set of u with respect to $X \cap Y$ is non-trivial. The line xu thus obtained from the support C of $X \cap Y$ is independent of the choice of $X \cap Y$. If we define ux to be the corner of C, then the set of all such supports, together with their corners define an STi-plane over the set of lines through x (the transversal partitions will follow from Lemma 4.7).*

Now consider again traces with respect to (∞). Suppose we are given 4 pairwise non-collinear points collinear to (∞) and such that the 4 respective projections onto (∞) do not lie on one circle of $\mathcal{P}_{(\infty)}$ (this defines a *general position* for 4 points collinear to one fixed point). We can take without loss of generality the points (0), $(0,0)$, (k,b) and (k^*, b^*), $k, k^* \in K_\sigma^{(2)}$, $b, b^* \in K$, with $k = (k_0, k_1)$ not proportional to $k^* = (k_0^*, k_1^*)$. If $p = (a, l, a', l', a'', l'', a''')$ defines a trace containing these 4 points, then its coordinates must satisfy $a = a''' = 0$, $b = l_0 k_1 + l_0'' k_0$ and $b^* = l_0 k_1^* + l_0'' k_0^*$. It is clear that under the given assumptions these equations define uniquely a pair (l_0, l_0''), and hence we have shown:

LEMMA 4.5 *Let x be any point of a perfect Ree octagon, then there is a unique trace containing 4 points in general position collinear to x.*

Suppose now two traces (with respect to (∞)) meet in exactly one point, say (0). Using (I6), one can check that, if one trace is defined by o, then the other must be defined by a point p with coordinates $(0, (0, l_1), a', l', a'', (0, l_1''), a''')$ with $a''' \neq 0$ and consequently it consists of the points (k, a'''), $k \in K_\sigma^{(2)}$. So the set of traces meeting $(\infty)^o$ in only (0) is a set of traces with trivial intersection. We call such a set a *pencil* of traces based at (0). We have shown:

LEMMA 4.6 *Let x be any point of a perfect Ree octagon and let X be a trace with respect to x. For every point $y \in X$, there exists a unique pencil of traces based at y and containing X.*

Next, we look at intersections of pencils. Let $\mathcal{E}_\infty$ and $\mathcal{E}_l$ be two pencils of traces based at respectively (0) and $(0,0)$, both containing the trace $(\infty)^o$. Put $u = (0,0,0,0,0,0,a''')$ and $v = (a,0,0,0,0,0,0)$. Then $(\infty)^u$ and $(\infty)^v$ are arbitrary elements of $\mathcal{E}_\infty$ and $\mathcal{E}_l$. They meet on the line $[k]$, $k \in K_\sigma^{(2)}$, if and only if $a''' = aN(k)$ (by (I6) again). Hence their intersection is non-trivial and the support is a circle defined by $N(k) = a'''/a =$ constant. This shows that the set of supports of the intersections is the transversal partition with extremities the lines $[\infty]$ and $[0]$. Hence the lemma:

LEMMA 4.7 *Let x be any point of a perfect Ree octagon and consider, with respect to x, two pencils (based at resp. y_1 and y_2) sharing a common trace Y. Then the set of supports of all intersections of elements of one pencil with elements of the other pencil (excluding Y) is the transversal partition in $\mathcal{P}_\S$ with extremities xy_1 and xy_2.*

As a consequence we have:

COROLLARY 4.8 *There do not exist three distinct traces (with respect to a fixed point) in $O_R(K,\sigma)$ with pairwise trivial intersection and not contained in a pencil.*

Now we can define the following geometry $O_R(K,\sigma)_x$ for any point x of $O_R(K,\sigma)$. The points are of two types :

  (i) the traces with respect to x,

  (ii) the points collinear to x in $O_R(K,\sigma)$, including x itself.

The lines are also of two types :

  (a) the pencils of traces (with respect to x),

  (b) the lines of $O_R(K,\sigma)$ through x.

The incidence between points of type (i) (resp. type (ii)) and lines of type (a) (resp. type (b)) is containment (resp. the incidence in $O_R(K,\sigma)$). No point of type (i) is incident with a line of type (b). A point of type (ii) is incident with a line of type (a) if the pencil in question is based at the point in question. Incidence in $O_R(K,\sigma)_x$ will be denoted by I_x.

PROPOSITION 4.9 *The geometry $O_R(K,\sigma)_x$ as defined above is the generalized quadrangle of Tits-type $T_3(\mathcal{O}_{ST})$.*

PROOF. We give a geometric proof using the lemmas above. A group-theoretic proof or another algebraic one is also possible. For instance, one can coordinatize $O_R(K,\sigma)_x$ to identify it. Or one could determine the automorphism group of this geometry inside the stabilizer of a point of the automorphism group of $O_R(K,\sigma)$. One would find an affine group with a Suzuki group acting.

We first show that $O_R(K,\sigma)_x$ is a generalized quadrangle. Let Π and Λ be a point and a line of $O_R(K,\sigma)_x$ which are not incident. We have to show that there is a unique point Π' I_x Λ, and a unique line Λ' I_x Π such that Π' I_x Λ'. Suppose first that Π is of type (ii) and Λ is of type (b). Then

Π I_x M I_x x I_x Λ, where M is the line in $O_R(K,\sigma)$ joining Π and x, and this path is unique. Suppose now Π is of type (i) and Λ is of type (b). Then there is a unique point p in $O_R(K,\sigma)$ incident with the line Λ and contained in the trace Π. There is also a unique pencil Λ' of traces based at p and containing Π, by Lemma 4.6. Again we have Π I_x Λ' I_x p I_x Λ and no other such path exists. Next, let Π be of type (ii) and Λ of type (a). Let Λ be based at p. If $\Pi = x$, then Π I_x xp I_x p I_x Λ. Suppose now $\Pi = y$ is distinct from x. If y is collinear to p in $O_R(K,\sigma)$, then clearly Π I_x xy I_x p I_x Λ; if not, then Π I_x Λ^* I_x X I_x Λ, where X is the member of the pencil Λ containing y, and Λ^* is the pencil of traces based at y and containing X. Finally, suppose Π is of type (i) and Λ is of type (a). Let Λ be based at y. If $y \in \Pi$, then Π I_x Λ^* I_x y I_x Λ, where Λ^* is the pencil of traces based at y and containing Π. If $y \notin \Pi$, then by Corollary 4.8, there exists at most one member of the pencil Λ meeting the trace Π trivially. We now show that there exists at least one such element of Λ. Let X be any member of Λ. Suppose that X meets Π non-trivially (otherwise we are done) and let C be the support of the intersection $\overline{C}$. So C is a circle in $\mathcal{P}_{\S}$. By Lemma 4.3, there exists a unique transversal partition of $\mathcal{P}_{\S}$ containing C and having xy as one of its extremities. Let M (a line in $O_R(K,\sigma)$ through x) be the other extremity and let p be the unique point incident with M and lying on the trace Π. Let Λ^* be the pencil of traces containing Π and based at p, and Y be the unique member of Λ^* containing y. let Λ^{**} be the pencil of traces containing Y and based at y. Then by Lemma 4.7, there exists a trace $Z \in \Lambda^{**}$ meeting Π in $\overline{C}$, hence $\Lambda^{**} = \Lambda$, $Y \in \Lambda$ and $Y \cap \Pi = \{p\}$. So Π I_x λ^* I_x Y I_x Λ. We leave it to the reader to check that no other such paths exist and hence $O_R(K,\sigma)$ is a generalized quadrangle.

We still have to show that $O_R(K,\sigma)_x$ is isomorphic to $T_3(\mathcal{O}_{ST})$. If K is a finite field of order 2^{2e+1}, it follows from PAYNE & THAS [3],5.3.1 that $O_R(2^{2e+1})_x$ is isomorphic to some $T_3(\mathcal{O})$, $\mathcal{O}$ an ovoid in $PG(3,q)$ since the point x of $O_R(2^{2e+1})_x$ is a 3-regular point (this can be verified easily). From the first part of the proof of 5.3.1 of *loc.cit.*, it follows that $\mathcal{O}$ is the Suzuki-Tits ovoid $\mathcal{O}_{ST}$. In the infinite case, one has essentially the same proof, replacing some counting arguments in 5.3.1 of *loc.cit.* by arguments using the lemma's of this section. Alternatively, an algebraic proof goes as follows.

The map

$$x^{\perp} \to x^{\perp} : (k, b) \mapsto (k, b + A''' + AN(k) + L_0 k_1 + L_0'' k_0)$$

defines an automorphism of $\mathcal{P}_{(\infty)}$ (this follows from (I6)). The group Ω of all such maps is isomorphic to $K \times K \times K \times K, +$, where the above element corresponds to (A''', A, L_0, L_0''). But Ω acts regularly on the set of traces (this is immediately verified), hence we can put the structure of the affine space $AG(4, K)$ on the set of traces in $(\infty)^{\perp}$ (induced by Ω). One can easily check that the set of lines of $O_R(K, \sigma)_x$ of type (a) are all lines of that affine space of certain parallel classes, and these parallel classes determine exactly a Suzuki-Tits ovoid at infinity.

REMARK. In the finite case, the above proof simplifies. Indeed, a counting argument replaces the use of Lemma 4.3, Lemma 4.7 and Corollary 4.8.

5   The metasymplectic space $\mathcal{M}(\mathcal{K})$

Dual traces.
One can also ask what the geometry of the *dual* traces look like (defined dually). From the relation (I1), we can deduce the following. Let L be a line of $O_R(K, \sigma)$, x_1 and x_2 two different points on L and L_1, L_2 two lines incident with x_1 resp. x_2, not equal to L. Let C_i, $i = 1, 2$, be the circle in $\mathcal{P}_{\S)}$ with corner L and containing L_i. If M is a line, varying over the set of all lines whose (dual) trace with respect to L contains an element of C_i for $i = 1, 2$, then the set of lines contained in the trace of M and incident with any point x of L varies over a circle C in $\mathcal{P}_{\S}$ with corner L. We call the set of lines of traces with respect to L of all such lines M a *Suzuki regulus*. These sets will play an important role in the reconstruction of the ambient metasymplectic space for $O_R(K, \sigma)$.

One further property of traces.
Let x be any point in $O_R(K, \sigma)$ and let p be any point opposite x. Let $\mathcal{G}$ be the set of points y opposite x such that $x^y = x^p$. We define a graph on

$\mathcal{G}$ by the rule: two points of G are adjacent if they are not opposite each other. One can show that two such points have distance 6 in the octagon and the unique middle element of the shortest path joining them has distance 5 to x. The graph $\mathcal{G}$ is not connected, in fact, it has exactly $|K|$ connected components. A connected component of $\mathcal{G}$ will be called a *trace direction*.

The reconstruction.
We briefly sketch how one can now reconstruct the ambient metasymplectic space $\mathcal{M}(\mathcal{K})$. The points of $\mathcal{M}(\mathcal{K})$ are of three types. Type (I) consist of the points of the octagon $O_R(K, \sigma)$ itself. The points of type (II) are the Suzuki reguli. A point of type (III) is a trace direction. The lines and planes must then be defined using the properties of traces listed in this paper. The hyperlines then follow rather easily, as well as the polarity. This will be proved in detail elsewhere.

References

[1] G. HANSSENS and H. VAN MALDEGHEM, A new look at the classical generalized quadrangles, *Ars. Combin.* **24** (1987), 199 – 210.

[2] M. JOSWIG and H. VAN MALDEGHEM, An essay on the Ree Octagons, *preprint*.

[3] S. E. PAYNE and J. A. THAS, Finite generalized quadrangles, Pitman (1984).

[4] M. RONAN, A geometric characterization of Moufang Hexagons, *Inventiones Math.* **57** (1980), 227 – 262.

[5] J. SARLI, The geometry of root subgroups in Ree groups of type $^2F_4$, *Geom. Ded.* **26** (1988), 1 – 28.

[6] J. TITS, Sur la trialité et certains groupes qui s' en déduisent, *Publ. Math. IHES* **2** (1959), 14 – 60.

[7] J. TITS, Ovoïdes et groupes de Suzuki, *Arch. Math.* **13** (1962), 187 – 198.

[8] J. TITS, Buildings of spherical type and finite BN-pairs, Springer, *Lecture Notes Math.* **386** (1974).

[9] J. TITS, Endliche Spiegelungsgruppen, die als Weylgruppen auftreten, *Inventiones Math.* **43** (1977), 283 – 295.

[10] J. TITS, Moufang octagons and the Ree groups of type $^2F_4$, *Am. J. Math.* **105** (1983), 539 – 594.

[11] H. VAN MALDEGHEM, A configurational characterization of the Moufang generalized polygons, *European J. Combin.* **11** (1990), 362 – 372.

Address of the Author :

Department of Pure Mathematics and Computer Algebra
University Gent
Galglaan 2
B – 9000 Gent
BELGIUM

Small Rank Exceptional Hurwitz Groups

Gunter Malle *, Mathematisches Institut und IWR,
Im Neuenheimer Feld 368, D - 69120 Heidelberg.

A group G is called a $(2,3)$-group if it can be generated by an involution and an element of order 3. This is equivalent to saying that G is a factor group of $\mathrm{PSL}_2(\mathbb{Z})$, which is the free product of two cyclic groups of order two and three. Thus every $(2,3)$-group corresponds to a normal (in general: non-congruence) subgroup of $\mathrm{SL}_2(\mathbb{Z})$. There has recently been some interest in determining the $(2,3)$-generated finite simple groups (see for example [3,13,14]). In this paper we consider the exceptional groups of Lie type ${}^3D_4(q)$ and ${}^2F_4(2^{2n+1})$.

Theorem 1: *The groups ${}^3D_4(q)$ and ${}^2F_4(2^{2n+1})'$ are factor groups of the modular group $\mathrm{PSL}_2(\mathbb{Z})$. Here q is any prime power.*

A finite group G is called a *Hurwitz group* if it is a factor group of

$$G_{2,3,7} := \langle \sigma_1, \sigma_2, \sigma_3 \mid \sigma_1^2 = \sigma_2^3 = \sigma_3^7 = \sigma_1\sigma_2\sigma_3 = 1 \rangle.$$

The interest in such groups stems from the fact that they can be represented as a group of $84(\gamma-1)$ automorphisms of a Riemann surface of genus γ. Hurwitz showed that this is the biggest possible value for the order of a group acting on a surface of genus $\gamma \geq 2$. For an introduction to the present knowledge on Hurwitz groups see the recent overview [1] or also [3]. Clearly a Hurwitz group is $(2,3)$-generated. In [8] we showed that the groups $G_2(q)$ for $q \geq 5$ and the groups ${}^2G_2(3^{2n+1})$ are Hurwitz. Here we extend this result to the Steinberg triality groups ${}^3D_4(q)$ and to some of the Ree groups ${}^2F_4(2^{2n+1})$. Together with the results in [8] this completely determines the Hurwitz groups among simple exceptional groups of Lie type of rank at most 2.

Theorem 2:

(a) *The group ${}^3D_4(q)$, $q = p^n$, is a Hurwitz group if and only if $p \neq 3$, $q \neq 4$.*

(b) *The group ${}^2F_4(2^{2n+1})'$ is a Hurwitz group if and only if $n \equiv 1 \pmod 3$.*

---

* The author gratefully acknowledges financial support by the Deutsche Forschungsgemeinschaft

(This in particular implies Theorem 1 for the groups $^3D_4(q)$, $p \neq 3$, $q \neq 4$, and also for $^2F_4(2^{2n+1})$ with $n \equiv 1 \pmod 3$.)

The method of proof is similar to the one employed in [8]. To show $(2,3)$-generation (resp. $(2,3,7)$-generation) of G we first calculate a structure constant from the ordinary character table of G. Then information on the maximal subgroups of G is used to rule out the possibility that all such $(2,3)$-pairs (resp. $(2,3,7)$-triples) generate proper subgroups. Since this was already described in [8] in some detail, we will not repeat the precise description.

We remark that the calculation of structure constants was carried out with the computer algebra system CHEVIE [5], which contains the generic character table of the groups $^3D_4(q)$ (as first published in [4]) and of $^2F_4(2^{2n+1})$ (computed from the results in [7]).

1 The groups $^3D_4(q)$

In this section we prove that the Steinberg triality groups $G := {}^3D_4(q)$, $q = p^n$, $p \neq 3$, $q \neq 4$, are Hurwitz groups, and that the remaining groups are $(2,3)$-generated. For this we first have to know the classes of involutions, of 3-elements and of 7-elements in G. The unipotent classes of G were determined by Spaltenstein [12], while the semisimple ones can be found in [4]. With these references one easily checks the following facts:

Lemma 1: (a) If $p \neq 2$ then G has a single class of involutions, with centralizer order $q^4(q^2-1)(q^6-1)$.

(b) If $p \neq 3$ then G has two classes of elements of order three, with centralizer orders $q^3(q^2-1)(q^3-\epsilon)(q^2+\epsilon q+1)$, resp. $q^3(q^6-1)(q-\epsilon)$, where $q \equiv \epsilon \pmod 3$, $\epsilon \in \{1,-1\}$.

(c) If $p \neq 7$ then G has a single class of regular elements of order 7.

Proof : The first two assertions readily follow from the tables in [4]. If $q \equiv 2,3,4$ or $5 \pmod 7$, then a 7-Sylow subgroup of G is contained in a maximal torus T of G of order $(q^2+q+1)^2$ resp. $(q^2-q+1)^2$. These tori have a relative Weyl group of order 24, and from the action given in [4, Table 1.1], it follows that there exists a single regular class in T containing elements of order 7. If $q \equiv \pm 1 \pmod 7$, the 7-Sylow subgroups lie in tori

T of order $(q^3 - 1)(q - 1)$ resp. $(q^3 + 1)(q + 1)$. Of the 48 elements of order 7, 36 lie in subgroups centralizing a root subgroup. The remaining 12 elements are fused into a single class of regular elements by the Weyl group $W = S_3 \times Z_2$. Alternatively, this follows from the embedding $G_2(q) < {}^3D_4(q)$ and an application of Proposition 1 in [8]. ∎

If $p \neq 2$ we write $2A$ for the class of involutions, $3A$ and $3B$ for the two classes of 3-elements (where $3A$-elements have centralizer order $q^3(q + \epsilon)(q^3 - \epsilon)^2)$ if $p \neq 3$, and $7A$ for the class of regular 7-elements if $p \neq 7$. The strategy is now to prove that there exist triples $\sigma_1 \in 2A$, $\sigma_2 \in 3A$, $\sigma_3 \in 7A$ with $\sigma_1\sigma_2\sigma_3 = 1$ generating G. For this we need a further preparatory lemma.

Lemma 2: *Let $(\sigma_1, \sigma_2, \sigma_3)$ be a $(2, 3, 7)$-system of ${}^3D_4(q)$ with $\sigma_1\sigma_2\sigma_3 = 1$. If $2 \nmid q$, or $2|q$ and $\sigma_3 \in 7A$, then $\langle \sigma_1, \sigma_2, \sigma_3 \rangle$ is not contained in a maximal parabolic subgroup.*

Proof: By [6] the maximal parabolic subgroups of ${}^3D_4(q)$ have the structure $P_1 = [q^9] : (SL_2(q^3) \circ Z_{q-1}).d$ respectively $P_2 = [q^{11}] : (Z_{q^3-1} \circ SL_2(q)).d$, where $d = \gcd(2, q - 1)$. Here $[n]$ denotes a group of order n of unspecified structure, as in the Atlas notation [2]. Since a Hurwitz group is necessarily perfect, if $\langle \sigma_1, \sigma_2, \sigma_3 \rangle$ is contained in one of those parabolic subgroups, then already in the commutator subgroups $P_1' = [q^9] : SL_2(q^3)$ respectively $P_2' = [q^{11}] : SL_2(q)$, and we have $q \neq 2, 3$. Now first assume that $2 \nmid q$. Then we use that the factors of the above commutator subgroups by their unipotent radicals would also have to possess a $(2, 3, 7)$-system. This is a contradiction, as $SL_2(q^3)$ and $SL_2(q)$ both contain a single, central involution, which can thus not lie in a $(2, 3, 7)$-system.

If $p = 2$ and $\sigma_3 \in 7A$, then we may first exclude P_1', as this is the centralizer of a p-element, hence in particular can not contain regular semisimple elements. In the second case, a 7-Sylow subgroup of P_2' is necessarily cyclic, so any 7-element is conjugate to one in the complement $SL_2(q)$. But this complement centralizes a p-element lying in a short root subgroup, hence its semisimple elements are again not regular. ∎

Let $\mathbf{C} = (C_1, C_2, C_3)$ be a class vector of the finite group H. We write $n_H(\mathbf{C})$ or simply $n(\mathbf{C})$ for the normalized structure constant of $\mathbf{C}$ (see for example [8]), and $\ell(\mathbf{C})$ for the number of generating triples in $\mathbf{C}$ with product 1 modulo conjugation. See [8] for the connection between $n(\mathbf{C})$ and $\ell(\mathbf{C})$.

Proposition 1: *Theorem 2(a) holds for $p \neq 2, 3, 7$ and the class vector $\mathbf{C} = (2A, 3A, 7A)$.*

Proof: The normalized structure constant $n(\mathbf{C})$ of $\mathbf{C} = (2A, 3A, 7A)$ can be calculated from the ordinary character table of G. This calculation is most easily done with the computer algebra system CHEVIE [5], which contains the generic ordinary character table of the groups $^3D_4(q)$. The results are collected in Table 1. (They depend on the congruence classes of q modulo 3 and 7.)

Table 1: The $(2, 3A, 7A)$ normalized structure constants in $^3D_4(p^n)$, $p \neq 3, 7$.

$q \equiv$	1 (mod 7)	-1 (mod 7)	2,4 (mod 7)	3,5 (mod 7)
1 (mod 3)	$q^2 + 1$	$q^2 - 2q + 1$	$q^2 - 3q + 1$	$q^2 + q + 1$
-1 (mod 3)	$q^2 + 2q + 1$	$q^2 + 1$	$q^2 - q + 1$	$q^2 + 3q + 1$

Note that this shows $\frac{1}{2}q^2 < n(\mathbf{C}) < 2q^2$ for $p \neq 2, 3, 7$, $q \neq 5$. It now remains to check that $\ell(\mathbf{C}) \neq 0$, i.e., not all of the structure constant can be accounted for by the contribution of proper subgroups. For this we make use of the list of maximal subgroups of G in [6]. In Table 2 we list the maximal subgroups of G which might contribute to the $(2A, 3A, 7A)$-structure constant, i.e., which are non-solvable and have order potentially divisible by 42. (We have omitted the parabolic subgroups, since these can not contain a triple from $\mathbf{C}$ by Lemma 2.)

Table 2: Some maximal subgroups of $^3D_4(q)$, $q = p^n$, p odd.

K	occurs for	$n_K(2,3,7)$
$G_2(q)$	always	0
$PGL_3(q)$	$q \equiv 1$ (mod 3)	≤ 6
$PGU_3(q)$	$q \equiv -1$ (mod 3)	≤ 6
$(SL_2(q^3) \circ SL_2(q)) \cdot 2$	always	≤ 18
$((q^2 + q + 1) \circ SL_3(q)).f_+.2$	always	≤ 3
$((q^2 - q + 1) \circ SU_3(q)).f_-.2$	always	≤ 3
$^3D_4(q_0)$	$q = q_0^t$, $3 \neq t \in \mathbb{P}$	$< 2q_0^2$

Here $f_\epsilon := \gcd(3, q^2 + \epsilon q + 1)$.

The last column gives an upper bound for the contribution to $n(\mathbf{C})$ possibly coming from this class of maximal subgroups. This is obtained as follows.

In $G_2(q)$ there are two classes of 3-elements, with centralizers $SL_3(q)$ and $SL_2(q)(q-1)$ if $q \equiv 1 \pmod 3$, respectively $SU_3(q)$ and $SU_2(q)(q+1)$ if $q \equiv -1 \pmod 3$. Of these, the first one fuses into the class $3A$ of G. But the structure constant of G_2 involving this first class is seen to vanish using CHEVIE. The contribution from the remaining groups can also be computed from the generic character tables [5], respectively from Table 1 for $^3D_4(q_0)$.

First assume $q = 5$. Then $n(\mathbf{C}) = 40$ by Table 1, while the proper subgroups contribute at most 30 by Table 2. If $q = 25$ one also checks that $\ell(\mathbf{C}) \geq 1$. For $q = p \geq 11$ we have $n(\mathbf{C}) \geq 1/2q^2 \geq 60$, which is larger than the structure constants coming from proper subgroups. Finally, if $q = p^n$, $n \geq 2$, we write $n = \prod_{i=1}^r t_i^{e_i}$, so that the subfield subgroups $^3D_4(q_0)$ contribute at most

$$\sum_{i=1}^r 2p^{2n/t_i} \leq \sum_{i=1}^r 2p^n = 2rp^n.$$

Thus there remains at least

$$\ell(\mathbf{C}) \geq \frac{1}{2}p^{2n} - 2rp^n - 30 = \frac{1}{2}p^n(p^n - 4r - \epsilon) \geq 1,$$

with some $\epsilon < 1$ since $p^n \geq 121$, for the number $\ell(\mathbf{C})$ of generating Hurwitz-triples. This achieves the proof.   ∎

For $p = 7$ let $7U$ be the class of regular unipotent elements, which by the description of the elementary divisors of unipotent elements in the overgroup $D_n(q^3) = O_8^+(q^3)$ of $^3D_4(q)$ have order 7.

Proposition 2:   *Theorem 2(a) holds for $p = 7$ and the class vector $\mathbf{C} = (2A, 3A, 7U)$.*

Proof: The normalized structure constant of $\mathbf{C}$, computed with CHEVIE, is displayed in Table 3.

Table 3: The $(2A, 3A, 7U)$ normalized structure constants in $^3D_4(7^n)$.

$q \equiv$	$1 \pmod 3$	$-1 \pmod 3$
	$q(q-1)$	$q(q+1)$

The maximal subgroups which could possibly contribute to this structure constant are contained in Table 1. Again, the corresponding structure constant in $G_2(q)$ vanishes. It is an easy exercise to check that the contribution from subgroups is always strictly smaller than $n(\mathbf{C})$.   ∎

Proposition 3: *Theorems 2(a) and 1 hold for $^3D_4(3^n)$.*

Proof: We first show that all $(2,3,7)$-triples of G generate proper subgroups. By Lemma 1(a), G has a single class of involutions. The (unipotent) classes containing elements of order 3 are A_1, $3A_1$, A_2' and A_2'' in the notation of [12]. In Table 4 we tabulate those $(2,3,7)$-structure constants which are at least 1; the remaining ones can not come from generating triples anyway. In particular, it turns out that only the regular class $7A$ of 7-elements is involved in possible candidates.

Table 4: The $(2A,3,7A)$ normalized structure constants in $^3D_4(3^n)$.

$q \equiv$	1 (mod 7)	-1 (mod 7)	$2,4$ (mod 7)	$3,5$ (mod 7)
$3A_1$	1	1	1	1
A_2'	$\frac{1}{2}q(q+2) - \delta_+$	$\frac{1}{2}q(q-2)$	$\frac{1}{2}q(q-1) - \delta_+$	$\frac{1}{2}q(q+1)$
A_2''	$\frac{1}{2}q(q+2)$	$\frac{1}{2}q(q-2) - \delta_-$	$\frac{1}{2}q(q-1)$	$\frac{1}{2}q(q+1) - \delta_-$

Here $\delta_+ := (q^2 + q)/(q^2 + q + 1)$ and $\delta_- := (q^2 - q)/(q^2 - q + 1)$. The interesting maximal subgroups are listed in Table 1. It turns out that the corresponding $(2,3,7)$-structure constants in the maximal subgroup $G_2(q)$ are the same as those displayed in Table 4, up to the summands $-\delta_\epsilon$. We now compare the centralizers of $(2,3,7A)$-triples in $G_2(q)$ and $^3D_4(q)$. If the centralizer increases in $^3D_4(q)$, then by Table 2 the Hurwitz triple must lie inside $((q^2 + \epsilon q + 1) \times L_3^\epsilon(q)).2$, where $q^3 \equiv \epsilon$ (mod 7), since $7A$-elements are regular. (Here we write $L_3^+(q) := L_3(q)$, $L_3^-(q) := U_3(q)$.) But the $(2,3,7)$-structure constant of $L_3^\epsilon(q).2$ equals 1 if $q^3 \equiv \epsilon$ (mod 7). Thus passing from $G_2(q)$ to $^3D_4(q)$, the centralizer increases for precisely one Hurwitz triple, by $q^2 + \epsilon q + 1$, so the contribution to the structure constant decreases by

$$1 - \frac{1}{q^2 + \epsilon q + 1} = \delta_\epsilon.$$

Hence all Hurwitz triples in Table 4 come from $G_2(q)$, and G can not be a Hurwitz group.

To prove $(2,3)$-generation, we calculate $n(2A, A_2', [s_{14}]) = q(q-1)/2$, where s_{14} is a representative of the regular elements in the maximal torus of order $q^4 - q^2 + 1$. By [6], any such triple generates G. ∎

For $p = 2$ let $2U$ be the larger of the two classes of involutions of G, denoted by $3A_1$ in [12].

Proposition 4: *Theorems 2(a) and 1 hold for* $^3D_4(2^n)$ *and the class vector* $\mathbf{C} = (2U, 3A, 7A)$.

Proof : The structure constants in this case turn out to be the same as for $p \neq 2, 3, 7$, so are also given by Table 1. The maximal subgroups which could possibly contribute to $n(\mathbf{C})$ are as follows (see [6]):

Table 5: Some maximal subgroups of $^3D_4(2^n)$.

K	occurs for	$n_K(2,3,7)$
$G_2(q)$	always	0
$PGL_3(q)$	$q \equiv 1 \pmod 3$	≤ 3
$PGU_3(q)$	$2 < q \equiv -1 \pmod 3$	≤ 3
$L_2(q^3) \times L_2(q)$	always	≤ 10
$((q^2 + q + 1) \circ SL_3(q)).f_+.2$	always	≤ 3
$((q^2 - q + 1) \circ SU_3(q)).f_-.2$	always	≤ 3
$^3D_4(q_0)$	$q = q_0^t, \quad 3 \neq l \in \mathbb{P}$	$< 2q_0^2$

If $q = 2$ we have $n(\mathbf{C}) = 3$, but none of the maximal subgroups contains elements from $2U, 3A$ and $7A$ [2]. For $q \geq 8$, the usual estimate works. Now let $q = 4$. Then $n(\mathbf{C}) = 5$, of which 3 already comes from the subgroup $^3D_4(2)$. It is then clear that $\mathbf{C}$ contains no generating triples, since the outer automorphism group of order 6 would have to act faithfully on those. (In fact, the remaining two triples lie already inside the maximal subgroup $PGL_3(4)$.) The structure constant $n(2U, 3B, 7A)$ with the second class of 3-elements $3B$ equals 16 and is the only other $(2,3,7)$-constant not smaller than 1. But in $G_2(4)$ we have $n_{G_2}(2B, 3B, 7A) = 16$, originating from J_2 and $L_2(13)$ [8, §6]. Since these two subgroups have trivial centralizer in G as well, they account for the total structure constant in G. It follows that $^3D_4(4)$ is not a Hurwitz group. But $n(2U, 3A, 241A) = 13$, where $241A$ is a class of elements of order $241 = q^4 - q^2 + 1$, and it is easily verified that this gives a $(2,3)$-generation of $^3D_4(4)$.   ∎

2 The groups $^2F_4(q^2)$

In this section we study $G := {}^2F_4(q^2)$, $q^2 = 2^{2n+1}$ with $n \geq 0$. The order of G is $q^{24}(q^2 - 1)(q^6 + 1)(q^8 - 1)(q^{12} + 1)$ and hence divisible by 7 if and only if

$3|(2n+1)$, so $n \equiv 1 \pmod 3$. We take the notation for Steinberg generators and for representatives of conjugacy classes of G from [10]. Let $2B$ be the class of the unipotent element $\alpha_{10}(1)$, with centralizer order $q^{20}(q^4-1)$, $3A$ the unique class of elements of order three with representative t_4 and centralizer order $q^6(q^4-1)(q^6+1)$, and $7A$ a class of elements of order 7 with representative t_2 and centralizer order $q^2(q^2-1)(q^4-1)$.

Proposition 5: *Theorem 2(b) holds with the class vector* $\mathbf{C} = (2B, 3A, 7A)$.

Proof : It was already noted above that $|{}^2F_4(2^{2n+1})|$ with $n \not\equiv 1 \pmod 3$ is not divisible by 7, hence can not be a Hurwitz group. So now let us assume $n \equiv 1 \pmod 3$. From the CHEVIE table [5] of $G = {}^2F_4(q^2)$ one calculates the structure constant $n(\mathbf{C}) = q^4 + q^2 + 1 + \frac{1}{q^4-1}$. Next we have to determine the contribution coming from proper subgroups. The non-solvable maximal subgroups of G with order divisible by 3 are the following [9]:

Table 6: Some maximal subgroups of ${}^2F_4(q^2)$, $q^2 = 2^{2n+1}$.

K	occurs for	$n_K(2,3,7)$
$[q^{22}]\colon (L_2(q^2) \times (q^2-1))$	always	$q^2 + 1 + \frac{1}{q^2-1}$
$SU_3(q^2)\colon 2$	always	≤ 3
$PGU_3(q^2)\colon 2$	always	≤ 3
$Sp_4(q^2)\colon 2$	always	1
${}^2F_4(q_0^2)$	$q^2 = q_0^{2t}, \quad t \in \mathbb{P}$	$< \frac{4}{3}q_0^4$

The result follows from the upper bounds for $n_K(2,3,7)$ in this table by a calculation as for ${}^3D_4(q)$. So it remains to deduce these bounds.

For the subgroups $SU_3(q^2)\colon 2$, $PGU_3(q^2)\colon 2$ and $Sp_4(q^2)\colon 2$, this follows from the generic character tables if we keep in mind that a $(2,3,7)$-triple has to lie already in the commutator subgroup. For the subfield groups ${}^2F_4(q_0^2)$ the upper bound is obtained by induction, since the classes $2B$, $3A$ and $7A$ of any subfield group fuse into the same classes in G. We are left with the parabolic subgroup $P_a := [q^{22}]\colon (L_2(q^2) \times (q^2-1))$. Again, a Hurwitz triple must lie already inside $\tilde{P}_a := [q^{22}]\colon L_2(q^2)$. We determine explicitly all Hurwitz triples in that group and thus compute the contribution to $n(\mathbf{C})$.

For this it is necessary to work with the Steinberg generators of G as given for example in [10]. In this notation we have

$$\tilde{P}_a = U_1 U_2 U_4 \cdots U_{12} \cdot \langle U_3, r_a \rangle,$$

where the $U_i := \{\alpha_i(t) \mid t \in \mathbb{F}_{q^2}\}$ are as in [11] and r_a is a fundamental generator of the Weyl group. The Levi factor $L_a = \langle U_3, r_a \rangle$ of $\tilde{P}_a$ is isomorphic to the factor group of $\tilde{P}_a$ modulo its unipotent radical $R_u(\tilde{P}_a) = U_1 U_2 U_4 \cdots U_{12}$. Clearly if $(\sigma_1, \sigma_2, \sigma_3)$ is a Hurwitz triple of $\tilde{P}_a$, then so is its image in $\tilde{P}_a / R_u(\tilde{P}_a)$. But the Hurwitz triples in $L_a \cong L_2(q^2)$ can easily be classified: they are all conjugate in L_a to a triple of the form

$$(\alpha_3(\gamma) r_a \alpha_3(1), \; \alpha_3(1) r_a, \; \alpha_3(\gamma))$$

with $\gamma^3 + \gamma + 1 = 0$ (i.e., γ is an element of order 7 in $\mathbb{F}_8 \leq \mathbb{F}_{q^2}$). We have to count the number of liftings of this to Hurwitz triples of $\tilde{P}_a$. Since the 7-Sylow subgroup of $\tilde{P}_a$ is cyclic, we may moreover assume that the lifting of σ_1 is again $\alpha_3(\gamma) r_a \alpha_3(1)$. So we look for $u \in R_u(\tilde{P}_a) = U_1 U_2 U_4 \cdots U_{12}$ such that

$$(\alpha_3(1) r_a u)^3 = 1 \quad \text{and} \quad (\alpha_3(\gamma) u)^2 = 1. \qquad (*)$$

By the description of the involution classes in P_a in [11, (3.3)] we first deduce that $u \in U_4 \cdots U_{12}$. We distinguish two cases according to whether $u \in U_5 \cdots U_{12}$ or not. In the first case, let

$$u = \alpha_5(c_1)\alpha_6(c_2)\alpha_7(c_3)\alpha_8(c_4)\alpha_9(c_5)\alpha_{10}(c_6)\alpha_{11}(c_7)\alpha_{12}(c_8).$$

The conditions $(*)$, evaluated modulo the normal subgroup $U_7 U_9 \cdots U_{12}$ yield $c_1 = c_2 = c_4$ and $c_1(\gamma + \gamma^{2\theta}) = 0$, where θ is the automorphism of $\mathbb{F}_{q^2}$ with $2\theta^2 = 1$ (so θ is exponentiation by 2^n). But $\gamma + \gamma^{2\theta} = 0$ implies $\gamma = 1$, which is not the case, so we have $c_1 = c_2 = c_4 = 0$. Repeating the calculation with these new values then yields

$$u = \alpha_9(c_5)\alpha_{10}(c_5^{2\theta})\alpha_{12}(c_8), \qquad \text{with} \quad c_5, c_8 \in \mathbb{F}_{q^2}.$$

Since we may still conjugate this by the centralizer of L_a (which is isomorphic to $U_{10}.H_{10}$, with a subgroup H_{10} of order $q^2 - 1$ of the maximally split torus inside P_a), this gives a contribution of $q^2/(q^2 - 1)$ to $n_{P_a}(2,3,7)$.

Now assume

$$u = \alpha_4(c_0)\alpha_5(c_1)\alpha_6(c_2)\alpha_7(c_3)\alpha_8(c_4)\alpha_9(c_5)\alpha_{10}(c_6)\alpha_{11}(c_7)\alpha_{12}(c_8)$$

with $c_0 \neq 0$, where after conjugation with $h \in H_{10}$ we can take $c_0 = 1$. Calculation modulo $U_7 U_9 \cdots U_{12}$ then gives

$$c_1 = \frac{1 + \gamma^{2\theta}}{\gamma + \gamma^{2\theta}}, \quad c_2 = c_1 + 1, \quad c_4 = c_1$$

(note that we already saw that $\gamma + \gamma^{2\theta} \neq 0$). The complete computation shows that for u to satisfy $(*)$ we need

$$c_3 = \frac{(\gamma+1)(\gamma^{2\theta}+1)}{(\gamma+\gamma^{2\theta})(\gamma^2+\gamma^{2\theta})}, \quad c_6 = c_1^2 + c_1 + c_3^{2\theta} + c_5 + c_5^{2\theta},$$

$$c_7 = \gamma^{-1}(c_1^{2\theta+1} + c_1c_3\gamma^{2\theta} + c_3\gamma^{2\theta} + c_1c_3 + c_1c_5 + c_3^{2\theta} + c_5^{2\theta}).$$

Thus there exist q^4 such triples, corresponding to the choices of $c_5, c_8 \in \mathbb{F}_{q^2}$. Under conjugation with U_{10}, this leads to a contribution of q^2 to $n(\mathbf{C})$. This completes the proof of the table and hence of the proposition. ∎

To prove $(2,3)$-generation for arbitray n we just alter the third class in the above class vector $\mathbf{C}$. Namely, let C_T be the class in G of a regular semisimple element of order $q^4+\sqrt{2}q^3+q^2+\sqrt{2}q+1$, hence a generator of a cyclic maximal torus T of G of the same order, denoted by t_{17} in [10].

Proposition 6: *Theorem 1 holds for* $^2F_4(2^{2n+1})'$ *with the class vector* $\mathbf{C} = (2B, 3A, C_T)$.

Proof: Again with CHEVIE one computes the structure constant to be

$$n(\mathbf{C}) = q^3(q^5 - \sqrt{2}q^4 + 3q - \sqrt{2}).$$

Among the maximal subgroups of G, only the normalizer of the torus T contains elements with order divisible by a primitive prime divisor of $q^4 + \sqrt{2}q^3 + q^2 + \sqrt{2}q + 1$ [9]. This normalizer is an extension of T by the cyclic group of order 12 and thus clearly solvable, while the group generated by elements from $\mathbf{C}$ must be perfect. Hence G is $(2,3)$-generated. ∎

REFERENCES

[1]   M. D. E. CONDER, Hurwitz groups: a brief survey, *Bull. Amer. Math. Soc.* **23** (1990), 359–370.

[2]   J. H. CONWAY, R. T. CURTIS, S. P. NORTON, R. A. PARKER, R. A. WILSON, *Atlas of finite groups*, Clarendon Press 1985, Oxford.

[3]   L. DI MARTINO, N. A. VAVILOV, $(2,3)$-generation of $SL(n,q)$. I, II. Preprint Univ. degli Studi Milano (1992).

[4]   D. I. DERIZIOTIS, G. O. MICHLER, Character table and blocks of finite simple triality groups $^3D_4(q)$, *Trans. Am. Math. Soc.* **303** (1987), 39–70.

[5] M. GECK, G. HISS, F. LÜBECK, G. MALLE, G. PFEIFFER, CHEVIE
     —generic character tables of finite groups of Lie type, Hecke algebras
     and Weyl groups, IWR-Preprint 93-62, Universität Heidelberg.

[6] P. KLEIDMAN, The maximal subgroups of the Steinberg triality groups
     $^3D_4(q)$ and of their automorphism groups, J. Algebra 115 (1988), 182–
     199.

[7] G. MALLE, Die unipotenten Charaktere von $^2F_4(q^2)$, Comm. Algebra
     18 (1990), 2361–2381.

[8] G. MALLE, Hurwitz groups and $G_2(q)$, Canad. Math. Bull. 33 (1990),
     349–357.

[9] G. MALLE, The maximal subgroups of $^2F_4(q^2)$, J. Algebra 139 (1991),
     52–69.

[10] K. SHINODA, The conjugacy classes of the finite Ree groups of type
     (F_4), J. Fac. Sci. Univ. Tokyo 22 (1975), 1–15.

[11] K. SHINODA, A characterization of odd order extensions of the Ree
     groups $^2F_4(q)$, J. Fac. Sci. Univ. Tokyo 22 (1975), 79–102.

[12] N. SPALTENSTEIN, Caractéres unipotents de $^3D_4(\mathbb{F}_q)$, Comment. Math.
     Helvetici 57 (1982), 676–691.

[13] M.C. TAMBURINI, J.S. WILSON, N. GAVIOLI, On the (2,3)-generation
     of some classical groups, to appear in J. Algebra.

[14] A.J. WOLDAR, On Hurwitz generation and genus action of sporadic
     groups, Illinois J. Math. 33 (1989), 416–437.

THE DIRECT SUM PROBLEM FOR CHAMBER SYSTEMS

Antonio Pasini

1 Introduction

The Direct Sum Theorem for geometries states that a geometry belonging to a disconnected diagram is the direct sum of subgeometries corresponding to the connected components of that diagram. No analogous statement holds for chamber systems in general.

This situation has some uncomfortable consequences. For instance, we cannot reduce a classification problem for a class of chamber systems to cases with connected diagram, except when we have previously proved that the Direct Sum Theorem holds for the chamber systems of that class. Or, if we apply the celebrated criterion by Tits on rank 3 residues of spherical type to see if a given chamber system C belonging to a Coxeter diagram is covered by a building, we should check if the residues of C corresponding to disconnected rank 3 subdiagrams split as direct sums of subsystems of rank 1 or 2.

Unfortunately, some of the authors who have written on chamber systems seem to have been not awared of these problems. It would be stupid making a list of those who occasionaly said something wrong because of this oversight. I am not going to do that. Rather, I want to show that this situation is not really so bad as it might look. To support my optimistic opinion, I will show that in some important cases the counterexamples to the statement of the Direct Sum Theorem are quite sporadic, so that things can be kept under control in those cases. In particular, I want to say clearly that nothing wrong is contained in the literature on transitive locally classical chamber systems, apart from a few claims occasionaly made as "side-remarks" by some authors, which can be dropped with no substantial consequences for the rest of what those authors said.

This paper is organized as follows. In Section 2 we give a brief survey of the general theory of chamber systems. Direct sums of chamber systems are defined in Section 3. In Section 4 we describe the conterexamples that we know to the statement of the Direct Sum Theorem for chamber systems. The remaining sections are devoted to an investigation of transitive locally finite

chamber systems belonging to Coxeter diagrams (sections 5 and 6). We will
not study the thin case in this paper, in spite of its undoubtable interest. The
reader may see [26] for it.

As chambers systems are families of equivalence relations satisfying certain
properties (see the next section), it will be useful to have stated some notation
for equivalence relations. Given a nonempty set C, we denote by Ω the
trivial equivalence relation on C, namely the largest equivalence relation on
C, having C as its unique equivalence class. The identity relation $=$ will be
denoted by $\mathcal{U}$. Given an equivalence relation Φ on C and an element $x \in C$, we
denote the equivalence class of Φ containing x by $[x]\Phi$. Given two equivalence
relations Φ and Ψ on C, $\Phi \vee \Psi$ is the least equivalence relation containing both
Φ and Ψ, whereas $\Phi \cdot \Psi$ is the relation defined by the following clause: given
any two elements $x, y \in C$, $(x, y) \in \Phi \cdot \Psi$ iff $x\Phi z$ and $z\Psi y$ for some $z \in C$.
We will often write $\Phi\Psi$ for $\Phi \cdot \Psi$. If $\Phi \subseteq \Psi$, then Ψ/Φ denotes the quotient
of Ψ by Φ, defined on C/Φ by the following clause: $([x]\Phi, [y]\Phi) \in (\Psi/\Phi)$ iff
$x\Psi y$.

We follow the notation of [9] for finite groups. We presume the reader is
familiar with the language of diagrams for geometries and we are not going
to repeat the definition of geometries here. The reader may consult [27] for
that. We only warn that we assume geometries to be residually connected
and firm by definition, as in [27].

2   A Survey of Chamber Systems

2.1   Definitions

A *chamber system* over a finite set of *types* I is a pair $\mathcal{C} = (C, (\Phi_i)_{i \in I})$ where
C is a nonempty set, whose elements are called *chambers*, and $(\Phi_i)_{i \in I}$ is a
family of equivalence relations on C with the following properties:

(C1) $\bigvee_{i \in I} \Phi_i = \Omega$;
(C2) $\Phi_i \cap \Phi_j = \mathcal{U}$ for any two distinct types $i, j \in I$;
(C3) $|[x]\Phi_i| \geq 2$ for every type $i \in I$ and every chamber $x \in C$.

This definition of chamber systems is a bit more restrictive than other ones
that can be found in the literature (compare [48], [29], [14]). However, it is
general enough to cover all interesting examples, or almost all of them.

Properties (C1) and (C3) are often called the *connectedness* property and
the *firmness* property, respectively. The positive integer $n = |I|$ is called the
rank of $\mathcal{C}$. The relation Φ_i is called the *i-adjacency* relation. Two chambers
are said to be *adjacent* if they are *i*-adjacent for some $i \in I$.

We set $\Phi_\emptyset = \mho$ and $\Phi_J = \bigvee_{j \in J} \Phi_j$ for every nonempty subset J of I. Given a chamber x and a subset J of I, the pair $([x]\Phi_J, J)$ is called a *cell* of *type* J (of *cotype* $I - J$). We also write $[x]\Phi_J$ for $([x]\Phi_J, J)$ (thus identifying a cell with its set of chambers) when this abbreviation does not cause any confusion. $|J|$ and $n - |J|$ are respectively the *rank* and the *corank* of $([x]\Phi_J, J)$. Cells of rank 1 are called *panels*. Trivially, if X is a cell of type $J \neq \emptyset$ and Φ_i^X is the restriction of Φ_i to X, then $\mathcal{C}_X = (X, (\Phi_j^X)_{j \in J})$ is a chamber system over the set of types J and it is called a *residue* of $\mathcal{C}$. The type and the rank of X are the *type* and the *rank* of $\mathcal{C}_X$.

A chamber system $\mathcal{C}$ is said to be *tight* if some of its cells of corank 1 contains all chambers. Note that no chamber system of rank 1 or 2 is tight (by (C3) and (C2)).

A chamber system can also be viewed as a coloured graph: its edges are the pairs of distinct adjacent chambers and an edge $\{x, y\}$ gets the colour i if x and y are i-adjacent (note that every edge has just one colour, by (C2)). An *automorphism* of a chamber system $\mathcal{C}$ is just a colour-preserving automorphism of the graph $\mathcal{C}$. A chamber system $\mathcal{C}$ (a subgroup G of the automorphism group $Aut(\mathcal{C})$ of $\mathcal{C}$) is said to be *transitive* if $Aut(\mathcal{C})$ (respectively, G) is transitive on the set of chambers of $\mathcal{C}$. A *morphism* (in particular, an *isomorphism*) of chamber systems over the same set of types is a colour-preserving morphism (isomorphism) of graphs.

Given a transitive chamber system $\mathcal{C}$, let G be a transitive subgroup of $Aut(\mathcal{C})$. Given a chamber $c \in \mathcal{C}$, let B be stabilizer of c in G and, for every $i \in I$, let P_i be the stabilizer in G of the panel $[c]\Phi_i$. The following hold:

(P1) $G = \langle P_i \rangle_{i \in I}$;

(P2) $P_i \cap P_j = B$ for any two distinct types $i, j \in I$;

(P3) $B \neq P_i$ for all $i \in I$;

(P4) $\bigcap_{g \in G} B^g = 1$.

(P1) follows from (C1) and from the transitivity of G. (P2) and (P3) correspond to (C2) and (C3) respectively. (P4) holds because G, being an automorphism group of $\mathcal{C}$, acts faithfully on the set of chambers of $\mathcal{C}$.

B is called the *Borel subgroup* of G (relative to c) and, for every type $i \in I$, the subgroup P_i is the (*minimal*) *parabolic subgroup* of G of *type* i (relative to c). Given a proper nonempty subset J of I, the subgroup $P_J = \langle P_j \rangle_{j \in J}$ is the stabilizer in G of the cell of type J containing c. We call it the *parabolic subgroup* of G of *type* J relative to c. $|J|$ is the *rank* of P_J. We denote the elementwise stabilizer of $[c]\Phi_J$ in P_J by K_J and we call it the *kernel* of P_J. Thus, P_J / K_J is the action of P_J on $[c]\Phi_J$. Clearly, P_J / K_J is a transitive automorphism group of the residue $\mathcal{C}_{[c]\Phi_J}$ and (P1)-(P4) hold for the family $(P_j / K_J)_{j \in J}$ of subgroups of P_J / K_J, with I and B replaced by J and B / K_J

respectively.

We denote $(P_i)_{i \in I}$ by $\mathcal{P}_c(G, \mathcal{C})$. Note that, if d is another chamber of $\mathcal{C}$, then $\mathcal{P}_d(G, \mathcal{C})$ and $\mathcal{P}_c(G, \mathcal{G})$ are conjugated in G. Thus, as far as we are interested in $\mathcal{P}_c(G, \mathcal{C})$ 'modulo conjugation', we can write $\mathcal{P}(G, \mathcal{C})$ for $\mathcal{P}_c(G, \mathcal{G})$, dropping the subscript c.

Conversely, let B and $\mathcal{P} = (P_i)_{i \in I}$ be a subgroup of a group G and a family of subgroups of G satisfying (P1)-(P4). Then we can construct a chamber system $\mathcal{C}(\mathcal{P})$ as follows. Take the right cosets in G of B as chambers and for every $i \in I$ define the i-adjacency relation Φ_i by the following clause: given $f, g \in G$, fB and gB are i-adjacent iff $g^{-1}f \in P_i$. The group G, acting on the right cosets of B by left multiplication, is a transitive subgroup of $Aut(\mathcal{C}(\mathcal{P}))$ and we have $\mathcal{P}(G, \mathcal{C}(\mathcal{P})) = \mathcal{P}$. On the other hand, if $\mathcal{C}$ is a transitive chamber system and G is a transitive subgroup of $Aut(\mathcal{C})$, then $\mathcal{C}(\mathcal{P}(G, \mathcal{C})) \cong \mathcal{C}$.

2.2 Chamber Systems and Geometries

Given a chamber system $\mathcal{C} = (C, (\Phi)_{i \in I})$, we can construct a graph $\Gamma(\mathcal{C})$ from $\mathcal{C}$ taking the cells of $\mathcal{C}$ of corank 1 as vertices and declaring that two distinct cells X, Y of corank 1 are adjacent in $\Gamma(\mathcal{C})$ precisely when $X \cap Y \neq \emptyset$. It is clear that $\Gamma(\mathcal{C})$ is a connected n-partite graph (with $n = |I|$), the classes of the n-partition being the sets of cells of $\mathcal{C}$ of corank 1 of the same cotype. This graph is called the *incidence graph* of $\mathcal{C}$.

In order to recover $\mathcal{C}$ from $\Gamma(\mathcal{C})$, we should be able to interpret the cells of $\mathcal{C}$ as cliques of $\Gamma(\mathcal{C})$ (in particular, the chambers of $\mathcal{C}$ as maximal cliques of $\Gamma(\mathcal{C})$) in such a way that a cell is the intersection of all cells of $\mathcal{C}$ of corank 1 containing it and every clique of $\Gamma(\mathcal{C})$ is the set of cells of $\mathcal{C}$ of corank 1 containing some given cell of $\mathcal{C}$. We can do that if and only if the following hold (compare [14]):

(G1) $\Phi_J = \bigcap_{j \notin J} \Phi_{I-\{j\}}$ for every $J \subseteq I$;

(G2) $\Phi_J \cap (\Phi_{I-\{i\}} \cdot \Phi_{I-\{j\}}) = (\Phi_J \cap \Phi_{I-\{i\}}) \cdot (\Phi_J \cap \Phi_{I-\{j\}})$ for any two distinct types $i, j \in I$ and every subset J of I containing i and j.

If (G1) and (G2) hold in $\mathcal{C}$, then $\Gamma(\mathcal{C})$ is a geometry of rank n, in the meaning of [27]. In that case we say that $\mathcal{C}$ is *geometric* and its incidence graph $\Gamma(\mathcal{C})$ is called the *geometry* of $\mathcal{C}$.

Conversely, given a geometry Γ over a set of types I, it is well known that the chambers (i.e. maximal flags) of Γ form a chamber system $\mathcal{C}(\Gamma)$, two chambers of Γ being i-adjacent when they intersect in a flag of cotype i (see [27], Chapter 1). Clearly, $\mathcal{C}(\Gamma)$ is geometric and $\Gamma(\mathcal{C}(\Gamma)) \cong \Gamma$. On the other hand, $\mathcal{C}(\Gamma(\mathcal{C})) \cong \mathcal{C}$ for every geometric chamber system $\mathcal{C}$.

It is not difficult to check that (G1) is equivalent to the following condition (considered by Meixner and Timmesfeld in [22]):

(G1') $\quad \Phi_J \cap \Phi_K = \Phi_{J \cap K}$ for all $J, K \subseteq I$.

Note that (G1') fails to hold in tight chamber systems. Thus, tight chamber systems are non-geometric.

Trivially, all chamber systems of rank 2 are geometric. The geometries of rank 2 we will most frequently consider in this paper are generalized m-gons. We do not recall their definition here, presuming that the reader is familiar with them. We only remark that the following relation characterizes generalized digons: $\Phi_1 \Phi_2 = \Phi_2 \Phi_1$. Thus, a pair $\mathcal{P} = (P_1, P_2)$ of proper subgroups of a given group G satisfying (P1)-(P4) defines a generalized digon precisely when $P_1 P_2 = P_2 P_1 = G$.

2.3 Diagrams, Orders and Local Properties

Since all chamber systems of rank 2 are geometric, diagrams can be defined for chamber systems just as for geometries. We are not going to recall the definition of diagrams here, assuming that the reader is familiar with them. We only state a few conventions and fix some notation.

We denote the connected Coxeter diagrams of spherical type and rank $n \geq 3$ by A_n, C_n, D_n, F_4, E_6, E_7, E_8, H_3, H_4, as usual. We denote the diagram representing the class of generalized m-gons by $I_2(m)$, as in [48]. Thus, $I_2(3) = A_2$ and $I_2(4) = C_2$. We denote the diagram of rank 1 by A_1.

We recall that a Coxeter diagram can also be viewed as a symmetric matrix $M = (m_{ij})_{i,j \in I}$ where, for $i \neq j$, m_{ij} is an integer ≥ 2 or the symbol ∞, whereas $m_{ii} = 1$ for all $i \in I$. For $i \neq j$, the entry m_{ij} is called the *weight* of the edge $\{i, j\}$ of the Coxeter diagram represented by M.

A graph can be associated to every diagram $\mathbf{D}$, by joining two distinct types i and j when the class of rank 2 geometries associated in $\mathbf{D}$ to the stroke $\{i, j\}$ is not the class of generalized digons. This graph will be called the *underlying graph* of $\mathbf{D}$. If $\mathcal{C}$ is a chamber system belonging to $\mathbf{D}$, then the underlying graph of $\mathbf{D}$ is called the *diagram graph* of $\mathcal{C}$ (in [6]: the *basic diagram* of $\mathcal{C}$). Thinking of underlying graphs, we can extend to diagrams the terminology currently used for simple graphs: *connected, disconnected, connected components, string, tree, circuit, complete graph,...* For instance, we say that a diagram is *trivial* if its underlying graph has no edges.

If a diagram $\mathbf{D}$ splits into m connected components $\mathbf{D}_1$, $\mathbf{D}_2$,..., $\mathbf{D}_m$, then we write $\mathbf{D} = \mathbf{D}_1 + \mathbf{D}_2 + ... + \mathbf{D}_m$. For instance, $A_1 + A_1$ is a name for the class of generalized digons (we recall that A_1 is the diagram of rank 1), $A_1 + A_1 + A_1$ is the trivial diagram of rank 3

• • •

and $A_1 + I_2(m)$ is the following disconnected Coxeter diagram

$$(m)$$
• •———•

A chamber system $\mathcal{C}$ admits *order q* at some type i if all panels of $\mathcal{C}$ of type i have size $q + 1$. If $\mathcal{C}$ admits the same order q at every type, then we say that it has *uniform order q*. Following [7], we write orders below and types above the nodes of a diagram, when possible.

A chamber system $\mathcal{C}$ is said to be *thin* if it admits uniform order 1. On the other hand, if all panels of $\mathcal{C}$ have size ≥ 3, then $\mathcal{C}$ is *thick*. A chamber system is *locally finite* if all its cells of rank 2 are finite.

We recall that finite thick generalized m-gons exist only for $m = 2, 3, 4, 6$ or 8 (see [11]). A finite thick generalized m-gon with $2 < m$ is said to be *classical* if it arises from the natural BN-pair of a Chevalley group of rank 2 (possibly of twisted type). Thus, finite classical generalized m-gons exist only for $m = 3, 4, 6$ or 8. A locally finite thick chamber system $\mathcal{C}$ belonging to a Coxeter diagram **D** is *locally classical* if, for every edge $\{i, j\}$ of **D** of weight $m_{ij} \geq 3$, the residues of $\mathcal{C}$ of type $\{i, j\}$ are classical generalized m_{ij}-gons. Note that all locally classical chamber systems are locally finite and thick.

2.4 Covers and Quotients

According to [29] and [48] (see also [14]), given two chamber systems $\mathcal{C}$ and $\mathcal{C}'$ of rank n over the same set of types I and a morphism $f : \mathcal{C} \longrightarrow \mathcal{C}'$, we say that f is an *m-covering* for some $m = 1, 2, ..., n - 1$ if, for every chamber x of $\mathcal{C}$ and every $J \subseteq I$ with $|J| = m$, f maps the cell X of $\mathcal{C}$ of type J containing x onto the cell of $\mathcal{C}'$ of type J containing $f(x)$ and induces an isomorphism from the residue $\mathcal{C}_X$ of X in $\mathcal{C}$ to the residue $\mathcal{C}'_{f(X)}$ of $f(X)$ in $\mathcal{C}'$. If there is an m-covering $f : \mathcal{C} \longrightarrow \mathcal{C}'$, then $\mathcal{C}$ is called an *m-cover* of $\mathcal{C}'$ and $\mathcal{C}'$ is an *m-quotient* of $\mathcal{C}$.

Given chamber systems $\mathcal{C}$ and $\tilde{\mathcal{C}}$ of rank n over a same set of types I, we say that $\tilde{\mathcal{C}}$ is the *universal m-cover* of $\mathcal{C}$ for some positive integer $m < n$ if there is an m-covering $f : \tilde{\mathcal{C}} \longrightarrow \mathcal{C}$ such that, for every m-covering $g : \mathcal{C}' \longrightarrow \mathcal{C}$, there is just one m-covering $h : \tilde{\mathcal{C}} \longrightarrow \mathcal{C}'$ with $hf = g$. An m-covering $f : \tilde{\mathcal{C}} \longrightarrow \mathcal{C}$ as above is said to be *universal*.

The existence problem for universal m-covers has been solved by Ronan in [29] (see also Tits [48]):

Theorem 2.1 (Ronan) *Every chamber system of rank n admits a universal m-cover, for every $m = 1, 2, ..., n - 1$.*

It is clear that the universal m-cover of a chamber system is uniquely determined up to isomorphisms. A chamber system is said to be *m-simply connected* if it is its own universal m-cover (equivalently, if it is the universal m-cover of some chamber system).

m-Coverings and universal m-covers are defined for geometries in the same way as for chamber systems. However, there is no general analogue of Theorem 2.1 for geometries. Thus, when we search for the universal m-cover of a given geometry Γ, the only strategy we can follow in general is to study the universal m-cover $\tilde{C}$ of the chamber system $C = C(\Gamma)$ of Γ (which exists, by Theorem 2.1) checking if it is geometric. If it is geometric, then its geometry $\Gamma(\tilde{C})$ is the universal m-cover of Γ.

Unfortunately, we do not know if the universal m-cover of every geometric chamber system of rank n is geometric when $2 \leq m < n - 1$ (even if no counterexamples are known). When $m = n - 1$ things go better. Indeed:

Proposition 2.2 *Every $n - 1$-cover of a geometric chamber system of rank n is geometric.*

(See [25], Proposition 5; also [27], 12.5.1.) Thus, every geometry of rank n admits an universal $n - 1$-cover. In particular, every geometry of rank 3 admits an universal 2-cover.

Buildings are perhaps the most important family of geometric chamber systems belonging to Coxeter diagrams. The following celebrated theorem of Tits [48] gives us a "local" caracterization of those chamber systems belonging to a Coxeter diagram that are 2-covered by buildings:

Theorem 2.3 (Tits) *Let C be a chamber system belonging to a Coxeter diagram $\mathbf{D}$ over a set of types I and assume that, for every subset J of I of size 3 such that the diagram $\mathbf{D}_J$ induced by $\mathbf{D}$ on J is spherical, every residue of C of type J is 2-covered by a building. Then the universal 2-cover of C is a building.*

Therefore:

Corollary 2.4 *All buildings of rank ≥ 3 are 2-simply connected.*

The following also holds in the finite thick case:

Theorem 2.5 *Finite thick buildings with connected diagram of rank ≥ 3 do not admit any transitive proper 2-quotients.*

Proof. Any 2-quotient of a building C is obtained as a quotient C/X of C by a subgroup X of $G = Aut(C)$ acting semi-regularly on the set of chambers of C (see [29]). Furthermore, $Aut(C/X) \cong N_G(X)/X$ and C/X is transitive if and only if $N_G(X)$ is transitive on $\mathcal{C}$ (see [29]). On the other hand, if $\mathcal{C}$ is finite and thick and belongs to a connected diagram of rank ≥ 3, then the only subgroup X of G semi-regular on the set of chambers of $\mathcal{C}$ and with $N_G(X)$ transitive on $\mathcal{C}$ is the trivial one (Seitz [31]). $\square$

The connectedness of the diagram is essential in Theorem 2.5, as it will be clear from the examples of Section 4. Note also that only transitive proper quotients are forbidden by Theorem 2.5. Non-transitive proper quotients exist. We only mention one of them, discovered by Timmesfeld [46]. Let $\Gamma = PG(3,2)$. Its automorphism group $G = L_3(2)$ admits a subgroup $X = Z_5$ acting semi-regularly on the set of chambers of Γ. That subgroup defines a 2-quotient $\mathcal{C}(\Gamma)/X$ of the chamber system $\mathcal{C}(\Gamma)$ of Γ. According to Theorem 2.5, that quotient is not transitive. Actually, $N_G(X) = D_{10} \times 3$, which is not transitive on $\mathcal{C}(\Gamma)$. Note that $\mathcal{C}(\Gamma)/X$ is not even geometric. Indeed all geometries of type A_n, D_n and E_6 are buildings ([48], Proposition 6; [39], Lemma 3.2; [4]). Furthermore, the following has been proved in [4]:

Theorem 2.6 (Brouwer-Cohen) *Finite thick buildings of rank $n \geq 3$ do not admit any geometric proper 2-quotients.*

3   Direct Sums and Reducibility

3.1   Truncations

Given a geometry Γ over a set of types I and a proper nonempty subset J of I, let V_J be the set of elements of Γ of type $j \in J$. The *truncation* of Γ over J (*J-truncation* of Γ, for short) is the geometry $tr_J(\Gamma)$ over the set of types J induced by Γ on V_J. If $\mathcal{C} = (C, (\Phi_i)_{i \in I})$ is the chamber system of Γ, then the chamber system of $tr_J(\Gamma)$ can be recovered from $\mathcal{C}$ as follows: the quotient C/Φ_{I-J} corresponds to the set of chambers of $tr_J(\mathcal{C})$ and, for every $j \in J$, we can take $\Phi_{(I-J)\cup\{j\}}/\Phi_{I-J}$ as the j-adjacency relation. This construction can be done in general for every chamber system $\mathcal{C}$. It gives us a chamber system provided that both the following hold:

(T1) $\Phi_{(I-J)\cup\{j\}} \cap \Phi_{(I-J)\cup\{k\}} = \Phi_{I-J}$ for any two distinct types $j, k \in J$;
(T2)   for every $j \in J$, the equivalence classes of $\Phi_{(I-J)\cup\{j\}}/\Phi_{I-J}$ have size at least 2.

Clearly, both (T1) and (T2) hold if $\mathcal{C}$ is geometric. If (T1) and (T2) hold, then the chamber system $tr_J(\mathcal{C}) = (C/\Phi_{I-J}, (\Phi_{(I-J)\cup\{j\}}/\Phi_{I-J})_{j \in J})$ will be called the *truncation* of $\mathcal{C}$ over J (also *J-truncation* of $\mathcal{C}$, for short).

3.2   Direct Sums

Given a geometry Γ over a set of types I and a partition of I in two nonempty disjoint subsets J and K, we say that Γ is the *direct sum* $tr_J(\Gamma) \oplus tr_K(\Gamma)$ of its J- and K-truncations if all elements of Γ of type $j \in J$ are incident with all elements of Γ of type $k \in K$. More generally, Γ is the direct sum $\Gamma_1 \oplus \Gamma_2$ of two disjoint geometries Γ_1, Γ_2 over the sets of types J and K respectively if it is obtained by assembling Γ_1 and Γ_2 in the most trivial way, choosing the trivial incidence relation to relate elements of Γ_1 to elements of Γ_2. If $\Gamma = \Gamma_1 \oplus \Gamma_2$ with Γ_1 and Γ_2 as above, then $\Gamma_1 \cong tr_J(\Gamma)$ and $\Gamma_2 \cong tr_K(\Gamma)$ and, for every flag F of Γ of type K (of type J), the geometry Γ_1 (respectively, Γ_2) is isomorphic with the residue Γ_F of F.

Let $\mathcal{C}$, $\mathcal{C}_1$, $\mathcal{C}_2$ be the chamber systems of Γ, Γ_1 and Γ_2, respectively. Thus, $\mathcal{C}_1$ and $\mathcal{C}_2$ are the truncations of $\mathcal{C}$ over J and K, respectively. The relation $\Gamma = \Gamma_1 \oplus \Gamma_2$ is characterized by the following properties in $\mathcal{C}$ (note that the first one just says that all residues of type $\{j, k\}$ with $j \in J$ and $k \in K$ are generalized digons):

(R1) $\Phi_j \Phi_k = \Phi_k \Phi_j$ for all $j \in J$ and $k \in K$;

(R2) $\Phi_J \cap \Phi_K = \mho$.

Therefore, the set C of chambers of $\mathcal{C}$ can be identified with the direct product of the sets of chambers of $\mathcal{C}_1$ and $\mathcal{C}_2$, representing $x \in C$ as $([x]\Phi_K, [x]\Phi_J)$. For every $j \in J$, the j-adjacency relation Φ_j of $\mathcal{C}$ corresponds to the pair of equivalence relations $((\Phi_{K \cup \{j\}})/\Phi_K, \mho_2)$, where $\mho_2$ is the identity relation on the set of chambers of $\mathcal{C}_2$. The k-adjacency relations with $k \in K$ can be represented in a similar way. The fact that $tr_J(\Gamma) \cong \Gamma_F$ for every flag F of Γ of type K can now be rephrased as follows: we have $tr_J(\mathcal{C}) \cong \mathcal{C}_X$, for every cell X of $\mathcal{C}$ of type J.

Thus, direct sums of possibly non-geometric chamber systems can be defined as follows. Let $\mathcal{C}_1 = (C_1, (\Psi_j)_{j \in J})$ and $\mathcal{C}_2 = (C_2, (\Psi_k)_{k \in K})$ be chamber systems over mutually disjoint sets of types J and K. We define the *direct sum* $\mathcal{C} = \mathcal{C}_1 \oplus \mathcal{C}_2$ of $\mathcal{C}_1$ and $\mathcal{C}_2$ by taking $I = J \cup K$ as set of types, $C = C_1 \times C_2$ as set of chambers and the pairs $(\Psi_j, \mho_2)$, $(\mho_1, \Psi_k)$ as adjacency relations ($j \in J$, $k \in K$ and $\mho_i$ is the identity relation on C_i, $i = 1, 2$). Trivially, (R1) and (R2) hold in $\mathcal{C}$ for the partition $\{J, K\}$ of I. (T1) and (T2) also hold, $tr_J(\mathcal{C}) \cong \mathcal{C}_1 \cong \mathcal{C}_X$ for every cell X of $\mathcal{C}$ of type J and $tr_K(\mathcal{C}) \cong \mathcal{C}_2 \cong \mathcal{C}_Y$ for every cell Y of $\mathcal{C}$ of type K.

Clearly, $\mathcal{C}_1 \oplus \mathcal{C}_2$ is geometric if and only if both $\mathcal{C}_1$ and $\mathcal{C}_2$ are geometric. If this is the case, then $\Gamma(\mathcal{C}_1 \oplus \mathcal{C}_2) = \Gamma(\mathcal{C}_1) \oplus \Gamma(\mathcal{C}_2)$.

Conversely, let $\mathcal{C}$ be a chamber system over a set of types I and let $\{J, K\}$ be a partition of I in two disjoint nonempty subsets. Assume that (R1) and (R2) hold in $\mathcal{C}$ for the partition $\{J, K\}$. Then (T1) and (T2) also hold ([25],

Proposition 3). Thus we can consider the truncations of $\mathcal{C}$ over J and K. By (R1) and (R2), we have $\mathcal{C} \cong tr_J(\mathcal{C}) \oplus tr_K(\mathcal{C})$.

We have only considered direct sums of two geometries or chamber systems here. Needless to say, all the above can be rephrased for direct sums with more than two summands.

3.3   Reducibility

A chamber system is said to be *reducible* if it splits as a direct sum of some of its truncations. Otherwise, it is said to be *irreducible*. Clearly, every reducible chamber system $\mathcal{C}$ splits as the direct sum of a finite number of irreducible chamber systems and that splitting is unique, modulo permutations of the summands. The summands of that splitting are called the *irreducible components* of $\mathcal{C}$ (of Γ).

We have shown in the previous subsection that a chamber systems $\mathcal{C}$ over a set of types $I = J \cup K$, with $\{J, K\}$ a partition of I, splits as the direct sum of chamber systems over the sets of types J and K if and only if both (R1) and (R2) hold in $\mathcal{C}$ with respect to J and K. Thus we call (R1) and (R2) the *reducibility conditions*.

The first reducibility condition (R1) just states that the sets of types J and K are not joined by any path in the diagram graph of $\mathcal{C}$. The second reducibility condition (R2) is a special case of (G1') of Subsection 2.2, which holds in every geometric chamber system. Furthermore, every geometric chamber system admits truncations over every nonempty set of types and its truncations are geometric. Therefore:

Theorem 3.1 (Direct Sum Theorem) *The irreducible components of a geometric chamber system $\mathcal{C}$ are the truncations of $\mathcal{C}$ over the connected components of the diagram graph of $\mathcal{C}$.*

In particular:

Corollary 3.2 *A geometric chamber system is irreducible if and only if its diagram graph is connected.*

Unfortunately, the statement of Theorem 3.1 fails to hold for non-geometric chamber systems. We will give a number of conterexamples in Section 4, but the reason of that failure can be explained right now. The first reducibility condition (R1) is the only information we can get from the disconnectedness of a diagram, but (R1) is not sufficient to obtain splittings in direct sums: we also need (R2) for that. However, (R2) does not hold in non-geometric chamber systems, in general.

We say that a chamber sytem C with disconnected diagram graph $\mathbf{D}$ is *completely reducible* if C admits truncations over every connected component of $\mathbf{D}$ and these truncations are the irreducible components of C.

Given a chamber system C of rank $n \geq 3$ and a triple J of types such that the graph induced on J by the diagram graph of C is disconnected, we say that C is *reducible over* J if all residues of C of type J are reducible (hence completely reducible, since $|J| = 3$ and since reducible chamber systems of rank 2 are generalized digons). We say that C is *locally reducible* if it is reducible over every triple of types J as above.

Note that all reducible chamber systems of rank 3 are geometric and all geometries belonging to disconnected Coxeter diagrams of rank 3 are buildings. Thus, by Theorem 2.3 we obtain the following

Proposition 3.3 *Let C be a locally reducible chamber system belonging to a Coxeter diagram. Then the universal 2-cover of C is a building if and only if all residues of C of type A_3, C_3 and H_3 are 2-quotients of buildings.*

3.4   Covers of Direct Sums

Let C be a chamber system of rank n over a set of types I, let $\{I_1, I_2, ..., I_r\}$ be a partition of I such that C admits truncation over each of $I_1, I_2, ..., I_r$ and let $C = \bigoplus_{i=1}^{r} C_i$, where C_i is the I_i-truncation of C $(i = 1, 2, ..., r)$. Given an integer $m = 2, 3, ..., n - 1$, if $m < |I_i|$, then $\tilde{C}_i$ will denote the universal m-cover of C_i. If $m \geq |I_i|$, then we set $\tilde{C}_i = C_i$. Let $\tilde{C}$ be the universal m-cover of C. Then the following holds ([25], Proposition 6; also [27], 12.5.2):

Proposition 3.4 $\tilde{C} = \bigoplus_{i=1}^{r} \tilde{C}_i$.

The next corollaries easily follow from this proposition:

Corollary 3.5 *Let $C = \bigoplus_{i=1}^{r} C_i$ and assume that, for every $i = 1, 2, ..., r$, either C_i is m-simply connected or it has rank $\leq m$. Then C is m-simply connected.*

Corollary 3.6 *Every direct sum of chamber systems of rank 1 or 2 is 2-simply connected.*

Corollary 3.7 *Let C be a geometric chamber system of rank n with disconnected diagram graph $\mathbf{D}$ and let $m = 2, 3, ...$ or $n - 1$. Then the universal m-cover of C is the direct sum of the universal m-covers of the truncations of C over the connected components of $\mathbf{D}$ (with the convention that a chamber system of rank $\leq m$ should be regarded as its own universal m-cover).*

Note that Proposition 3.4 only states that, if a chamber system C is reducible, then its universal m-cover is also reducible. It does not say that, if C is reducible, then every m-cover of C is reducible (but I do not know any counterexamples).

4   Counterexamples

4.1   Ronan Systems

Given a family $\mathcal{P} = (P_i)_{i=1}^n$ of $n \geq 3$ proper subgroups of a group G, let $B = \bigcap_{i=1}^n P_i$. Assume that $\mathcal{P}$ satisfies (P2)-(P4) of Subsection 2.1 and the following condition, stronger than (P1):

(P1*) $\langle P_i, P_j \rangle = G$ for any two distinct indices $i, j = 1, 2, ..., n$.

By (P1*), the chamber system $\mathcal{C}(\mathcal{P})$ defined by $\mathcal{P}$ is tight (hence it is not geometric). We call $\mathcal{P}$ a *Ronan system*, after Ronan [30], who used a system like that (actually, three copies of Z_3 in $G = Frob(21)$) to produce a tight chamber system of type $\tilde{A}_2$ (see also [15], Example 3.3(1)). We will use Ronan systems to construct irreducible transitive chamber systems with disconnected diagrams.

4.2   Examples from Frobenius Groups

Let G be a Frobenius group and let F be its Frobenius kernel. Take $P_1 = F$ and choose $P_2, P_3,..., P_n$ among the complements of F in G. Then $\mathcal{P} = (P_i)_{i=1}^n$ is a Ronan system, with $B = 1$. The diagram of $\mathcal{C}(\mathcal{P})$ is disconnected. Indeed the type 1 is not joined to any other type in that diagram. However, $\mathcal{C}(\mathcal{P})$ is irreducible.

For instance, if $G = Frob(9.73)$, $Frob(21)$ or $S_3(= Frob(6))$ and $n = 3$, then $\mathcal{C}(\mathcal{P})$ belongs to the Coxeter diagram $A_1 + A_2$ and has orders $(72, 8, 8)$, $(6, 2, 2)$ or $(2, 1, 1)$ respectively:

$$
\begin{array}{ccc}
1 & 2 & 3 \\
\bullet & \bullet\!\!-\!\!-\!\!-\!\!-\!\!\bullet \\
72 & 8 & 8
\end{array}
$$

$$
\begin{array}{ccc}
1 & 2 & 3 \\
\bullet & \bullet\!\!-\!\!-\!\!-\!\!-\!\!\bullet \\
6 & 2 & 2
\end{array}
$$

$$
\begin{array}{ccc}
1 & 2 & 3 \\
\bullet & \bullet\!\!-\!\!-\!\!-\!\!-\!\!\bullet \\
2 & 1 & 1
\end{array}
$$

G stabilizes the (unique) residue of type $\{2, 3\}$ and acts faithfully on it. If $G = Frob(9.73)$ or $Frob(21)$, then $\mathcal{C}(\mathcal{P})$ is locally classical.

4.3   Examples with $G = X^2{:}B$

Let A be a split extension $A = X{:}B$ of a group X by a group B, let $Y = X_1 \times X_2$ be the direct product of two copies X_1, X_2 of X and let $G = Y{:}B$ be the split extension of Y by B, with B acting on X_1 and X_2 as on X. let $\mathcal{F}$ be a finite family of isomorphisms from $X_1{:}B$ to $X_2{:}B$ such that $f(b) = b$ for all $f \in \mathcal{F}$ and all $b \in B$ and $f(x) \neq g(x)$ for every choice of distinct members $f, g \in \mathcal{F}$ and all $x \in X_1 - \{1\}$. Set $P_i = X_i{:}B$ for $i = 1, 2$ and, for every $f \in \mathcal{F}$, define $P_f = \{(x, f(x))b \mid x \in X_1, b \in B\}$. We obtain a Ronan system $\mathcal{P}$ of rank n $(= 2+|\mathcal{F}|)$. The diagram of $\mathcal{C}(\mathcal{P})$ is trivial, but $\mathcal{C}(\mathcal{P})$ is irreducible.

Needless to say, we might also take $B = 1$ and $G = Y$ in the above.

Examples of rank 3 can be obtained from any split extension $A = X{:}B$ by taking any isomorphism f from $X_1{:}B$ to $X_2{:}B$ and $\mathcal{F} = \{f\}$. For instance, let $X = Z_3$, $B = Z_2$ and $A = S_3 = X{:}B$. Then $G = 3^2{:}2$ is a subgroup of the alternating group A_6. We can find an involution $\varphi \in S_6 - A_6$ normalizing G and B and interchanging X_1 with X_2. We set $\mathcal{F} = \{f\}$, with f the outer automorphism of G induced by φ and $\mathcal{P} = (P_1, P_2, P_3)$ with $P_i = X_i{:}B$ $(i = 1, 2)$ and $P_3 = \{(x, f(x))b \mid x \in X_1, b \in B\}$. The diagram of $\mathcal{C}(\mathcal{P})$ is the trivial diagram $A_1 + A_1 + A_1$, with uniform order 2:

$$\overset{\bullet}{2} \qquad \overset{\bullet}{2} \qquad \overset{\bullet}{2}$$

However, $\mathcal{C}(\mathcal{P})$ is irreducible. We call $\mathcal{C}(\mathcal{P})$ the $3^2{:}2$-*chamber system*.

Remark. Every chamber system can be viewed as a partial plane with a parallelism relation. The points are the chambers, the lines are the panels (two lines meet in at most one point by (C2)) and two lines are parallel when they are panels of the same adjacency relation. Thus, every net of degree $n \geq 3$ is an irreducible chamber system with trivial diagram. The examples described in this subsection are indeed of this kind.

4.4   Covers of the Previous Examples

Let $\mathcal{P}$ be as in the previous subsections. We will show that the universal 2-cover of $\mathcal{C}(\mathcal{P})$ is reducible.

In the examples of Subsection 4.2 we have $B = 1$ and at least one of the minimal parabolics, say P_1, is normal in G. Thus, we can construct a suitable split extension $\tilde{G} = N{:}G$ with $N \cong P_1$ and we can choose a family $\tilde{\mathcal{P}} = (\tilde{P}_i)_{i=1}^n$ of subgroups of $\tilde{G}$ with $\tilde{P}_1 = N$ and $\tilde{P}_i = P_i \leq G$ for $i = 2, 3, ..., n$. The homomorphism $f : \tilde{G} \longrightarrow G$ mapping (x, y) onto xy $(x \in N, y \in G)$ induces a 2-covering from $\mathcal{C}(\tilde{\mathcal{P}})$ to $\mathcal{C}(\mathcal{P})$. The reducibility condition (R2) holds in $\mathcal{C}(\tilde{\mathcal{P}})$ for the partition $\{\{1\}, \{2, ..., n\}\}$ of the set of types $\{1, 2, ..., n\}$.

Hence the lack of connections from the node 1 to the rest of the diagram now corresponds to a splitting of $\mathcal{C}(\tilde{\mathcal{P}})$ as the direct sum of its truncations over the sets of types $\{1\}$ and $\{2, ..., n-1\}$. Iterating this construction if necessary, we eventually reach a 2-cover of $\mathcal{C}(\mathcal{P})$ which splits as a direct sum of irreducible components corresponding to the connected components of the diagram of $\mathcal{C}(\mathcal{P})$. Therefore the universal 2-cover of $\mathcal{C}(\mathcal{P})$ is reducible, by Proposition 3.4.

Turning to the examples of 4.3, let us set $I = \{1, 2\} \cup \mathcal{F}$ and $X_i = P_i \cap Y$ for $i \in I$. Let $\tilde{G} = \tilde{Y}:B$ be the split extension of the direct product $\tilde{Y} = \prod_{i \in I} X_i$ by B, with B acting on X_i as in P_i. Clearly, Y and G can be viewed as subgroups of $\tilde{Y}$ and $\tilde{G}$ respectively. For every $i \in I$, the subgroup P_i of G can be identified with its copy $X_i:B$ in $\tilde{G}$. Hence (P1)-(P4) hold in the family $\tilde{\mathcal{P}} = (P_i)_{i \in I}$ of subgroups of $\tilde{G}$. Clearly, the chamber system $\mathcal{C}(\tilde{\mathcal{P}})$ defined by $\tilde{\mathcal{P}}$ is the direct sum of $n = |I|$ geometries of rank 1. Hence it is 2-simply connected, by Corollary 3.6.

It is not difficult to prove that there is some subgroup K of $\tilde{Y}$ isomorphic to a product of $n - 2$ copies of X and such that $K \cap (X_i \times X_j) = 1$ for every choice of distinct types $i, j \in I$. Clearly B normalizes all subgroups of $\tilde{Y}$. Hence $K \trianglelefteq \tilde{G}$. It is easily seen that $G \cong \tilde{G}/K$. Thus, we can identify G with $\tilde{G}/K$. The canonical projection $\pi : \tilde{G} \longrightarrow \tilde{G}/K = G$ induces the identity on P_i for every $i \in I$ and does not identify any two of these subgroups. Therefore π induces a 2-covering $f_\pi : \mathcal{C}(\tilde{\mathcal{P}}) \longrightarrow \mathcal{C}(\mathcal{P})$, which is universal because $\mathcal{C}(\tilde{\mathcal{P}})$ is 2-simply connected.

Remark. Some examples of 2-simply connected irreducible chamber systems with disconnected diagram have been constructed by Tits ([48], 6.1.6(b)), by some kind of 'free construction'. Those examples are neither transitive nor finite.

4.5   Some Non-Tight Examples

The previous examples of irreducible chamber systems with disconnected diagram are tight. However, non-tight irreducible chamber systems with disconnected diagram are easy to construct.

Let $\mathcal{P} = (P_i)_{i \in I}$ be a family of subgroups of a group G satisfying (P1)-(P4) of Subsection 2.1. Given a family $(H_i)_{i \in I}$ of non-trivial groups, let us set $\hat{G} = G \times \prod_{i \in I} H_i$ and $\hat{P}_i = P_i \times H_i$. Then $\hat{P}_i$ can be wieved as a subgroup of $\hat{G}$ and (P1)-(P4) hold in $\hat{\mathcal{P}} = (\hat{P}_i)_{i \in I}$. The chamber system $\mathcal{C}(\hat{\mathcal{P}})$ is non-tight and it has the same diagram graph as $\mathcal{C}(\mathcal{P})$. Note that $\mathcal{C}(\hat{\mathcal{P}})$ is geometric if and only if $\mathcal{C}(\mathcal{P})$ is geometric. Let the diagram graph of $\mathcal{C}(\mathcal{P})$ (hence of $\mathcal{C}(\hat{\mathcal{P}})$) be disconnected. Then $\mathcal{C}(\hat{\mathcal{C}})$ is reducible if and only if $\mathcal{C}(\mathcal{P})$ is reducible.

Let $\mathcal{C}(\mathcal{P})$ be tight, irreducible but with a disconnected diagram (as in the examples of the previous subsections). Then $\mathcal{C}(\widehat{\mathcal{P}})$ is irreducible with the same (disconnected) diagram graph as $\mathcal{C}$, but it is not tight.

4.6   The 3^2 : 2-Chamber System, the Wester Chamber System and the A_7-Geometry

The 3^2:2-chamber system (defined in Subsection 4.3) appears as a rank 3 residue in the following chamber system, discovered by Wester [52].

Let S be a set of seven objects and let G be the alternating group A_7, in its natural action on S. Let Π^+ be a model of the projective plane $PG(2,2)$ built on the seven points of S. Given $a \in S$, let $m = \{a,b,c\}$ be a line of Π^+ on a. The transposition $(b,c) \in S_7$ does not preserve Π^+, but it maps Π^+ onto another model Π^- of $PG(2,2)$ for which m is still a line. Let P_0 be the stabilizer of Π^+ and Π^- in G. It is not difficult to check that P_0 also stabilizes a. We denote the stabilizers in G of m and Π^+ and of m and Π^- by P_+ and P_-, respectively. The stabilizer of a and m in G will be denoted by P_1. The quadruple $\mathcal{P} = (P_+, P_-, P_0, P_1)$ satisfies (P1)-(P4), with $B = D_8$. The chamber system $\mathcal{C}(\mathcal{P})$ belongs to the Coxeter diagram of affine type $\widetilde{B}_3$, with uniform order 2:

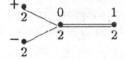

$P_J = G = A_7$ for every triple J of types other than $\{+,-,1\}$ and $P_{\{+,-,1\}} = 2^2 \cdot (3^2{:}2)$, with kernel $K_{\{+,-,1\}} = 2^2$ and $P_{\{+,-,1\}}/K_{\{+,-,1\}} = 3^2{:}2$.

$\mathcal{C}(\mathcal{P})$ has 35 residues of type $\{+,-,1\}$, but just one residue of each of the types $\{+,0,1\}$, $\{-,0,1\}$ and $\{+,0,-\}$. Hence it is tight. The residues of type $\{+,-,1\}$ are isomorphic to the 3^2:2-chamber system. The residue of type $\{+,0,-\}$ is isomorphic with the chamber system of $PG(3,2)$ and $P_{\{+,0,-\}} = A_7$ acts faithfully on it. The residues of type $\{+,0,1\}$ and $\{-,0,1\}$ belong to the Coxeter diagram C_3. They are models of the so-called A_7-geometry.

We call $\mathcal{C}(\mathcal{P})$ the *Wester chamber system*, after its discoverer. The Wester chamber system also appears as a rank 4 residue in two chamber systems of rank 6 for the groups $O_6^-(3)$ and $U_3(5)$ (see [52] and [23]; also [19], pages 54 and 55).

5   Coxeter Diagrams of Rank 3

5.1   Some Reducibility Theorems

In this subsection $\mathcal{C} = (C, (\Phi_1, \Phi_2, \Phi_3))$ is a transitive thick locally finite chamber system of rank 3 belonging to the disconnected Coxeter diagram $A_1 + I_2(m)$

$$\begin{array}{ccc} 1 & 2 \ (m) \ 3 \\ \bullet & \bullet\!\!-\!\!-\!\!\bullet \end{array}$$

where 1, 2, 3 are the types. In particular, when $m = 2, 3$ or 4 we have the following diagrams:

$$\bullet \quad \bullet \quad \bullet \qquad\qquad (m = 2)$$

$$\bullet \quad \bullet\!\!-\!\!-\!\!\bullet \qquad\qquad (m = 3)$$

$$\bullet \quad \bullet\!\!=\!\!=\!\!\bullet \qquad\qquad (m = 4)$$

Note that $m = 2, 3, 4, 6$ or 8 by [11], since we have assumed that $\mathcal{C}$ is locally finite and thick. The residues of type $\{1\}$ and $\{2,3\}$ containing a given chamber c of $\mathcal{C}$ will be denoted by C_1 and $C_{2,3}$ respectively. P_1, P_2, P_3 and B are the minimal parabolic subgroups and the Borel subgroup of $Aut(\mathcal{C})$ relative to c. According to the notation of 2.1, $P_{2,3} = \langle P_2, P_3 \rangle$ is the stabilizer of $C_{2,3}$ and $K_{2,3}$ is its kernel.

Theorem 5.1 *Let $m = 3$. Then $\mathcal{C}$ is reducible, except possibly if $P_{2,3}/K_{2,3}$ is a Frobenius group of order $(q + 1)(q^2 + q + 1)$, sharply flag-transitive on the projective plane $C_{2,3}$ (of order q), and $(P_1 \cap P_{2,3})/K_{2,3} = Z_{q^2+q+1}$.*

Proof. Set $X = P_1 \cap P_{2,3}$. Let $\mathcal{C}$ be irreducible. Then X properly contains B (compare (R2) of Subsection 3.1). By a theorem of Kantor [16] one of the following holds:

(i) q is a prime power, $C_{2,3} = PG(2, q)$ and $P_{2,3}/K_{2,3} \geq L_3(q)$;

(ii) $q^2 + q + 1$ is prime, q is even and $P_{2,3}/K_{2,3}$ is a Frobenius group of order $(q + 1)(q^2 + q + 1)$, acting sharply flag-transitively on the projective plane $C_{2,3}$.

Let (i) hold. Then $X/K_{2,3} \geq (P_i/K_{2,3}) \cap L_3(q)$ for some $i = 2$ or 3. However, this contradicts (P2) of 2.1 (indeed $X \leq P_1$).

Thus (ii) only survives. We still must prove that $(P_1 \cap P_{2,3})/K_{2,3} = Z_{q^2+q+1}$. Either $X/K_{2,3}$ is a non-trivial subgroup of some complement of the Frobenius kernel F of the Frobenius group $P_{2,3}/K_{2,3}$, or $X/K_{2,3} = F$. However, $(X/K_{2,3})(P_i/K_{2,3}) = (P_i/K_{2,3})(X/K_{2,3})$ for $i = 2, 3$, because $P_1 P_i =$

$P_i P_1$ for $i = 2, 3$. On the other hand, this condition does not hold if $X/K_{2,3}$ is contained in some complement of F. Therefore $X/K_{2,3} = F$. $\square$

Theorem 5.2 *Let* $m \geq 3$ *and let* $\mathcal{C}$ *be locally classical. Then* $\mathcal{C}$ *is reducible, except possibly when* $m = 3$ *and* $P_{2,3}/K_{2,3} = Frob((q + 1)(q^2 + q + 1))$, *with* $q = 2$ *or* 8.

Proof. When $m = 3$ the statement follows from Theorem 5.1 and from a well known theorem of Higman and McLaughlin [13]. Let $m > 3$ and let $\mathcal{C}$ be irreducible, if possible. As in the proof of Theorem 5.1, $X = P_1 \cap P_{2,3} > B$. Furthermore, one of the following holds, by a well known theorem of Seitz [31] (see also [14], Theorem C.7.1):

(1) $P_{2,3}/K_{2,3}$ contains the Chavalley group naturally associated to the generalized m-gon $\mathcal{C}_{2,3}$;
(2) $\mathcal{C}_{2,3} = Q_4(2)$ and $P_{2,3}/K_{2,3} = A_6 = S_4(2)'$;
(3) $\mathcal{C}_{2,3} = Q_4(3)$ and $P_{2,3}/K_{2,3} = 2^4{:}S_5$ or $2^4{:}A_5$ or $2^4{:}Frob(20)$;
(4) $\mathcal{C}_{2,3} = Q_5^-(3)$ and $P_{2,3}/K_{2,3} = L_3(4).2$ or $L_3(4).2^2$;
(5) $\mathcal{C}_{2,3}$ is the generalized hexagon for $G_2(2)$ and $P_{2,3}/K_{2,3} = G_2(2)'$;
(6) $\mathcal{C}_{2,3}$ is the generalized octagon for $^2F_4(2)$ and $P_{2,3}/K_{2,3} = {}^2F_4(2)'$.

In case (1) a contradiction is obtained as in (i) of the proof of Theorem 5.1. Let (2) occur. Then $B/K_{2,3}$ is the Sylow 2-subgroup of A_6 and $X/K_{2,3}$ is contained in the stabilizer of a point or a line of $\mathcal{C}_{2,3}$, as $X \geq B$ and $X \neq P_{2,3}$ by (P2) of 2.1. However, $XP_i = P_iX$ and $X \cap P_i = B$ for $i = 2, 3$. It is not difficult to check that no subgroup X of $P_{2,3} = K_{2,3}.A_6$ can have all the above properties.

Assume we are in case (3) with $P_{2,3}/K_{2,3} = V{:}A_5$, $V = Vect(4, 2)$. Then $B/K_{2,3} = Z_2 \times Z_3$ and $P_i/K_{2,3} = 3.D_8$, $P_j/K_{2,3} = \langle e \rangle{:}A_4$, with e a nonzero vector of V, $(i, j) = (2, 3)$ or $(3, 2)$. Clearly, $e \in X/K_{2,3}$, since e is the involution of $B/K_{2,3}$. If $X/K_{2,3}$ contains an element g of order 5, then it contains all of V, since g displaces e and $X/K_{2,3}$ also contains an element of order 3 fixing e, inherited from $B/K_{2,3}$. This forces $X = P_{2,3}$, which contradicts (P2) of 2.1. Therefore $|X/K_{2,3}|$ divides $2^6 3$. We have $X \cap P_i = B$ by (P2) and $XP_i = P_iX$ ($i = 2, 3$). Therefore 2^5 does not divide $|X/K_{2,3}|$. Hence $|X| = 2^h 3$ for some $h = 2, 3$ or 4. It is now easy to check that no subgroup X of $P_{2,3}$ can have all those properties.

The case of $P_{2,3}/K_{2,3} = 2^4{:}S_5$ can be ruled out in the same way. Let $P_{2,3}/K_{2,3} = 2^4{:}Frob(20)$. We now have $B/K_{2,3} = Z_2$, $P_i/K_{2,3} = D_8$ and $P_j/K_{2,3} = Z_2 \times Z_4$ for $(i, j) = (2, 3)$ or $(3, 2)$). An analysis as above shows that $|X/K_{2,3}| = 10$. This implies $X/K_{2,3} = B.5 = 2 \times 5$. However, no subgroup like this esists in $2^4{:}Frob(20)$.

Let us consider (4) with $P_{2,3}/K_{2,3} = L_3(4).2$. We have $B/K_{2,3} = 3^2{:}2$, $P_i/K_{2,3} = 3^2{:}Q_8.2$ and $P_j/K_j = A_6$, for $(i, j) = (2, 3)$ or $(3, 2)$. The subgroup

$P_i/K_{2,3}$ is maximal in $P_{2,3}/K_{2,3}$, $i = 2,3$. Therefore $XP_i = P_iX = P_{2,3}$, $i = 2,3$. Since $X \cap P_i = B$ by (P2), $|X|.|P_i|=|P_{2,3}|.|B|$ for $i = 2,3$. Therefore $|P_i|=|P_j|$ ($i,j = 2,3$), which is false. The case of $P_{2,3}/K_{2,3} = L_3(4).2^2$ can be ruled out in a similar way.

An argument similar to the above can also be used to rule out (6). Finally, assume we are in case (5). For $(i,j) = (2,3)$ or $(3,2)$, we have $P_i/K_{2,3} = 4.S_4$ and $P_j = 4^2:S_3$. These are maximal subgroups of $P_{2,3}/K_{2,3}$ of index 63. Since $XP_i = P_iX$ and $X \cap P_i = B$ for $i = 2,3$, $|P_{2,3}|.|B|= 3$. However, $G_2(2)'$ does not admit any subgroup of index 3.

Thus, each of the cases (1)-(6) leads to a contradiction. $\square$

Examples for the irreducible cases of theorems 5.1 and 5.2 have been given in Subsection 4.2, with $q = 2$ and 8. Let q be the order of a projective plane as in the irreducible case of Theorem 5.1, with $q \neq 2,8$. A theorem of Feit [10] gives us more information on this case: 3 divides $q + 1$, 8 divides q, but q is not a power of 2. Furthermore $q > 14,000,000$.

In the next theorem we assume that the parabolic subgroups of $Aut(C)$ satisfy the following for some prime $p > 1$:

(1) P_{ij} is finite for all $i,j = 1,2,3$ and B is the normalizer in P_{ij} of a Sylow p-subgroup of P_{ij}.
(2) For every $i = 1,2,3$, $O^{p'}(P_i/O_p(P_i))$ is either a perfect central extension of a rank 1 Chevalley group in characteristic p (possibly of twisted type) or the dihedral group D_{10} (of index 2 in $Sz(2) = Frob(20)$; $p = 2$ in this case).
(3) for $i = 2, 3$, $O^{p'}(P_{1,i}/O_p(P_{1,i}))$ is either a direct product $X_1 \times X_i$ with $X_k = (P_k \cap O^{p'}(P_{1,i}))/O_p(P_{1,i})$ for $k = 1,i$, or the subgroup $3^2:2$ of the alternating group A_6, obtained by intersecting A_6 with the stabilizer in S_6 of two disjoint triples of objects ($p = 2$ in this case). The same condition holds for $O^{p'}(P_{2,3}/O_p(P_{2,3}))$ when $m = 2$.
(4) If $m > 2$, then $O^{p'}(P_{2,3}/O_p(P_{2,3}))$ is either a perfect central extension of a rank 2 Chevalley group in characteristic p, possibly of twisted type, or one of $S_4(2)'$, $G_2(2)'$ or $^2F_4(2)'$ ($p = 2$ if any of these three cases occurs).

(We have mentioned perfect central extensions in (2) and (4); we recall that a central extension $X = Z.Y$ is said to be perfect if $Z \leq X'$.) The next theorem generalizes a result implicit in Corollary 5.9 of [39]:

Theorem 5.3 *Let* (1) − (4) *hold for some prime* $p > 1$. *Then either* C *is reducible or* $m = 2$, $p = 2$ *and* C *is the* $3^2{:}2$-*chamber system* (*defined in Subsection* 4.3).

Proof. Let C be irreducible. Then $m = 2$ by Theorem 5.2. We can assume without loss that $P_i = O^{p'}(P_i)B$, for all $i = 1,2,3$ (see [39], (4.5)). Let

$O^{p'}(P_{2,3}/O_p(P_{2,3}))$ be a direct product $X_2 \times X_3$ as in (3), if possible. Then $O^{p'}(P_{2,3})B = P_2P_3$, the subgroups P_2, P_3 normalize each other and

$$(B \cap O^{p'}(P_{2,3}))/O_p(P_{2,3}) = Y_2 \times Y_3 = (O^{p'}(P_2) \cap O^{p'}(P_3))/O_p(P_{2,3})$$

for suitable subgroups Y_2, Y_3 of X_2 and X_3 (recall that $O_p(P_{2,3}) \leq B$, by (1)). As $\mathcal{C}$ is irreducible, $P_1 \cap P_2P_3$ properly contains B. Hence there are elements $x \in A_2$, $y \in A_3$ such that $xy \in ((P_1 \cap O^{p'}(P_{2,3}))/O_p(P_{2,3})) - (Y_1 \times Y_2)$. As $xy \notin Y_2 \times Y_3$, either $x \notin Y_2$ or $y \notin Y_3$. We can assume without loss that $x \notin Y_2$. Let P be the unique Sylow p-subgroup of B (see (1)). $B \cap O^{p'}(P_{2,3})$ is the normalizer of P in $O^{p'}(P_{2,3})$. Let x normalize Y_2, if possible. Then, as x also centralizes Y_3, every representative x' of x in $O^{p'}(P_{2,3})$ normalizes the normalizer $B \cap O^{p'}(P_{2,3})$ of P in $O^{p'}(P_{2,3})$, hence it belongs to $B \cap O^{p'}(P_{2,3})$, contrary to the choice of $x \in X_2 - Y_2$. Therefore x does not normalize Y_2. That is, there is an element $z \in Y_2$ such that $zxz^{-1}x^{-1} \notin Y_2$. Therefore, and since z centralizes y:

$$zxyzy^{-1}x^{-1} = zxyz^{-1}y^{-1}x^{-1} \in (P_1 \cap O^{p'}(P_{2,3}))/O_p(P_{2,3}) \cap (X_2 - Y_2).$$

This implies that $P_1 \cap P_2$ properly contains B, contrary to (P2). Hence, for every choice of distinct types $i, j = 1, 2, 3$, none of the groups $O^{p'}(P_{i,j}/O_p(P_{i,j}))$ is a direct product. Therefore all of them are isomorphic to the subgroup $3^2 : 2 = (S_3 \times S_3) \cap A_6$ of A_6, we have $p = 2$ and $O^{p'}(P_i/O_p(P_i)) = S_3$ for $i = 1, 2, 3$.

Since $\mathcal{C}$ is irreducible, $P_1 \cap P_{2,3}$ properly contains B. Hence there is an element x of $((P_1 \cap O^{2'}(P_{2,3}))/O_2(P_{2,3})) - (S/O_2(P_{2,3}))$ of order 3 or 6, with S the Sylow 2-subgroup of $P_{2,3}$. We can generate all of P_1 by a representative of x and by S. Therefore $P_1 \leq P_{2,3}$. By (P4), $O_2(P_{2,3}) = 1$. It is now clear that $\mathcal{C}$ is the 3^2:2-chamber system. $\square$

Properties like (1)-(4) are considered by Timmesfeld in [39] to define parabolic systems. However, Timmesfeld assumed non-tightness in his definition of parabolic systems. In Corollary 4.9 of [39], Timmesfeld considered a parabolic system $\mathcal{P}$ of arbitrary rank in a group G. He also assumed that, for every connected component J of the diagram $\mathbf{D}$ of $\mathcal{C}(\mathcal{P})$, $O^{p'}(P_J/O_p(P_J))$ is a perfect central extension $X = Z.Y$ of a simple Chevalley group Y in characteristic p by a p'-group Z. Then he could prove that G splits as the central product of its parabolic subgroups corresponding to the connected components of $\mathbf{D}$. Needless to say, that result does not apply to the 3^2:2-chamber system. Indeed the 3^2:2-chamber system is tight, whence its parabolics do not form a parabolic system in the meaning of [39]. Furthermore, $O^{p'}(P_i/O_p(P_i)) = S_3$ for each minimal parabolic P_i of the 3^2:2-chamber system, contrary to the hypotheses of Corollary 5.9 of [39].

Since the 3^2:2-chamber system is tight, every transitive chamber system having it as a rank 3 residue is tight. Note also that, if C is as in Theorem 5.1 or 5.2, then $P_3 \leq P_1P_2$. This inclusion forces C to be tight. Thus:

Corollary 5.4 *Let C be as in theorems* 5.1, 5.2 *or* 5.3, *but let C be non-tight. Then C is reducible.*

Problem. The 3^2:2-chamber system has a reducible universal 2-cover (see Subsection 4.4). Can we prove the same for any of the irreducible chamber system allowed by Theorem 5.1 ?

5.2   On Geometric Chamber Systems of Rank 3

We will draw some consequences from the previous theorems in Section 6. For the rest of this section we discuss some conditions sufficent for a chamber systems of rank 3 to be geometric. We will need them later.

Meixner and Timmesfeld [22] have proved that, if the diagram graph of a chamber system C is a string, then C is geometric if and only if it satifies (G1'). The following is a consequence of that theorem:

Lemma 5.5 *Let $C = (C, (\Phi_1, \Phi_2, \Phi_3))$ be a chamber system of rank 3 with $\Phi_1\Phi_3 = \Phi_3\Phi_1$, $\Phi_{1,2} \cap \Phi_{i,3} = \Phi_i$ for $i = 1, 2$ and $\Phi_i\Phi_2 \neq \Phi_2\Phi_i$ for $i = 1, 3$. Then C is geometric.*

(See [25], Corollary 13.) Henceforth we assume that C is a transitive chamber system admitting finite orders s, s, t with $s > 1$ and belonging to the diagram A_3 or C_3

$$(A_3) \quad \underset{s}{\overset{1}{\bullet}} \!\!\!-\!\!\!-\!\!\!- \underset{s}{\overset{2}{\bullet}} \!\!\!-\!\!\!-\!\!\!- \underset{s}{\overset{3}{\bullet}}$$

$$(C_3) \quad \underset{s}{\overset{1}{\bullet}} \!\!\!-\!\!\!-\!\!\!- \underset{s}{\overset{2}{\bullet}} \!\!\!=\!\!\!=\!\!\!= \underset{t}{\overset{3}{\bullet}}$$

(1, 2, 3 are the types.)

Theorem 5.6 *Let C be as above. Then C is one of the following:*

(i) a 3-dimensional projective geometry;
(ii) a rank 3 polar space;
(iii) the A_7-geometry (see Subsection 4.6);
(iv) an (unknown) geometric chamber system of type C_3 with s even, t odd, $14.10^6 < s < t$ and $P_{1,2}/K_{1,2} = Frob((s + 1)(s^2 + s + 1))$;
(v) an (unknown) non-geometric chamber system of type C_3 with $2s < t \leq s^2 - s$ and $1 + s + s^2$ dividing $(1 + t)(t - s - 1)$.

Proof. Let $\mathcal{C}$ be non-geometric. By Lemma 5.5, $P_{i,j} \cap P_{i,3} > P_i$, with $(i,j) = (1,2)$ or $(2,1)$. By [16], one of the following holds:

(a) s is a prime power, $\mathcal{C}_{1,2} = PG(2,s)$ and $P_{1,2}/K_{1,2} \geq L_3(s)$;
(b) s is even, $1 + s + s^2$ is prime and $P_{1,2}/K_{1,2} = Frob((1 + s)(1 + s + s^2))$ acting sharply flag-transitively on the projective plane $\mathcal{C}_{1,2}$.

In both cases P_h is maximal in $P_{1,2}$, for $h = 1, 2$. Hence $P_{i,j} \cap P_{i,3} = P_{1,2}$, since $P_{i,j} \cap P_{i,3} > P_i$ ($i = 1$ or 2). Therefore $P_{1,2} \leq P_{i,3}$. Since $P_j \leq P_{1,2}$ with $\{i,j\} = \{1,2\}$, $P_j \leq P_{i,3}$. Whence $P_{j,3} \leq P_{i,3}$ and:

(*) $|P_{i,3}|/|P_3| = (|P_{i,3}|/|P_{j,3}|) \cdot (|P_{j,3}|/|P_3|)$.

However, $|P_{1,3}|/|P_3| = 1 + s$ and $|P_{2,3}|/|P_3| = 1 + s + s^2$ or $(1 + s)(1 + st)$ according to whether $m = 3$ or 4. Substituting in (*), we see that a contradiction is avoided only if $(i,j) = (2,1)$ and $m = 4$. In this case we obtain $P_{1,2} \leq P_{2,3}$ and $|P_{2,3}|/|P_{1,3}| = 1 + st$. However, $|P_{2,3}|/|P_2| = (|P_{2,3}|/|P_{1,2}|) \cdot (|P_{1,2}|/|P_2|)$. Hence $1 + s + s^2$ divides $(1 + t)(1 + st)$. By standard manipulations, we obtain that $1 + s + s^2$ also divides $1 + t^3$. Therefore $1 + s + s^2$ divides $d(1 + t)$, with $d = g.c.d.(1 - t + t^2, 1 + st, 1 + s + s^2)$. It is not difficult to see that d divides $s + 1 - t$. Therefore $1 + s + s^2$ divides $(1 + t)(t - s - 1)$ and it is easily seen that this forces $s^2 \neq t > s$, and $t \leq s^2 - s$ (see [28]). Thus, $\mathcal{C}$ is as in (v).

Let $\mathcal{C}$ be geometric. If its diagram is A_3, then $\mathcal{C}$ is a projective geometry by Proposition 6 of [48]. If the diagram is C_3 and $\mathcal{C}$ is neither a polar space nor the A_7-geometry, then (iv) holds (see [53]). $\square$

Note that none of the known finite thick generalized quadrangles admits orders s, t satisfying relations as in (iv) or (v) of the previous theorem. In particular, $\mathcal{C}$ cannot be locally classical in those cases. Therefore:

Corollary 5.7 *Let $\mathcal{C}$ be locally classical. Then $\mathcal{C}$ is one of the following:*

(i) a 3-dimensional projective geometry;
(ii) a rank 3 polar space;
(iii) the A_7-geometry.

The previous statement had been obtained by Aschbacher [2] for geometries in the early eighties. A short time later, Timmesfeld [41] generalized Aschbacher's theorem to chamber systems.

6   Coxeter Diagrams of Rank ≥ 3

Throughout this section $\mathcal{C}$ is a transitive locally finite thick chamber system belonging to a Coxeter diagram **D** of rank $n \geq 3$ with set of types I and G is a transitive subgroup of $Aut(\mathcal{C})$. As in Subsection 2.1, $\mathcal{P}(G,\mathcal{C}) = (P_i)_{i \in I}$ is the system of parabolic subgrpups of G defined by a chamber c of $\mathcal{C}$ and B is the stabilizer of c. Given a subset J of I, $\mathcal{C}_J$ is the residue of $\mathcal{C}$ of type J containing c, $P_J = \langle P_j \rangle_{j \in J}$ is the stabilizer of $\mathcal{C}_J$ in G and K_J is the kernel of P_J on $\mathcal{C}_J$. The diagram induced by $\mathbf{D}$ on J will be denoted by $\mathbf{D}_J$.

6.1   Some Lemmas

Lemma 6.1 *Let $\mathcal{C}$ be locally classical, let $\mathbf{D}$ be connected and B finite. Let J be a triple of types such that $\mathbf{D}_J$ is disconnected but $\mathcal{C}_J$ is irreducible. Then $\mathbf{D}_J$ is trivial and one of the following holds:*

(i) $\mathcal{C}_J$ is the $3^2{:}2$-chamber system (see Subsection 4.3);
(ii) $\mathbf{D}$ is a bipartite graph with all edges of weight 4; furthermore, for any two types i, j joined in $\mathbf{D}$, the residues of type $\{i, j\}$ are isomorphic with the $S_4(3)$ generalized quadrangle or its dual and $P_{i,j}/K_{i,j} = 2^4{:}A_5$, $2^4{:}S_5$ or $2^4{:}Frob(20)$;
(iii) $\mathbf{D}$ is a star with all edges of weight 4

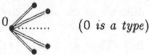
0 *(0 is a type)*

and, for every $i \in I - \{0\}$, the residue $\mathcal{C}_{0,i}$ is isomorphic with the $U_4(3)$ generalized quadrangle; $P_{0,i}/K_{0,i} = L_3(4).2$ or $L_3(4).2^2$.

Proof. This lemma is an easy consequence of the classification of transitive locally classical chamber systems with connected diagram and finite Borel subgroup, accomplished by Timmesfeld, Stroth and Meixner in a series of papers (Timmesfeld [39], [40], [41], [42], [43], [44], [45], Stroth [32], [33], [34], [35], [36], [37], [38], Meixner [18], [19], [20], [21]). In some of those papers some additional hypotheses were made. For instance, Stroth restricted his interest to non-tight chamber systems (sometimes, even to geometric chamber systems). Anyhow, a complete account of that classification, free from any additional hypotheses, is given by Meixner in [19]. We refer to that paper.

Let us prove Lemma 6.1. Let $\mathcal{C}$, G and J be as in the hypotheses of that lemma. It follows from [19] that (1)-(4) of Subsection 5.1 hold in $\mathcal{P}(P_J/K_J, \mathcal{C}_J)$ except when (ii) or (iii) occurs. If (1)-(4) of 5.1 hold, then (i) holds by Theorem 5.3. In this case $\mathbf{D}_J$ is trivial. Clearly, $\mathbf{D}_J$ is trivial also if (iii) holds. If (ii) holds, then $\mathbf{D}_J$ is trivial by Theorem 5.2. □

Let $\mathcal{C}$, $\mathbf{D}$ and B be as in Lemma 6.1. Assume furthermore that $\mathcal{C}$ is non-tight and that it is not as in (ii) or (iii) of Lemma 6.1. Then $\mathcal{C}$ is locally reducible (indeed $\mathcal{C}$ is tight if (i) of Lemma 6.1 holds for some triple J of types, because the $3^2{:}2$-chamber system is tight). Thus, the non-tightness assumption made by Stroth in many of his papers on the case of characteristic 2 is a way to keep off non locally reducible examples.

Corollary 6.2 *Let* C *be locally classical, let* D *be connected and let* B *be finite. Assume furthermore that none of the rank 3 residues of* C *is isomorphic with the* A_7*-geometry. Then the universal 2-cover of* C *is a building, except possibly when* C *is as in (ii) or (iii) of Lemma* 6.1.

Proof. Easy, by Corollary 5.7, Lemma 6.1 and Proposition 3.3, and because the universal 2-cover of the 3^2:2-chamber system is reducible (see 4.4). $\square$

Lemma 6.3 (Aschbacher) *Let* C *be locally classical and locally reducible and let* D *be a string of length* $n \geq 4$. *Then the none of the residues of* C *of rank 3 is isomorphic to the* A_7*-geometry.*

This is Theorem 2 of [2]. Actually, Achbacher stated that theorem only for geometries. However, the local reducibility is the only property of geometries he exploited in his proof.

Corollary 6.4 *Let* C *and* D *be as in the hypotheses of Lemma* 6.3. *Then the universal 2-cover of* C *is a building.*

Easy, by Lemma 6.3, Corollary 5.7 and Proposition 3.3.

Problem. I do not know if (ii) and (iii) of Lemma 6.1 are really exceptions for the statement of Theorem 6.2. The information available from [19] and [20] on those cases should be worked out in order to obtain some conclusions on the structure of P_J when D_J is trivial. I do not claim that this is extremely difficult to do. Nevertheless, it does not seem to be a trivial exercise.

6.2 Connected Spherical Diagrams

Theorem 6.5 *Let* D *be connected, of spherical type and rank* $n \geq 4$. *Then* C *is either a building or an (unknown) non-geometric chamber system of type* C_n *with* C_3 *residues as in (v) of Theorem* 5.6.

Proof. By Theorem 5.6, the residues of C of type A_3 are projective geometries and those of type C_3 are polar spaces or the A_7-geometry or non-geometric as in (v) of Theorem 5.6 (note that (iv) of Theorem 5.6 cannot occur for C_3 residues of transitive geometric chamber systems of type C_n with $n \geq 4$; see [17]).

   Assume that C is of type C_n with C_3 residues as in (v) of Theorem 5.6. Then $D = C_n$ or F_4. Let $D = F_4$, if possible. Only one of the two C_3-residues of $\mathcal{C}$ containg a given chamber can be as in (v) of Theorem 5.6. Therefore the other C_3-residue is either a polar space or the A_7-geometry. This forces C_2-residues of $\mathcal{C}$ to be classical, whereas C_2-residues cannot be classical in a chamber system as in (v) of Theorem 5.6. Hence $\mathbf{D} = C_n$.

Assume now that no C_3-residues of $\mathcal{C}$ are as in (v) of Theorem 5.6. Then $\mathcal{C}$ is locally classical. Furthermore, $\mathcal{C}$ is finite [51]. Thus we can apply Lemma 6.1, obtaining that $\mathcal{C}$ is locally reducible, except perhaps if some rank 3 residues of $\mathcal{C}$ were isomorphic with the $3^2{:}2$-chamber system. However, this possibility is ruled out by the following inductive argument.

Let a residue $\mathcal{C}_J$ be isomorphic with the $3^2{:}2$-chamber system, if possible. Then $n \geq 5$ and there is a set of types H containing two types $i, j \in J$ such that the diagram induced on H is A_m with $4 \leq m < n$. We have $\mathcal{C}_H \cong PG(m, q)$ for some prime power q, by induction, and $P_H/K_H \geq L_{m+1}(q)$, by a theorem of Higman [12]. Furthermore, $q = 2$ as $\mathcal{C}$ has order 2 at the types i and j and $P_{ij}/K_{ij} = 3^2{:}2$, as $\mathcal{C}_J$ is isomorphic with the $3^2{:}2$-chamber system. However $L_{m+1}(q)$ does not contain any parabolic subgroup P_{hk} of rank 2 and such that $P_{hk}/K_{hk} = 3^2{:}2$. A contradiction has been reached.

Therefore $\mathcal{C}$ is locally reducible. Hence all C_3 residues of $\mathcal{C}$ are polar spaces by Corollary 5.7 and Lemma 6.3. The universal 2-cover $\tilde{\mathcal{C}}$ of $\mathcal{C}$ is a building by Proposition 3.3. By Theorem 2.5, $\mathcal{C} = \tilde{\mathcal{C}}$. $\square$

By Theorem 6.5 and Corollary 5.7 we easily get the following well known result (see [41]; also [40]).

Corollary 6.6 *Let* $\mathbf{D}$ *be connected, of spherical type and rank* $n \geq 3$ *and let* $\mathcal{C}$ *be locally classical. Then* $\mathcal{C}$ *is either a building or the* A_7-*geometry.*

6.3   Affine Diagrams

Lemma 6.7 *Let* $\mathcal{C}$ *be locally classical and assume the following:*

(i) for every triple of types J *such that* $\mathbf{D}_J$ *has just one edge, there is a set of types* H *such that* $|H| \geq 3$, $|H \cap J| \geq 2$ *and* $\mathbf{D}_H$ *is a connected spherical diagram;*

(ii) for every triple of types J *with* $\mathbf{D}_J$ *trivial, there are sets of types* H *and* H' *such that* $H \cup H' \supseteq J$, $H \cap H' \cap J \neq \emptyset$ *and both* $\mathbf{D}_H$ *and* $\mathbf{D}_{H'}$ *are connected spherical diagrams;*

(iii) none of the residues of $\mathcal{C}$ *of rank 4 is isomorphic with the Wester chamber system (defined in 4.6);*

Then $\mathcal{C}$ *is locally reducible*

Proof. Let $J = \{i, j, k\}$ be a triple of types with $\mathbf{D}_J$ disconnected and let $\mathcal{C}_J$ be any of the residues of $\mathcal{C}$ of type J. Let $\mathcal{C}_J$ be irreducible, if possible.

We first assume that $\mathbf{D}_J$ is non-trivial. By Theorem 5.2, P_i/K_i, P_j/K_j and P_k/K_k are cyclic groups. On the other hand, this is impossible for at least two of those types, by (i), by Theorem 6.5 and by a theorem of Seitz [31]. We have reached a contradiction. Thus, $\mathbf{D}_J$ is trivial.

By (i), there is a set H of types containing at least two of the types of J and such that $\mathbf{D}_H$ is connected and spherical. By Corollary 6.6, by [31] and (ii), P_H/K_H is either a classical group or the alternating group A_7. In the latter case, $\mathcal{C}_H$ is either $PG(3,2)$ or the A_7-geometry. In any case, $\mathcal{C}_H$ is geometric.

Let P_H/K_H be classical, if possible. We can assume without loss that $i,j \in H$. Let $X = P_k \cap P_{ij}$. As $\mathcal{C}_J$ is irreducible, X properly contains B. Therefore $X = P_h$ for some $h \in H$, as P_H/K_H is a classical group. On the other hand, X is different from any of P_i and P_j, by (P2). Hence $X = P_h$ for some $h \in H - \{i,j\}$. However, $P_h \not\leq P_{ij}$, because $\mathcal{C}_H$ is geometric. We have reached a contradiction.

Therefore, if H is a set of types containing two of the types i, j, k and such that $\mathbf{D}_H$ is connected and spherical, then H has size 3, $P_H/K_H = A_7$ and $\mathcal{C}_H$ is either $PG(3,2)$ or the A_7-geometry. $\mathcal{C}$ has order 2 at each of the types i, j, k, each of the groups P_i/K_i, P_j/K_j, P_k/K_k is isomorphic with S_3 and each of the groups P_{ij}/K_{ij}, P_{jk}/K_{jk}, P_{ki}/K_{ki} is isomorphic with the subgroup $3^2{:}2$ of A_6 described in Subsection 4.3. Note that this group has just three subgroups isomorphic with S_3 and containing a given 2-subgroup.

We must rule out this case. As $\mathcal{C}_J$ is irreducible, $X = P_k \cap P_{ij}$ properly contains B. Furthermore, X does not contain any of P_i, P_j, by (P2). Therefore X/K_{ij} is one of the three subgroups of P_{ij}/K_{ij} isomorphic with S_3 and containing B/K_{ij} ($= Z_2$). P_i/K_{ij} and P_j/K_{ij} are the other two subgroups with those properties. On the other hand, B has index 3 in P_k, as $\mathcal{C}$ has order 2 at k. Hence $X = P_k$. It is now clear that $\mathcal{C}_J$ is the $3^2{:}2$-chamber system.

Let H contain i and j with $P_H/K_H = A_7$ and $\mathcal{C}_H$ isomorphic either to $PG(3,2)$ or to the A_7-geometry. We have already proved that $P_k \leq P_i P_j$. Hence $P_{ij} = P_J$, $K_H \leq K_k$ and $P_k \leq P_H$. Therefore $K_{H \cup \{k\}} = K_H$, $P_{H \cup \{k\}}/K_{H \cup \{k\}} = P_H/K_H = A_7$ and $P_J/K_H = P_{ij}/K_H = 2^2 \cdot (3^2.2)$. The groups P_i/K_H, P_j/K_H and P_k/K_H are the subgroups of A_7 called P_+, P_- and P_2 in Subsection 4.6 and $\mathcal{C}_{H \cup \{k\}}$ is the Wester chamber system. However, (iii) forbids this. $\square$

We assume that $\mathbf{D}$ is of affine type in the next theorem. We presume that the reader knows the list of affine Coxeter diagrams (anyhow, he can find it in [3], VI, 4.3).

Theorem 6.8 *Let $\mathcal{C}$ be locally classical and let $\mathbf{D}$ be of affine type. Then either the universal 2-cover of $\mathcal{C}$ is a building or $\mathcal{C}$ is the Wester chamber system.*

Proof. By Theorem 2.3, we can assume that $\mathcal{C}$ has rank $n \geq 4$. Let $\mathcal{C}$ be not the Wester chamber system. Then it is easy to check that (i), (ii) and (iii) of Lemma 6.5 hold in $\mathbf{D}$ and $\mathcal{C}$. Hence $\mathcal{C}$ is locally reducible (note that we cannot use Lemma 6.1 now, as we do not know if B is finite).

By Corollary 6.6 and Lemma 6.3, all residues of $\mathcal{C}$ of type A_3 are projective geometries and all residues of $\mathcal{C}$ of type C_3 are polar spaces, except possibly when $\mathbf{D} = \tilde{B}_3$ with uniform order 2 (in this case, some C_3-residues might be isomorphic to the A_7-geometry). Thus, if $\mathbf{D} \neq \tilde{B}_3$, then the universal 2-cover of $\mathcal{C}$ is a building, by Proposition 3.3.

Let $\mathbf{D} = \tilde{B}_3$. If none of the rank 3 residues of $\mathcal{C}$ is isomorphic with the A_7-geometry, then $\mathcal{C}$ is a 2-quotient of a building by Proposition 3.3. Thus, we assume that some residues of $\mathcal{C}$ are isomorphic with the A_7-geometry. Stroth [36] (Lemma 3.3) has proved that a locally classical transitive chamber system $\mathcal{C}$ of type $\tilde{B}_2$, admitting the A_7-geometry as a residue, must be tight. However, a stronger conclusion is implicit in his proof: namely, that $\mathcal{C}$ is just the Wester chamber system. (We warn that Stroth assumes the finiteness of B in [36], but he does not exploit that hypothesis in the proof of Lemma 3.3 of [36].) □

6.4   A Result on Disconnected Diagrams

Lemma 6.9 *Let $\mathcal{C}$ be locally classical and assume the following:*

(i) for every triple of types J with $\mathbf{D}_J$ disconnected, there is a set of types H such that $|H| \geq 3$, $|H \cap J| \geq 2$ and $\mathbf{D}_H$ is a connected spherical diagram;
(ii) for every triple of types J such that $\mathbf{D}_J = A_3$ or C_3, P_J / K_J is not the alternating group A_7.

Then $\mathcal{C}$ is locally reducible.

The proof is similar to that of Lemma 6.7. We leave it for the reader. The following thoerem can be proved by an argument quite similar to that employed for Theorem 6.8, but using Lemma 6.9 instead of Lemma 6.7.

Theorem 6.10 *Let $\mathcal{C}$ be locally classical, of rank ≥ 4, and let $\mathbf{D}$ have just two connected components J_1, J_2, of spherical or affine type. Assume also that, for every triple of types J with $\mathbf{D}_J = A_3$ or C_3, P_J / K_J is not the alternating group A_7. Then the universal 2-cover of $\mathcal{C}$ is a building.*

Remark. (i) of Lemma 6.9 forces $\mathbf{D}$ to admit at most two connected components. Note that (ii) of Lemma 6.7 is rather stronger. It implies that $\mathbf{D}$ is either connected or the disjoint union of two complete graphs.

Problem. What can we say when $\mathbf{D}$ has more than two connected components ?

References

[1] M. Aschbacher, *Flag structures on Tits geometries*, Geom. Dedicata, 14 (1983), 21-32.

[2] M. Aschbacher, *Finite geometries of type C_3 with flag-transitive automorphism groups*, Geom. Dedicata, 16 (1984), 195-200.

[3] N. Bourbaki, "Groupes et Algebres de Lie", chps. 4, 5 and 6, Actu. Sci. Ind. 1337, Hermann, 1968.

[4] A. Brouwer and A. Cohen, *Some reamrks on Tits' geometries*, Indag. Math. 45 (1983), 393-402.

[5] F. Buekenhout, *Diagrams for geometries and groups*, J. Comb. Th. A 27 (1979), 121-171.

[6] F. Buekenhout, *The basic diagram of a geometry*, in "Geometries and Groups", L.N. 893, Springer (1981), 1-29.

[7] F. Buekenhout, *Diagram geometries for sporadic groups*, in Finite Groups Coming of Age, A.M.S. series Contemp. Math. 45 (1985), 1-32.

[8] F. Buekenhout and A. Pasini, *Finite diagram geometries extending buildings*, Chapter 22 of "Handbook of Incidence Geometry" (F.Buekenhout ed.), to appear.

[9] J. Conway, R. Curtis, S. Norton, R. Parker and R. Wilson, "Atlas of Finite Groups", Oxford Univ. Press, 1985.

[10] W. Feit, *Finite projective planes and a question about primes*, to appear.

[11] W. Feit and D. Higman, *The nonexistence of certain generalized polygons*, J.Algebra, 1 (1964), 114-131.

[12] D. Higman, *Flag-transitive collineation groups of finite projective spaces*, Ill. J. Math., 6 (1962), 434-446.

[13] D. Higman and J. McLaughlin, *Geometric ABA-groups*, Ill. J. Math., 5 (1961), 385-397.

[14] W.Kantor, *Generalized polygons, SCABs and GABs*, in "Buildings and the Geometry of Diagrams", L.N. 1181, Springer (1986), 79-158.

[15] W. Kantor, *Some locally finite flag-transitive buildings*, European J. Comb., 8 (1987), 429-436.

[16] W. Kantor, *Primitive groups of odd degree and an application to finite projective planes*, J.Algebra, 106 (1987), 14-45.

[17] G. Lunardon and A. Pasini, *Finite C_n geometries (a survey)*, Note di Matematica, 10 (1990), 1-35.

[18] T. Meixner, *Klassische Tits kammersysteme mit einer transitiven automorphismengruppen*, Mitteilungen Math. Sem. Giessen, 174 (1986).

[19] T. Meixner, *Groups acting transitively on locally finite Tits chamber systems*, Finite Geometries, Buildings and Related Topics (W. Kantor, R. Liebler, S. Payne and E. Shult eds.), Oxford Univ. P. (1990), 45-65.

[20] T. Meixner, *Locally finite Tits chamber systems with transitive group of automorphism in characteristic* 3, Geom. Dedicata 35 (1990), 13-30.

[21] T. Meixner, *Parabolic systems: the $GF(3)$-case*, to appear in Trans. A.M.S.

[22] T. Meixner and F. Timmesfeld, *Chamber systems with string diagrams*, Geom. Dedicata, 15 (1983), 115-123.

[23] T. Meixner and M. Wester, *Some locally finite buildings derived from Kantor's 2-adic groups*, Comm. Alg. 14 (1986), 389-410.

[24] R. Niles, *BN-pairs and finite groups with parabolic-type subgroups*, J. Algebra, 75 (1982), 484-494.

[25] A. Pasini, *Reducibility of chamber systems and factorization of parabolic systems*, to appear in the Proceedings of the International Conference of Algebra held in Barnaul (1991).

[26] A. Pasini, *On thin chamber systems*, to appear in Rivista di Matematica Pura ed Applicata.

[27] A. Pasini, An Introduction to Diagram Geometry, Oxford Univ. Press, to appear.

[28] S. Payne and J. Thas, "Finite Generalized Quadrangles", Pitman, Boston, 1984.

[29] M. Ronan, *Coverings and automorphisms of chamber systems*, European J. Comb., 1 (1980), 259-269.

[30] M. Ronan, *Triangle geometries*, J. Comb. Th. A, 37 (1984), 294-319.

[31] G. Seitz, Flag-transitive subgroups of Chevalley groups, Ann. Math., 97 (1973), 27-56.

[32] G. Stroth, *Geometries of type M related to A₇*, Geom. Dedicata **20** (1986), 265-293.

[33] G. Stroth, *Some sporadic geometries having C_3-residues of type A_7*, Geom. Dedicata **23** (1987), 215-219.

[34] G. Stroth, *One node extensions of buildings*, Geom. Dedicata, **25** (1988), 71-120.

[35] G. Stroth, *A local classification of some finite classical Tits geometries and chamber systems of characteristic $\neq 3$*, Geom. Dedicata **28** (1988), 93-106.

[36] G. Stroth, *The nonexistence of certain Tits geometries with affine diagrams*, Geom. Dedicata **28** (1988), 278-319.

[37] G. Stroth, *Some geometry for McL*, Comm. Alg. **17** (1989), 2825-2833.

[38] G. Stroth, *Chamber systems, geometries and parabolic systems whose diagram contains only bonds of strength 1 and 2*, Invent. Math. **102** (1990).

[39] F. Timmesfeld, *Tits geometries and parabolic systems in finitely generated groups; I, II*, Math. Zeit. **1984** (1983), 377-396 and 449-487.

[40] F. Timmesfeld, *Tits chamber systems and finite group theory*, in Buildings and the Geometry of Diagrams (L.A. Rosati ed.), L.N. **1181**, Springer (1986), 249-269.

[41] F. Timmesfeld, *Tits geometries and revisionism of the classification of finite simple groups of characteristic 2 type*, Proc. Rutgers Group Theoretic Year, Cambridge U.P. (1984), 229-242.

[42] F. Timmesfeld, *On the "non-existence" of flag-transitive triangle geometries*, Coxeter Festschrift (I), Mitt. Math. Sem. Giessen **163** (1984), 19-44.

[43] F. Timmesfeld, *Tits geometries and parabolic systems of rank 3*, J. Algebra **96** (1985), 442-478.

[44] F. Timmesfeld, *Locally finite classical Tits chamber systems of large order*, Invent. Math. **87** (1987), 603-641.

[45] F. Timmesfeld, *Classical locally finite Tits chamber systems of rank* 3, J. Algebra **124** (1989), 9-59.

[46] F. Timmesfeld, private communication.

[47] J. Tits, "Buildings of Spherical Type and Finite BN-pairs", L. N. 386, Springer (1974).

[48] J. Tits, *A local approach to buildings*, in "The Geometric Vein", Springer (1981), 519-547.

[49] J. Tits, *Buildings and groups amalgamations*, London Math. Soc. L. N. 121 (186), 110-127.

[50] J. Tits, *Ensembles ordonnes, immeubles et sommes amalgamees*, Bull. Soc. Math. Belgique, A38 (1986), 367-387.

[51] S. Tsaranov, *Representation and classification of Coxeter monoids*, European J. Comb. 11 (1990), 189-204.

[52] M. Wester, *Endliche fahnentransitive Tits geometrien und ihre universellen uberlagerungen*,. Mitt. Math. Sem. Giessen **170** (1985).

[53] S. Yoshiara and A. Pasini, *On flag-transitive anomalous C_3 geometries*, to appear.

A. Pasini
Dept. Math., Univ. Siena, Via del Capitano 15, SIENA (Italy)
PASINI@SIVAX.CINECA.IT

EMBEDDINGS AND HYPERPLANES OF LIE INCIDENCE GEOMETRIES

Ernest E. Shult

1. EMBEDDINGS OF POINT-LINE GEOMETRIES

Embeddings. Let $\Gamma = (\mathcal{P}, \mathcal{L})$ be a **point-line geometry**, that is, an incidence system of points $\mathcal{P}$ and lines $\mathcal{L}$, such that distinct lines possess distinct point-shadows, thus allowing lines to be viewed as sets of points. A **projective embedding** of point-line geometry Γ into the projective space $\mathbf{P}(V)$ of all proper subspaces of the vector space V is an injective mapping $e : \mathcal{P} \to$ projective points of $\mathbf{P}(V) = $ 1-spaces of V such that

(1) $e(L)$ is a projective line for each line L of $\mathcal{L}$, and

(2) the image points $e(\mathcal{P})$ span $\mathbf{P}(V)$.

Such an embedding is denoted by the symbol $e : \Gamma \to \mathbf{P}(V)$.

Morphisms of embeddings. Let $\tau : \mathbf{V} \to \mathbf{W}$ be a semilinear transformation of vector spaces. This induces a partial mapping of the corresponding projective spaces $\mathbf{P}(V)$ and $\mathbf{P}(W)$, sending points of $\mathbf{P}(V)$ not contained in $\ker\tau$ in $\mathbf{P}(V)$ to projective points of $\mathbf{P}(W)$. With some abuse of notation, we denote this by $\tau : \mathbf{P}(V) \to \mathbf{P}(W)$. If $e : \Gamma \to \mathbf{P}(V)$ is a projective embedding of the point-line geometry Γ, then composition with the partial map τ can yield a new embedding $e\tau$ if and only if

(1) τ is a surjective semilinear transformation, and

(2) For any points p and q of Γ, $\ker\tau$ meets the subspace $< e(p), e(q) >$ at the zero subspace of V.

In this case we call the transfer from embedding e to embedding $e\tau$, a **morphism of embeddings** and write $e \to e\tau$. In general if we insist that $e \to e'$ is a morphism of embeddings, it means that $e' = e\tau$ for an appropiate semilinear transformation τ. If $\ker\tau$ is the zero vector space, e and e' are said to be **equivalent embeddings**. Morphisms can be composed in the obvious way, and the collection of all projective embeddings of Γ along with all their morphisms forms a category $\mathcal{E}$.

Universal Embeddings. An embedding $u : \Gamma \to \mathbf{P}(V)$ is said to be **relatively universal** if and only if the existence of a morphism $w \to u$ in category $\mathcal{E}$ implies w is equivalent to u. An embedding u is said to be **universal for** e if and only if u is relatively universal and there is a morphism $\kappa : u \to e$ with the property that for any morphism $\phi : w \to e$, there exists a morphism $\phi' : u \to w$ such that κ is the composition of ϕ and ϕ'. Clearly any two embeddings universal for e are equivalent; we call any of them **the universal hull** of e. An important theorem of Mark Ronan states that every embedding possesses a universal hull (see [15]).

There is another entirely different notion of "universalness" for an embedding of a point-line geometry. We say that an embedding $u : \Gamma \to \mathbf{P}(V)$ is **absolutely universal** if and only if for any embedding e, there is in $\mathcal{E}$ a morphism $\phi_e : u \to e$.

The two notions can be compared as follows: An absolutely universal embedding is a source in the category $\mathcal{E}$. The universal hull of embedding e is a source in the induced subcategory $\mathcal{E}_e$ of all objects sending an arrow into e. But if there is a morphism $e \to f$, both e and f have equivalent universal hulls. Thus (using Ronan's Theorem) the relatively universal embeddings are the sources of the connected components of $\mathcal{E}$ viewed as a digraph.

Automorphisms of the geometry which lift to the embedded geometry. Suppose σ is an automorphism of the point-line geometry Γ and that $e : \Gamma \to \mathbf{P}(V)$ is an embedding of the geometry. Then the composition σe is also an embedding of Γ. In this way $G = \mathrm{Aut}(\Gamma)$ acts as a group of natural transformations of $\mathcal{E}$. Clearly the set

$$G_e = \{\sigma \in \mathrm{Aut}(\Gamma)|\ \sigma e \text{ is equivalent to } e\}$$

is a subgroup of $\mathrm{Aut}(G)$. Moreover, each element of this group, when applied to the embedded point-line geometry $(e(\mathcal{P}), e(\mathcal{L}))$ lifts to an automorphism of the ambient projective space $\mathbf{P}(V)$. In this way, one obtains an injective morphism of groups: $G_e \to \mathrm{P\Gamma L}(V)$.

It is easily seen from Ronan's homological construction of the universal hull $\hat{e}$ of embedding e, that G_e is a subgroup of $G_{\hat{e}}$.

If u is an absolutely universal embedding, then $G_u = \mathrm{Aut}(\Gamma)$— that is, all automorphisms of Γ lift to automorphisms of the ambient projective space in which u embedds it.

A question. Of course, in general, there is nothing to insure that an absolutely universal embedding of a geometry exists, or that a relatively

universal embedding of a geometry admit many automorphisms at all. Yet
it seems to be the case that the classical Lie incidence geometries possess essentially one relatively universal embedding as a point-line geometry.
And it virtually always seems to be the case that the relatively universal
embedding admits the full automorphism group G of the Lie incidence
geometry—that is, the embedding is into $\mathbf{P}(M)$ where M is a particular
G-module. We would like to find why this is so from first principles which
do not involve any knowledge of weight-modules or any other paraphenalia
of the representation theory of Lie-type groups.

2. HYPERPLANES OF POINT-LINE GEOMETRIES

A **subspace** of a point-line geometry $\Gamma = (\mathcal{P}, \mathcal{L})$ is a subset X of $\mathcal{P}$,
such that any line with at least two of its points in X is in fact contained
in X. The subspace X is **proper** if it is not all of $\mathcal{P}$. A **hyperplane**
is a proper subspace which meets each line non-trivially. We let $\mathcal{V}$ denote
the set of all hyperplanes of Γ. In order to endow $\mathcal{V}$ with some sort of
geometric structure we consider the following three hypotheses.

(Veldkamp Points Exist) *Every hyperplane is a maximal subspace of Γ.*

(Veldkamp Lines Exist) *If A, B and C are hyperplanes of Γ with A
not contained in C, but $A \cap B \subseteq C$, then $A \cap C \subseteq B$.*

(Velkamp Planes Exist) *If A, B, C, and D are hyperplanes of Γ with
$A \cap B \cap C \subseteq D$ but $A \cap B$ is not contained in D, then $A \cap B \cap D \subseteq C$*

By allowing some of the hyperplanes to be equal, is is easy to see that
each of the above hypotheses implies its predecessors. In fact the general
hypothesis is

(Veldkamp $(r-1)$-spaces Exist, $r \geq 0$) *For any hyperplanes $A_1, \ldots, A_{r+1}$
such that $A_1 \cap \ldots \cap A_r \subseteq A_{r+1}$ but $A_1 \cap \ldots \cap A_{r-1} \nsubseteq A_{r+1}$ then
$A_1 \cap \ldots \cap A_{r-1} \cap A_{r+1} \subseteq A_r$*

One may recognize these axioms as versions of the exchange axiom of a
dependance relation, but limited to sets of small cardinality (see [7]).

A well-known sufficient condition that Veldkamp points exist is that
the induced point-collinearity graph on the complement of a hyperplane be
connected. Similarly Velkamp lines exist if it can be shown that $A - (A \cap B)$
always has a connected collinearity graph for any two hyperplanes A and

B. Similar criteria apply in general, and in this way it can easily by shown that Velkamp $(r-1)$-spaces exist for non-degenerate polar spaces of rank r. (See [16] or [3] for all of the assertions of this paragraph.)

The importance of the hypothesis that Veldkamp lines exist is that in its presence $\mathcal{V}$ acquires the structure of a linear space called the **Veldkamp space**. Its points are the hyperplanes of $(\mathcal{P}, \mathcal{L})$, and its lines are the intersections of pairs of distinct hyperplanes, incidence being containment. That any line is uniquely determined by any two of its points is exactly the assertion that Veldkamp lines exist.

Another important hypothesis is the following:

(Teirlink's Condition (compare with [22])) *If A and B are distinct hyperplanes of $\Gamma = (\mathcal{P}, \mathcal{L})$ and p is a point of $\mathcal{P} - (A \cup B)$, then there is a unique hyperplane C of Γ containing $\{p\} \cup (A \cap B)$.*

We now have

THEOREM 1. ([19]) *Suppose Veldkamp planes exists and that Teirlink's condition holds for Γ. Then the Veldkamp space $\mathcal{V}$ is a projective space.*

REMARK: Note that whenever Velkamp lines exist, Teirlink's condition can be weakened by deleting the word "unique". The proof of this theorem appears in working notes kindly supplied to me by Professor A. Pasini, who, in generalizing beyond polar spaces the argument that $\mathcal{V}$ is a projective space saw that the hypothesis that Veldkamp planes exist was needed, though it was not so named at the time.

3. EMBEDDINGS AND HYPERPLANES.
Hyperplanes Arising from an Embedding. Let $e : \Gamma \to \mathbf{P}(V)$ be a projective embedding of the point-line geometry $\Gamma = (\mathcal{P}, \mathcal{L})$. Suppose $\mathbf{H}$ is a hyperplane of $\mathbf{P}(V)$. Then it is easy to see that the set

$$H(\mathbf{H}) := \{x \in \mathcal{P}| \ e(x) \in \mathbf{H}\}$$

is a hyperplane of Γ. A hyperplane H of Γ is said to **arise from the embedding** e if and only if it has the form $H = H(\mathbf{H})$ for some projective hyperplane $\mathbf{H}$ of the ambient target space $\langle e(\mathcal{P}) \rangle = \mathbf{P}(\mathcal{V})$. We denote the set of all hyperplanes which arise from the embedding e by the symbol $\mathcal{V}_e$. As usual $\mathcal{V}$ will denote the set of *all* hyperplanes of Γ.

THEOREM 2. $\mathcal{V}_e = \mathcal{V}$ *implies that* e *is a relatively universal embedding.*

THEOREM 3. *Suppose* $e : \Gamma \to \mathbf{P}(V)$ *is an embedding of the point-line geometry* Γ *for which Veldkamp lines exist. Then* $\mathcal{V}_e$ *is a projective subspace of the linear space* $\mathcal{V}$.

Embeddings Constructed from $\mathcal{V}$. Suppose Veldkamp lines exist so $\mathcal{V}$ is a linear space. Given a point p of the geometry Γ, let $\mathcal{V}_p = \{H \in \mathcal{V}|\ p \in H\}$. When is $\mathcal{V}_p$ a hyperplane of $\mathcal{V}$?

THEOREM 4. *If Veldkamp lines exist and Teirlink's condition holds for* Γ, *then* $\mathcal{V}_p$ *is a hyperplane of the linear space* $\mathcal{V}$.

It now follows that if Veldkamp lines exist, if Tierlink's condition holds, and if, for some reason, $\mathcal{V}$ is a projective space, then there is an embedding

$$\hat{u} : \Gamma \to \mathcal{V}_0^*$$

where $\mathcal{V}_0^*$ is the subspace of $\mathcal{V}^*$ generated by hyperplanes $\mathcal{V}_p$ as p ranges over $\mathcal{P}$. In particular, from Theorem 1, we have

COROLLARY 1. *If Veldkamp planes exist and Teirlink's condition holds for* Γ *then the projective embedding* $\hat{u}$ *exists.*

Historically this was the original approach taken by Veldkamp for polar spaces of rank at least three—as recast by Buekenhout and Cohen in their book ([3]). All of the results listed so far in this section are either well-known or have proofs easily lifted from those applied to polar spaces in [3], [4], [14] and [10].

Question: Is $\hat{u}$ relatively universal?

The Hypothesis (A). We consider here the following hypothesis:

(A): (i) *Veldkamp lines exist for* Γ.

(ii) *There exists an embedding* $e : \Gamma \to \mathbf{P}(V)$ *with* $\mathcal{V}_e = \mathcal{V}$ *and with* $dim(\mathbf{P}(V))$ *finite.*

The reader should note that (ii) does not imply (i) (Certain classical generalized quadrangles are the counterexamples.).

We have

THEOREM 5. *Hypothesis* (A) *implies Teirlink's condition.*

The assumption that $\mathcal{V}_e = \mathcal{V}$ in part (ii) of Hypothesis (A) implies $\mathcal{V} \simeq (\mathbf{P}(V)^*)$ as projective spaces. So in (ii), $\mathcal{V}$ is a projective space of finite dimension. By the remark preceeding the Corollary 1, $\hat{u} : \Gamma \to \mathcal{V}_0^*$ is an embedding of Γ into a finite dimensional projective space.

At this point it will be useful to unfold a few consequences of Hypothesis (A) in the form of a series of Lemmas whose proofs are sketched when necessary.

LEMMA 1. *Assume part (ii) of Hypothesis* (A) *only. Then the following are equivalent:*

(a) *Veldkamp lines exist.*

(b) *For any subspace* $\mathbf{R}$ *of codimension at most two in* $\mathbf{P}(V)$, $\mathbf{R} = \{e(\mathcal{P}) \cap \mathbf{R}\}$ — *i.e.* $\mathbf{R}$ *is spanned by the image points which it contains.*

REMARK: In general, if $e : \Gamma \to \mathbf{P}(V)$ satisfies $\mathcal{V}_e = \mathcal{V}$ and Veldkamp $(r-1)$-spaces exist for Γ, then any subspace $\mathbf{R}$ of codimension at most r in $\mathbf{P}(V)$ is generated by the points of $e(\mathcal{P})$ within it (see [16]).

NOTATION: If $e : \Gamma \to \mathbf{P}(V)$ is an embedding of Γ and x is a projective point of $\mathbf{P}(V)$, set

$$H_x = \{H \in \mathcal{V}|\ \mathbf{P}(\langle e(H)\rangle) \text{ contains } x\}$$

Note that x need not be in $e(\mathcal{P})$.

LEMMA 2. *Assume Hypothesis* (A). H_x *is a hyperplane of* $\mathcal{V}$

SKETCH OF PROOF: Let L be the Veldkamp line generated by hyperplanes A and B of Γ—i.e. $L = \{H \in \mathcal{V} \mid H \supseteq A \cap B\}$. Now as $\mathcal{V} = \mathcal{V}_e$

we see that $A = H(\mathbf{A})$ and $B = H(\mathbf{B})$ for hyperplanes $\mathbf{A}$ and $\mathbf{B}$ of $\mathbf{P}(V)$ and there exists a bijection sending L into the set of hyperplanes of $\mathbf{P}(V)$ which contain $\mathbf{A} \cap \mathbf{B}$. One or all of these contain x. Finally H_x is not all of V since there exists a hyperplane complement to x in $\mathbf{P}(V)$.

LEMMA 3. *Assume* (A). *The mapping*

$$\phi : \mathbf{P}(V) \to V^*$$

defined by sending x to H_x, and sending lines to the intersection of the H_y, y ranging over the points of the line, is an isomorphism of projective spaces. Moreover $\phi(\mathbf{P}(V)) = V_0^*$.

SKETCH OF PROOF: Since $V = V_e$, we have an isomorphism $\psi :$ $(\mathbf{P}(V))^* \to V$ which maps any hyperplane $\mathbf{A}$ of $\mathbf{P}(V)$ to $H(\mathbf{A})$, and the collection of all hyperplanes over any codimension 2 subspace of $\mathbf{P}(V)$, to the Veldkamp line of V defined by the ψ-image of any two hyperplanes of this collection. Thus ψ is an isomorphism of projective spaces. Also ψ takes the collection of all projective hyperplanes on x onto the set H_x: the former is an element of $(\mathbf{P}(V))^{**}$, while the latter is an element of V^*. Noting that $\mathbf{P}(V)$ has finite dimension, we obtain a dual isomorphism

$$\psi^* : V^* \to (\mathbf{P}(V))^{**}.$$

Then ϕ is the inverse of the composition of ψ^* with the canonical isomorphism $eval : (\mathbf{P}(V))^{**} \to \mathbf{P}(V)$, using finiteness of dimension once again.

LEMMA 4. *Assume* (A). *Then ϕ induces an equivalence $e \to \hat{u}$.*

COROLLARY 2. *Assume* (A). *Then all embeddings f for which $V_f = V$ are equivalent to e (or $\hat{u}$).*

Now suppose $e : \Gamma \to \mathbf{P}(V)$ is as in Hypothesis (A)(ii). Let σ be an automorphism of Γ and let $\sigma e : \Gamma \to \mathbf{P}(V)$ be the composition of σ and e. Then clearly $V = V_{\sigma e}$. If Veldkamp lines exist (so that the full Hypothesis (A) is in force), it follows from Corollary 2 that $\sigma e \simeq \hat{u} \simeq e$

since σe is relatively universal by Theorem 2. But to say that e and σe are equivalent embeddings is to say that there is a semilinear transformation $\bar{\sigma} : \mathbf{P}(V) \to \mathbf{P}(V)$ such that $e \circ \bar{\sigma} = \sigma e$. Thus

THEOREM 6. *Assume Hypothesis* (A) *relative to the embedding* e . *Then* e *admits the group* $G = \mathrm{Aut}(\Gamma)$, *so there is an embedding* $G \to P\Gamma L(V)$. *Thus* V *is automatically a* G -*module.*

This result does not prove that e is absolutely universal—that is, that the category $\mathcal{E}$ is connected as a digraph. It does , however, show that the connected component of $\mathcal{E}$ which contains e is invariant under the group of natural transformations induced by G .

But now assume Γ satisfies Hypothesis (A) relative to the embedding $e : \Gamma \to \mathbf{P}(V)$. Then by Lemma 3 we have the isomorphism $V_0^* = V^* \simeq \mathbf{P}(V)$ of projective spaces. More precisely,

(*)   every hyperplane of the projective space V has the form H_x for some $x \in \mathbf{P}(V)$, and if the hyperpane H_X contains $H_y \cap H_z$, for distinct $y, z \in \mathbf{P}(V)$, then x is on the line yz of $\mathbf{P}(V)$.

Suppose now, that $f : \Gamma \to \mathbf{P}(W)$ is any arbitrary embedding in $\mathcal{E}$. Then by Corollary 2, $\mathcal{V}_f$ is a subspace of the projective Veldkamp space $\mathcal{V}$. There is then a vector space $V_0 \simeq V^*$ and a subspace V_f of V such that $\mathcal{V} = \mathbf{P}(V_0)$, $\mathcal{V}_f = \mathbf{P}(V_f)$, $\mathcal{V}^* = \mathbf{P}(V_0^*)$ and $\mathcal{V}_f^* = \mathbf{P}(V_f^*)$.

Since $V_f \leq V$, there is an exact sequence

$$0 \to K \to V^* \xrightarrow{\tau} V_f \to 0$$

where K consists of all functionals of V which vanish on V_f , – that is, $\mathbf{P}(K)$ is all hyperplanes of $\mathbf{P}(V)$ which contain $\mathbf{P}(V_f)$. Then τ induces a mapping

$$\tau : \mathcal{V}^* - \mathbf{P}(K) \to \mathcal{V}_f^*.$$

We claim that τ defines a morphism $\hat{u} \to f$ in the category $\mathcal{E}$. For this purpose we need only show that for any two points p and q , in $\Gamma = (\mathcal{P}, \mathcal{L})$, the vector space $\langle \hat{u}(p), \hat{u}(q) \rangle$ is disjoint from K .

Suppose by way of contradiction, that $X \in \mathbf{P}(K) \cap \langle \hat{u}(p), \hat{u}(q) \rangle_{\mathcal{V}^*}$. Thus, as a collection of geometric hyperplanes of $\mathcal{V}$, X contains all hyperplanes of Γ arising from f and also contains all hyperplanes of $\hat{u}(p), \hat{u}(q)$, or, equivalently, the full set of hyperplanes of Γ containing the line $L = pq$.

Now by (*), X has the form H_x for some point x in the projective line $e(L)$ generated by $e(p)$ and $e(q)$. Let $\mathbf{U}$ be a projective hyperplane of $\mathbf{P}(W)$ containing $f(q)$ but not $f(p)$. Then $H = f^{-1}(\mathbf{U} \cap f(\mathcal{P}))$ is a geometric hyperplane of Γ belonging to $\mathcal{V}_f$. But since $\mathcal{V} = \mathcal{V}_e$, H arises from e, and so $H = e^{-1}(\mathbf{H} \cap e(\mathcal{P}))$ for some projective hyperplane $\mathbf{H}$ of $\mathbf{P}(V)$. Since $q \in H$, $e(q)$ is in $\mathbf{H}$. Similarly, p is not in H, whence $e(p)$ is not in $\mathbf{H}$. On the other hand, since $H \in \mathcal{V}_f$, H is a member of the collection $X = H_x$, so $x \in H_x$ as well. Thus x, $e(p)$ and $e(q)$ are points of the projective line $e(L) = e(p)e(q)$, and $\mathbf{H}$ is a hyperplane of $\mathbf{P}(V)$ which contains x and $e(q)$ but not $e(p)$. Since $e(p) \neq e(q)$ this forces $x = e(q)$. But by repeating the argument replacing $\mathbf{U}$ by a projective hyperplane $\mathbf{U}'$ of $\mathbf{P}(W)$ containing $f(p)$ but not $f(q)$, one deduces $x = e(p)$ as well. Clearly both possibilities are impossible, so no such X exists. It follows that τ induces a morphism of embeddings $\hat{u} \to u\tau$, where $\hat{u}\tau$ embedds Γ into $\mathcal{V}_f^*$.

It remains only to show that $\hat{u}\tau$ is equivalent to f. But the map $\tau : \mathcal{V}^* - \mathbf{P}(K) \to \mathcal{V}_f^*$ is effected by taking any hyperplane A^* of $\mathcal{V}$ which does not contain $\mathcal{V}_f$ and intersecting it with $\mathcal{V}_f$ to yield a hyperplane $A^* \cap \mathcal{V}_f$, an element of $\mathcal{V}_f^*$. Thus $\hat{u}\tau$ takes each point p to $H_p \cap \mathcal{V}_f$. On the other hand f is equivalent to $f \circ f^*$ (the composition of $f : \Gamma \to \mathbf{P}(W)$ and the projective-space isomorphism $f^* : \mathbf{P}(W) \to \mathcal{V}_f^*$), which takes each point x of $\mathbf{P}(W)$ to the collection

$$\{f^{-1}(\mathbf{A} \cap f(\mathcal{P})) | f(x) \in \mathbf{A} \in \mathbf{P}(W)^*\},$$

which, for $p \in \Gamma$ yields the same set $H_p \cap \mathcal{V}_f$.

Thus there is a morphism $\hat{u} \to f$, and as f was arbitrary, $\hat{u}$ is absolutely universal.

Thus:

THEOREM 7. *Assume Γ satisfies Hypothesis(A) with respect of the embedding e. Then e is absolutely universal.*

So the question raised in Section 1, whether embeddings of Lie incidence geometries must come from G-modules leads to two fundamental questions:

(1) When do Veldkamp lines exist for these incidence geometries?

(2) When is there an embedding e of the geometry such that all hyperplanes arise from it?

In the next two sections we review what is known concerning these two questions.

4. EXISTENCE OF VELDKAMP LINES AND PLANES.

Flats. For subsets $B, A_1, \ldots A_k$ in a family $\mathcal{F}$ of subsets of a set X, we say that B **depends on** $\{A_1, \ldots A_k\}$ if and only if $A_1 \cap \ldots \cap A_k \subseteq B$. This notion of dependence satisfies the first two laws of a dependence theory as expounded in Cohn ([7]): namely, the reflexive and transitive laws. What is missing that would give $\mathcal{F}$ the full strucure of a dependence theory or matroid is the exchange axiom. In fact if $X = \mathcal{P}$ and $\mathcal{F}$ is the collection $\mathcal{V}$ of all hyperplanes of $\Gamma = (\mathcal{P}, \mathcal{L})$, then the assertion that Veldkamp $(r-1)$-spaces exist is the assertion that

If C depends on $\{A_1, \ldots, A_{r-1}\}$ but not $\{A_1, \ldots, A_{r-2}\}$, then A_{r-1} depends on $\{A_1, \ldots, A_{r-2}, C\}$.

Let $I = H_1 \cap \ldots \cap H_k$ be an intersection of hyperplanes H_i of Γ . Then the set

$$\mathrm{fl}(I) := \{H \in \mathcal{V} | \ H \supseteq I\}$$

of all hyperplanes depending on I is called a **flat**.

It is easy to see that if Veldkamp lines exist, then a flat $\mathrm{fl}(I)$ is a subspace of the Veldkamp space $\mathcal{V}$. But even when $\mathcal{V}$ is a projective space (as when also $\mathcal{V} = \mathcal{V}_e$), $\mathrm{fl}(I)$ need *not* be the subspace *generated* by the Veldkamp points $H_1, \ldots H_k$, although it must contain the latter. In fact, the assertion that Veldkamp $(r-1)$-spaces exist is precisely the assertion that flats $\mathrm{fl}(J)$ *are* the subspaces generated by the Veldkamp points $H_1, \ldots H_k$, whenever the latter are $k \leq r$ independant hyperplanes whose intersection is J .

EXAMPLE: Veldkamp planes do not exist for the Grassmannians $\Gamma = \mathcal{G}_k(V)$ (see [19]). (If the Grassmannian is non-embeddable, $\mathcal{V}$ is a linear space, but is not even projective ([13]). If the Grassmannian is embeddable, the embedding $e : \Gamma \to \mathbf{P}(W)$ where W is the k-fold wedge product of the V's, is a G-admissable embedding, where $G = \mathrm{Aut}(\Gamma) = P\Gamma L(V)$, so under the isomorphism $\mathbf{P}(W)^* \to \mathcal{V}$, $\mathcal{V}$ becomes a projective G-module with a special G-invariant system of flats that includes all lines, but not all planes.

A special class of strong parapolar space. In order to obtain a fairly uniform proof that Veldkamp lines or planes exist, the following families of geometries were introduced in [16].

Let n be an integer greater than one. A point-line geometry $\Gamma = (\mathcal{P}, \mathcal{L})$ is in the class $\mathcal{E}_n$ if and only if it satisfies the following axioms:

(E_1) Γ is a connected Gamma space.

(E_2) Any geodesic path $(p_1, \ldots, p_k)$ in the point-collinearity graph $(\mathcal{P}, \sim)$ can be extended to one of length n; moreover $\operatorname{diam}(\mathcal{P}, \sim) = n$.

(E_3) For any point p, the set $\Delta^*_{n-1}(p)$ of points at distance at most $n-1$ from p in $(\mathcal{P}, \sim)$ forms a hyperplane of Γ.

(E_4) If p and q are two points of Γ at distance d in $(\mathcal{P}, \sim)$, $2 \leq d \leq n$, the convex closure $\langle p, q \rangle_\Gamma$ (i.e. the intersection of all subspaces S containing p and q which are convex subsets of $(\mathcal{P}, \sim)$) is a subgeometry belonging to $\mathcal{E}_d$.

Note that the class $\mathcal{E}_2$ is the class of non-degenerate polar spaces. Thus if p and q are two points at distance 2 in $(\mathcal{P}, \sim)$, then then $\langle p, q \rangle_\Gamma$ is a convex subspace which is a polar space—i.e. a **symplecton**. By (E_4) any distance-2 pair of points lies in some symplecton, so the geometries of $\mathcal{E}_n$ are strong parapolar spaces of point-diameter n.

KNOWN EXAMPLES

Example	symplectic rank	n
1. Polar spaces	≥ 2	2
2. Near Polygons with quads (see [21])	2	≥ 2
3. Grassmannians $A_{2n-1,n}(D)$	3	≥ 2
4. Half-spin geometries $D_{n,n}(F)$, n even	4	≥ 2
5. $E_{7,1}(F)$	6	3

In example 3, D is a division ring, while in examples 4 and 5, F is a field. Example 5 is a Lie incidence coset geometry whose points are the cosets of the maximal parabolic subgroup P of the algebraic group $E_7(F)$ associated with the terminal node of the longest arm of the Dynkin diagram.

Now define the family $\mathcal{F}_n$, $n \geq 2$ as follows: A point-line geometry $\Gamma = (\mathcal{P}, \mathcal{L})$ is in $\mathcal{F}_n$ if and only if

(F_1) diameter $(\mathcal{P}, \sim)$ is $n \geq 2$.
(F_2) If $d(x,y) = d \leq n$, then $\langle x,y \rangle_\Gamma \in \mathcal{E}_d$

We list below the known examples of these geometries beyond the members of $\mathcal{E}_n$ that are not polar spaces, near polygons with quads or a product geometry $\Gamma_1 \times \Gamma_2$, $\Gamma_i \in \mathcal{F}_{n(i)}$.

EXAMPLES OF GEOMETRIES IN $\mathcal{F}_n$

Example	symplectic rank	n
1. Grassmannians $A_{n,k}(D)$, $k \leq (n+1)/2$	3	k
2. Half-spin geometries $D_{n,n}(F)$	4	$[n/2]$
3. $E_{6,1}(F)$	5	2
4, $E_{7,1}(F)$	6	3

THEOREM 8.([16],[19]) *Let* $\Gamma \in \mathcal{F}_n$, $n \geq 2$ *and suppose* Γ *has symplectic rank at least* r .

(i) *If* $r \geq 3$, *Veldkamp lines exist.*
(ii) *If* $r \geq 4$, *Veldkamp planes exist.*

COROLLARY 3. *In the list of examples of geometries in* $\mathcal{F}_n$, *examples 1-4 have Veldkamp lines, while examples 2-4 have Veldkamp planes.*

REMARK: The proof of Theorem 8 immediately reduces to proving it for geometries Γ in $\mathcal{E}_n$ rather than $\mathcal{F}_n$. The result that Veldkamp $(r-1)$-spaces exist for polar spaces is proved in the book of Buekenhout and Cohen ([3])—although not exactly in these terms—and is used throughout the proof of this Theorem.

Strangely, there seems to be some obstacle to the proof of

CONJECTURE: *If* $\Gamma \in \mathcal{F}_n$ *has symplectic rank* $r \geq 4$, *then Veldkamp* $(r-1)$ *-spaces exist.*

Open Questions and Side-issues.

1. Prove the conjecture.

2. Are all geometries in $\mathcal{E}_n$ with symplectic rank 2 necessarily near polygons?

3. Aside from the classical examples (dual polar spaces) are there any geometries in $\mathcal{E}_n$, $n > 3$, with symplectic rank 2?

4. Are there any further examples of geometries in $\mathcal{E}_n$, with symplectic rank at least 3, aside from those listed above? Evidence suggesting the answer might be negative is offered by

Proposition (El-Atrash and Shult, unpublished, [12]) *Assume* Γ *is a member of* $\mathcal{F}_n$, $n \geq 2$ *having symplectic rank at least three.*

(i) *Suppose* $\Gamma \in \mathcal{E}_3$ *and that two symplecta never meet in exactly a single point. Then* Γ *is the Grassmanian* $A_{5,2}(D)$ *or* $E_{7,1}(F)$.

(ii) *Suppose* Γ *satisfies the weak hexagon property:*

(WH) *If* $(x_0, \ldots, x_5)$ *is a 6-circuit in* $(\mathcal{P}, \sim)$ *with* $d(x_i, x_{i+3}) = 3$ *for at least one subscript* i *(mod 3), then there exists a point* p *in* $x_0^\perp \cap x_2^\perp \cap x_4^\perp$ *or in* $x_1^\perp \cap x_3^\perp \cap x_5^\perp$.

Then Γ *is either a polar space, a Grassmannian, a half-spin geometry,* $E_{6,1}(F)$, *or* $E_{7,1}(F)$.

(The proof uses the fundamental characterization of Lie incidence geometries due to Cohen and Cooperstein ([6])

5. Do the other classical Lie incidence geometries of symplectic rank ≥ 3 (such as the polar Grassmannians, the metasymplectic spaces, the exceptional geometries $E_{6,4}, E_{7,7}$, or $E_{8,j}$, j indexing an end-node) possess Veldkamp lines?

5. WHEN DO HYPERPLANES ARISE FROM EMBEDDINGS?

We are concerned here with the second of the two questions raised at the end of Section 3: When is it true that $\mathcal{V} = \mathcal{V}_e$?

(Of course we must assume e is relatively universal (it is necessary by Theorem 2, and there is no loss in the assumption by Ronan's construction)).

In the case of the non-degenerate polar spaces, the answer is "always". This follows from the fact that the relatively universal embeddings of these geometries are completely known from the theorems of Buekenhout-LeFevre,

Dienst and Lemma 8.6 of Tits' book ([2], [11], [23]). All of these embeddings are the natural defining embeddings with an appropriate sesquilinear or pseudoquadratic form. Any hyperplane H is itself a (possibly degenerate) polar subspace, and if H is degenerate it arises from a hyperplane of the form $p^\perp$. Otherwise H is nondegenerate and by the quoted results the restriction of the embedding to H is also a "natural" one. The only known case where the embedded H still spans the target space for e requires e not to be universal.

For embeddings e of the other Lie incidence geometries, there are three methods for showing that $V = V_e$, which have been successful.

METHOD 1. (Circuitry) By results of Ronan ([15]), for a strong parapolar space, it is sufficient to show that for any hyperplane H of Γ, that in the subgraph of $(\mathcal{P}, \sim)$ induced on $\mathcal{P} - H$, every circuit C is a sum of triangles and 4-circuits. Since this property does not depend on the particular embedding, it means $V = V_e$ for *every* relatively universal embedding e. (It will then follow from Corollary 2 that if Veldkamp lines exist, e is absolutely universal without invoking Theorem 7.)

METHOD 2. (Inductive construction of a functional) If $e : \Gamma \to \mathbf{P}(V)$ is the embedding, we wish to show that for each hyperplane H of Γ, there exists a functional $h : V \to F$ of V such that $e \circ h$ vanishes on H but never vanishes on $\mathcal{P} - H$. Inductively there exists a family $\mathcal{S}$ of subgeometries of Γ belonging to a parameterized family of geometries containing Γ and such that the restriction of e to S in $\mathcal{S}$ is still relatively universal. Then for each $S \in \mathcal{S}$, there exists by induction a functional

$$h_S : \langle e(S) \rangle := W_S \to F$$

which vanishes on $e(S \cap H)$ but not on $e(S - H)$ (if the latter is empty, of course, $h_S = 0$). The problem is to patch the h_S together to produce the desired functional h. (To do this, there is a constant adjustment of the h_S by scalar multiples that involves further circuitry problems—this time on a graph with vertex set $\mathbf{P}(V)$.) This was the method used in [18].

METHOD 3. (The direct sum method) This method also uses induction in a different way. We assume as before that $e : \Gamma \to \mathbf{P}(V)$ is relatively universal. Again Γ is assumed to belong to a family of geometries parameterized by an integer-valued function τ. The theorem being proved asserts that for some function $\delta : \mathbf{Z} \to \mathbf{Z}$, if $\dim V \geq \delta(\tau(\Gamma))$, then (a) $\dim V = \delta(\tau(\Gamma))$ and (b) $\mathcal{V} = \mathcal{V}_e$. The geometries Γ must have the property that they always possess two subgeometries S_1 and S_2 in the same parameterized family satisfying

    (1) $\langle S_1, S_2 \rangle_\Gamma = \mathcal{P}$ (subspace generation in Γ), and

    (2) $\delta(\tau(S_1)) + \delta(\tau(S_2)) \leq \delta(\tau(\Gamma))$.

Induction on the S_i immediately yields the fact that $V = \langle e(S_1) \rangle \oplus \langle e(S_2) \rangle$ and that $\dim \langle e(S_i) \rangle_V = \delta(\tau(S_i))$. By induction, the hyperplane H can be assumed to meet each S_i properly at a hyperplane H_i of S_i. One sets X to be all points x of H such that $e(x)$ is not in the subspace $U := \langle e(H_1) \rangle \oplus \langle e(S_2) \rangle$ of codimension 2 in V. For $x, y \in X$, one writes

$$x \sim y \text{ if and only if } U \oplus \langle x \rangle = U \oplus \langle y \rangle.$$

The rest of the proof consists in showing that the graph $(X, \sim)$ is connected.

Summarizing known results, we have:

Geometry	Reference	Method
1. Projective spaces	Folklore	uniqueness
2. Polar spaces of rank ≥ 2	Buekenhout-LeFevre, Dienst, Tits	hard work
3. $E_{6,1}(F), D_{5,5}(F)$ and $A_{n,2}(F)$	Cooperstein and Shult [8], [9]	circuitry
4. $A_{n,k}, 2 \leq [(n+1)/2]$	Shult [18]	functional
5. Spin geometries (Dual polar spaces of type $\Omega(2n+1, F)$, $F = GF(q), q$ odd.)	Shult and Thas [20]	direct sum
6. Half spin geometries, $D_{n,n}(F)$	Shult [17]	direct sum

OPEN CASES:

    1.) Other dual polar spaces (types $\Omega^-(2n, F), \Omega(2n+1, F)$, with F infinite of odd characteristic, $n \geq 3$.

2.)   Other exceptional geometries: Metasymplectic spaces, $E_{6,4}, E_{7,7}$ and $E_{8,j}$ where j corresponds to a terminal node of the E_8 diagram.

3.)   Grassmann versions of the above (where points are cosets of a maximal parabolic subgroup corresponding to a non-terminal node).

6. EMBEDDINGS AND MODULES.

We put a few things together. We now know that Hypothesis (A) holds for the embeddable Grassmannians, the half-spin geometries, and the exceptional geometry $E_{6,1}$, where, in the respective cases V is the appropriate wedge product space, the half-spin module, or e is the universal hull of the embedding afforded by the module of dimension 27. Thus in each case the embedding is in fact $G = \mathrm{Aut}(\Gamma)$-admissable (which was known for these classical weight modules anyway) and, by Theorem 7, *all embeddings of these point-line geometries are morphic images of the classical one* (which was known to hold for the Grassmannians, half-spin geometries and odd-characteristic spin geometries, by a beautiful result of Albert Wells ([24])).

It would be especially nice to know whether all hyperplanes arise from an embedding in the case of the geometry $E_{7,1}$ where Veldkamp lines are already known to exist. Either part of Hypothesis (A), the existence of Veldkamp lines or the fact that $\mathcal{V} = \mathcal{V}_e$ for some embedding e, remains completely unknown for the parapolar spaces which are not strongly parapolar, such as $F_{4,1}$, $^2E_6$ (as a metasymplectic space), $E_{7,7}$, or $E_{8,j}$ with j marking a terminal node.

REFERENCES

[1] Buekenhout, F., Cooperstein's theory. *Simon Stevin*, (1983). **57**: 125-140.

[2] Buekenhout F. and LeFevre, C. Generalized quadrangles in projective spaces *Arch. Math.* (1983), **25**: 540-552.

[3] Buekenhout, F. and Cohen, A., *Diagram Geometries* To be published.

[4] Cohen, A. Point-line spaces related to buildings, in *Handbook of Incidence Geometry*, F. Buekenhout (Editor). To appear, 1993.

[5] Cohen, A. M. Point-line characterizations of buildings, in *Buildings and the Geometry of Diagrams: Como 1984*, L. Rosati, Editor. 1991, Oxford University Press, Springer Verlag, Berlin.

[6] Cohen, A. M. and Coopersein, B. N., A characterization of some geometries of exceptional Lie type. *Geom. Ded* (1983), **15**: 73-105.

[7] Cohn, P. M. *Algebra, vol. 2*, 1977, Wiley, London New York.

[8] Cooperstein, B.N. and Shult, E. E., Geometric hyperplanes of embeddable Lie incidence geometries, in *Advances in Finite Geometries and Designs*, J.W.P.Hirschfeld, D.R. Hughes and J.A. Thas Editors. 1991, Oxford University Press (Clarendon Press): Oxford. pp. 81-91.

[9] Cooperstein, B. N. and Shult, E. E.., Geometric hyperplanes of Lie incidence geometries. 1992, in manuscript.

[10] Cuypers, H., Johnson, P. and Pasini, A. Synthetic foundations of polar geometry. *Geom. Ded.* To appear.

[11] Dienst, K.J., Verallgemerte Vierecke in projectiven Raumen. *Arch. Math.*, (1980), **45**: 177-186.

[12] El-Atrash, M. and Shult, E., Characterizations of certain strong parapolar spaces, unpublished.

[13] Hall, J. I. and Shult, E. E., Geometric hyperplanes of non- embeddable Grassmannians, *Eur. J. Comb.*, (1993) **14**:29-35.

[14] Johnson, P. Personal communication regarding a forth-coming ms.

[15] Ronan, M.A., Embeddings and hyperplanes of discrete geometries, *Europ. J. Comb.*, (1987),**8**: 179-185.

[16] Shult, E.E., Veldkamp lines. Submitted.

[17] Shult, E.E., Geometric hyperplanes of the half-spin geometries. To appear in *Simon Steven*.

[18] Shult, E.E. Geometric hyperplanes of embeddable Grassmannians. *J. Alg.*(1992), **145**: 55-82.

[19] Shult, E.E., On Veldkamp planes, in *Finite Geometry and Combinatorics*, Lond. Math. Soc. Lec. Note Series no. 191, F. DeClerke *et al*, Editors, 1993, Cambridge U. Press. Cambridge, pp. 341-354.

[20] Shult, E.E., and Thas, J. A., Hyperplanes of dual polar spaces and the spin module. *Arch. Math* (1992), **59**: 610-623.

[21] Shult, E.E. and Yanushka, A. Near n-gons and line systems. *Geom. Ded.* (1980), **9**: 1-72.

[22] Teirlink, L. On projective and affine hyperplanes, *J. Combinatorial Theory (A)* (1980), **28**: 290-306.

[23] Tits, J. *Buildings of Spherical Type and Fine BN-pairs* Lecture Notes in Math, No.389, 1974, , New York-Berlin: Verlag Springer.

[24] Wells, A., Universal projective embeddings of the Grassmann, half-spinor, and dual orthogonal geometries, *Quart. J. Math. Oxford* (1983), **34**: 375-386.

INTERMEDIATE SUBGROUPS IN CHEVALLEY GROUPS

Nikolai Vavilov

Department of Mathematics and Mechanics
University of Sanct-Petersburg
Petrodvorets, 198904, Russia

and

Fakultät für Mathematik
Universität Bielefeld
33615, Bielefeld, Deutschland

Let G be a group and D be a subgroup of G. In this paper we are interested in the description of the lattice

$$\mathcal{L}(D, G) = \{H \mid D \leq H \leq G\}$$

of subgroups of G which contain D. We call subgroups $H \in \mathcal{L}(D, G)$ subgroups **intermediate** between D and G. We are primarily interested in the parametrization of intermediate subgroups and their behaviour with respect to set-theoretic and group-theoretic operations like intersection, conjugation, passage to normalizer of H or to normal closure of D in H, etc.

1991 *Mathematics Subject Classification.* 20G15, 20G35, 20H20, 20H25, 20D06, 20E15.

Key words and phrases. Chevalley groups, classical groups, lattice of subgroups, Aschbacher classes, maximal tori, semisimple subgroups, conjugacy theorems, abnormal subgroups, pronormal subgroups.

The author gratefully acknowledges support of the *Alexander von Humboldt-Stiftung* and *SFB 343 an der Universität Bielefeld* during the preparation of the present work.

We will speak about the structure of $\mathcal{L}(D, G)$ as opposed to the study of other classes of subgroups, like, for example, subgroups, generated by a given type of elements/subgroups (say, transvections, reflections, root elements, ...), defined by group theoretic properties (say, abelian, nilpotent, soluble, ...), ...

Little can be said about the structure of the lattice $\mathcal{L}(D, G)$ in general. In fact it is proven in [Tu] that *any* algebraic lattice is isomorphic to an interval in the subgroup lattice of an *infinite* group. This is why, to get sensible results, one has to specialize the class of groups considered. It turns out that in amazingly many examples arising in real life the lattice $\mathcal{L}(D, G)$ has a very transparent structure. There are some abstract conditions on a subgroup D, which are responsible for that (see § 10 and 14).

In our talk the group G will be a Chevalley group or its extension by diagonal automorphisms, usually over a field. Many results are only known for the classical groups, and some will be stated only for the general linear group to save space. We concentrate on results which are valid for *arbitrary* fields, not just finite or algebraically closed. This is why we speak only of *normal* types. While for a finite group theorist there is a good reason to consider also the twisted groups, for us they are no better than all other forms of semisimple algebraic groups.

Although we say something about generalizations to *commutative* rings, we do not discuss here results which *essentially* depend on the nature of the ground ring, or use some deeper ring theory. For the linear and unitary groups there are non-commutative generalizations of some of the results in the present paper, which depend on stability conditions, finiteness conditions, or non-commutative localizations. On the other side, for arithmetical rings there is a theory with many results which have no immediate counterparts in the theory we discuss. We leave all these things beyond the horizon. Some indications may be found in [HO], [PR], [Va16], [Va19], [Z2].

Several general remarks are in order here. First, one should note that many of the results we state below are easy – sometimes trivial – for finite fields. However for infinite fields they may be extremely difficult. People working in finite groups do not usually fully realize how much the finiteness condition – not even the Classification, but such trivial things as counting arguments or Sylow's theorem – make life easier. The most striking example of this is perhaps the theorem of O.King [Ki2] on the overgroups of diagonal subgroup in SL_2.

In some other cases the situation is strictly opposite: the results are easy to prove for infinite fields using Zariski topology and 'general position' arguments – which deliver no information whatsoever for the finite case. Sometimes it is easy to account for both the finite and the infinite case using *different* proofs for finite and infinite cases. However a 'natural' proof should cover *all* cases when a result holds by the same method. In many situations such natural proofs are still lacking.

Another point is that we describe *all* overgroups of a given group and not just the irreducible ones. Moreover we describe them up to inclusion – not just up to conjugacy. In many situations applying the classification of irreducible subgroups containing a long root unipotent element, a quadratic element or the like, from the very start we know the *irreducible* subgroups from $\mathcal{L}(D, G)$ *up to conjugacy*. However our work only starts rather than finishes at this point. One conjugacy class of such overgroups may lead to parametric families. Even the reduction of the general case to the irreducible one is not always trivial.

This survey complements [Kn], [V5], [V16], [V17], [V19], [Z1] – [Z3] rather than updates them. Many more details about the finite groups of Lie type as well as the corresponding algebraic groups, and especially on their maximal subgroups, may be found in [A1] – [A4], [KL1], [KL2], [Kn], [LSa2], [LSe1] – [LSe3], [S6], [S10] – [S15], [T]. A wealth of information concerning linear case may be found in [Z1] – [Z3]. Parabolic subgroups are discussed in [V5]; overgroups of tori in [V16], [V17]; overgroups of subsystem subgroups in [V16]. Consult [K3], [L2], [L6] for different views of the subject. In [V19] one can find general background on Chevalley groups over rings and many additional references. The works [BKH], [BV3], [HV], [Sv], [V14], [V15], [V18] also contain extensive lists of related references, not fully reproduced here.

§ 1. Notation

In this section we introduce the notation used in the rest of this paper. See [Bo], [C1], [H2], [St], [V19] for more details and further references.

Let Φ be a reduced root system of rank l, P be a lattice lying between the root lattice $Q(\Phi)$ and the weight lattice $P(\Phi)$. From this data one can construct an affine group scheme $G_P(\Phi, \ )$ over $\mathbb{Z}$, such that for any algebraically closed field K the value $G_P(\Phi, K)$ of this functor on K is the semisimple algebraic group over K corresponding to the pair Φ, P, which is called the *Chevalley-Demazure group scheme of type* (Φ, P). Let $T = T_P(\Phi, \ )$ be a split maximal torus of the Chevalley-Demazure group scheme $G_P(\Phi, \ )$. The values $G_P(\Phi, R)$ and $T_P(\Phi, R)$ of these functors on a commutative ring R with 1 ("the groups of rational points $G_P(\Phi, \ )$ and $T_P(\Phi, \ )$ with the coefficients in R") are called the *Chevalley group of type* (Φ, P) over R and its *split maximal torus* respectively. Usually we omit P in the notation and speak about a "*Chevalley group* $G = G(\Phi, R)$ *of type* Φ *over* R".

Let $L = L_{\mathbb{C}}$ be a complex semisimple Lie algebra of type Φ. Fix an order on Φ with Φ^+, Φ^- and $\Pi = \{\alpha_1, \ldots, \alpha_l\}$ being the sets of *positive, negative* and *fundamental roots* respectively. Let $\{e_\alpha, \alpha \in \Phi; h_\alpha, \alpha \in \Pi\}$ be a *Chevalley base* of L and $L_{\mathbb{Z}}$ be its the integral span. Now for an arbitrary commutative ring R we set $L_R = L_{\mathbb{Z}} \otimes_{\mathbb{Z}} R$.

Let now $\pi: L \to gl(V)$ be a representation of the complex semisimple Lie algebra $L = L_{\mathbb{C}}$ in a finite dimensional vector space V over $\mathbb{C}$. We often omit the symbol π in the action of G on V and for $x \in L$ and $u \in V$ write xu instead of $\pi(x)u$. Let $V_{\mathbb{Z}}$ be an *admissible* lattice in V (i.e. $V_{\mathbb{Z}}$ is invariant under the action of divided powers $e_\alpha^{(m)} = e_\alpha^m/m!$). Set $x_\alpha(\xi) = x_\alpha^\pi(\xi) = \exp(\xi\pi(e_\alpha))$, where $\alpha \in \Phi$, $\xi \in R$, and 'exp' is defined by the usual formula. For arbitrary commutative ring R we set $V_R = V_{\mathbb{Z}} \otimes R$. The group

$$E_\pi(\Phi, R) = \langle x_\alpha(\xi), \alpha \in \Phi, \xi \in R \rangle \leq GL(V_R).$$

is the *elementary Chevalley group of type* Φ *over* R in the representation π. Thus if the lattice of weights $P(\pi)$ of the representation π equals P, then the groups $E_P(\Phi, R)$ and $G_P(\Phi, R)$ from the very start appear together with their representation in the free R-module $V_R = V_{\mathbb{Z}} \otimes R$, which we denote by the same letter π.

The group G_π is called *simply connected* if $P(\pi) = P(\Phi)$ and *adjoint* if $P(\pi) = Q(\Phi)$. We write G_{sc} and G_{ad} for simply connected and adjoint groups respectively. For the cases $\Phi = E_8$, F_4 and G_2 one has $P(\Phi) = Q(\Phi)$ and thus $G_{sc} = G_{ad}$. However from our point of view these groups behave like simply connected groups of other types, rather than like the adjoint ones.

When $R = K$ is a field the Chevalley group $G_\pi(\Phi, K)$ is generated by $E_\pi(\Phi, K)$ and certain semisimple elements. In many important cases, notably, when K is an algebraically closed field, or when $G_\pi = G_{sc}$ is simply connected, one has $G_\pi(\Phi, K) = E_\pi(\Phi, K)$. However in general it is not true: adjoint groups are strictly larger than their elementary subgroups. For example, the adjoint group of type A_l coincides with $PGL(l+1, K)$, while its elementary subgroup coincides with $PSL(l+1, K)$. One can lift this distinction to the level of simply connected groups by considering 'diagonal extensions' of G_{sc}, the so called *extended Chevalley groups*, see [BMo], [V17]. Thus, for example, the extended (simply connected) group of type A_l coincides with $GL(l+1, K)$.

To any root $\alpha \in \Phi$ there correspond *unipotent root elements* $x_\alpha(\xi)$, $\xi \in K$. If we want to stress that these root unipotents correspond to a given choice of a split maximal torus we call them *elementary root unipotents*. In general any conjugate of an elementary root unipotent is called a *root unipotent*. It is *long* or *short* depending on whether the root α is long or short.

Let now $\alpha \in \Phi$ and $\epsilon \in K^*$. As usual we set $h_\alpha(\epsilon) = w_\alpha(\epsilon)w_\alpha(1)^{-1}$, where $w_\alpha(\epsilon) = x_\alpha(\epsilon)x_{-\alpha}(-\epsilon^{-1})x_\alpha(\epsilon)$. The elements $h_\alpha(\epsilon)$ – and their conjugates – are called *semisimple root elements*. They are called *long* or *short* depending on whether the root α is long or short.

For a fixed $\alpha \in \Phi$ the map $x_\alpha \colon \xi \mapsto x_\alpha(\xi)$ is a homomorphism of the additive group R^+ of R to the one-parameter subgroup $X_\alpha = \{x_\alpha(\xi) \mid \xi \in R\}$, which is called the *elementary unipotent root subgroup* corresponding to α. In fact x_α is an isomorphism of R^+ on X_α. When it does not lead to a confusion we omit the epithets "elementary" and "unipotent" and speak about *root elements* and *root subgroups*.

Let now $N = N(\Phi, K)$ be the subgroup of G, generated by $T = T(\Phi, K)$ and all the elements $w_\alpha(1)$, $\alpha \in \Phi$. It is very well known that if $|K| \geq 4$ the group N coincides with the normalizer of T in G (see [St]). The quotient group N/T is canonically isomorphic to the Weyl group W and for any $w \in W$ we fix a preimage n_w of w in N. Usually when we speak about subgroups containing T we tend to identify w and n_w.

Recall that we have fixed an order of Φ, with Φ^+ being the corresponding set of positive roots respectively. Set

$$U = U(\Phi, R) = \langle x_\alpha(\xi),\ \alpha \in \Phi^+,\ \xi \in R \rangle.$$

The product $B = \mathrm{B}(\Phi, R)$ of the groups T and U is called the *standard Borel subgroup* of G (corresponding to the given choice of T and Φ^+), U is the *unipotent radical of B*.

§ 2. ASCHBACHER CLASSES

A convenient framework for systematization of known results on intermediate subgroups is provided by Aschbacher classes $\mathcal{C}_1 - \mathcal{C}_8$. These classes were introduced in the study of maximal subgroups of the finite classical groups in [A1]. The groups from these classes are 'obvious' maximal subgroups of a finite classical group. The famous subgroup structure theorem of Aschbacher says that a maximal subgroup is either in one of these classes or in the class $\mathcal{S}$ consisting of almost simple groups in certain absolutely irreducible representations (subject to some further restrictions, see [A1], [KL2]).

The maximality of groups from Aschbacher classes was completely studied *modulo the Classification* by P.Kleidman and M.W.Liebeck, see [KL1], [KL2], where one can find all the details and many further references (of course, for many of these groups their maximality was established long before independently of the Classification).

Many of these classes have natural analogues also in exceptional groups. They arose in the subgroup structure theorem for the finite exceptional groups, proven by G.M.Seitz, M.W.Liebeck and others, compare [LSe1] and references there. See also [S12], [S13], [S15], [Te], [LSe3] for the general picture of maximal subgroups in exceptional groups.

Here we are interested in subgroups with non-trivial lattice of overgroups. Most of the large subgroups of Chevalley groups studied so far are obtained by dropping those additional conditions in the definition of Aschbacher classes, which guarantee maximality. We can also combine constructions of several Aschbacher classes and still get large subgroups. Here we informally describe how it works. Let V be a minimal representation of a group G of characteristic p, say the natural representation of a classical group. Then the classes $C_1 - C_8$ are *roughly* described as follows (see [A1] and [KL2] for the details):

C_1: Stabilizers of subspaces $U < V$;

C_2: Stabilizers of direct sum decompositions $V = U_1 \oplus \ldots \oplus U_t$ into <u>similar</u> summands;

C_3: Stabilizers of field extensions of <u>prime</u> degree;

C_4: Stabilizers of tensor decompositions $V = U_1 \otimes U_2$;

C_5: Stabilizers of subfields of <u>prime</u> degree;

C_6: Normalizers of some l-subgroups, $l \neq p$;[1]

C_7: Stabilizers of tensor decompositions $V = U_1 \otimes \ldots \otimes U_t$ into <u>similar</u> factors;

C_8: Classical subgroups.

It is clear what we should do to get large subgroups: we have to omit extra conditions (the underlined ones) introduced to enforce maximality. The classes C_1 and C_2 are fused and we get summand-wise stabilizers of direct sum decompositions $V = U_1 \oplus \ldots \oplus U_t$. In the classical cases these are basically (but not exclusively) subsystem subgroups. The classes C_4 and C_7 are fused and give us factor-wise stabilizers of tensor decompositions $V = U_1 \otimes \ldots \otimes U_t$. In the definitions of C_3 and C_5 we have only to drop the word 'prime'.

We can also combine procedures used in forming various Aschbacher classes to get other large subgroups. For example, we can fix a direct decomposition $V = U_1 \oplus \ldots \oplus U_t$ and then instead of taking the whole stabilizer of this decomposition consider field extension subgroups on each summand, etc.

In this paper we discuss at some length the classes $C_1 - C_3$ and their combinations (sections 3 – 18). This is a very interesting class of subgroups, which includes for example, maximal tori and subsystem subgroups. These

---

[1]It is not immediately clear what should be a correct analogue of the class C_6 for infinite fields (not to say rings) and we do not discuss this class any further.

classes are discussed for all Chevalley groups, or, at least, for all classical Chevalley groups.

Due to the lack of time (during the talk) and space (in this survey) we can not discuss the classes $C_4 + C_7$, $C_5 + C_8$ and S with the same amount of details. We restrict ourselves to stating a handful of results (sections 19 – 21) showing that the same patterns for distribution of intermediate subgroups apply also in these cases. We do not even attempt to properly describe the classes themselves for any group other than GL_n. There is of course a more serious reason for this: the definitive results for most of these classes in the natural generality are still not available. However the results are under way and the author hopes to be able to address this topic in later publications.

§ 3. STANDARD SUBGROUPS

In this section we describe what can be considered an analogue of the Aschbacher classes $C_1 + C_2$ for all Chevalley groups. Sections 4 – 13 will be dedicated to the analysis of overgroups of certain large groups from this class.

The groups U and U^- considered above are special cases of groups $E(S) = E(S, R)$ which we introduce now for any closed subset S in Φ. Recall that a subset S in Φ is called *closed* if for any two roots $\alpha, \beta \in S$ such that $\alpha + \beta \in \Phi$, one has $\alpha + \beta \in S$. Define $E(S) = E(S, R)$ as the subgroup generated by all the elementary root subgroups X_α, $\alpha \in S$, with respect to T:

$$E(S) = E(S, R) = \langle x_\alpha(\xi), \ \alpha \in S, \ \xi \in R \rangle.$$

Then U and U^- coincide with $E(\Phi^+)$ and $E(\Phi^-)$ respectively. The groups $E(S)$ are particularly important when the set S is *special* (alias *unipotent*), i.e. $S \cap (-S) = \emptyset$. In this case $E(S)$ is just the product of all X_α, $\alpha \in S$, in any fixed order. Set $G(S) = T(\Phi, R)E(S)$.

Suppose that $S \subseteq \Phi$ is any closed set of roots. Then S is the disjoint union of its *reductive* (alias *symmetric*) part S^r which consists of $\alpha \in S$ such that $-\alpha \in S$ and its *unipotent* part S^u which consists of $\alpha \in S$ such that $-\alpha \notin S$. The set S^r is a closed subsystem of roots while the set S^u is special. Moreover S^u is an *ideal* in S, i.e. if $\alpha \in S$, $\beta \in S^u$ and $\alpha + \beta \in \Phi$, then $\alpha + \beta \in S^u$. We avoid the common notations S^+ and S^- since we prefer to reserve them for $S^+ = S \cap \Phi^+$ and $S^- = S \cap \Phi^-$. It is easy to see that the group $G(S)$ is the semidirect product of the reductive subgroup $G(S^r)$ (a *Levi subgroup* of $G(S)$) and the unipotent subgroup $E(S^u)$ (the *unipotent radical* of $G(S)$) and analogously $E(S)$ is the semidirect product of $E(S^r)$ and $E(S^u)$.

Two sets of roots $S_1, S_2 \subseteq \Phi$ are called *conjugate* if there exists an element w of the Weyl group $W = W(\Phi)$ such that $wS_1 = S_2$. If the sets S_1 and S_2 are conjugate then there exists an $n \in N$ such that $nG(S_1)n^{-1} = G(S_2)$. We say that an element $w \in W$ normalizes S if $wS = S$. The set $X(S)$ of all $w \in W$ which normalize S is called the *Weyl normalizer* of S. It is clear that $X(S)$ contains the Weyl subgroup $W(S) = W(S^r)$. Moreover in the case when $S = S^r$ is a root subsystem $X(S)$ coincides with the normalizer of $W(S)$ in W (see [C2]). Denote by $N(S)$ the subgroup of G generated by $G(S)$ and n_w, $w \in X(S)$. A theorem of Tits (see [T2]) implies that almost always $N(S)$ coincides with the normalizer of $G(S)$ in G. The only cases when this equality may fail are those of the fields $K = \mathbb{F}_2, \mathbb{F}_3$.

§ 4. PARABOLIC SUBGROUPS

The first example (for many the *only* one) where one sees intermediate subgroups is the description of parabolic subgroups due to J.Tits [T1], [Bu].

Recall that we have fixed an order on the root system which determines Π, Φ^+ and Φ^-. A *standard parabolic* subset P is a closed set of roots containing Φ^+. The standard parabolic subsets are pair-wise non-conjugate and correspond bijectively to all of the subsets $J \subseteq \Pi$ of the fundamental system. Namely if $J \subseteq \Pi$ is such a subset, then we may define P_J to be the smallest closed set of roots containing Φ^+ and $-J$. The most important parabolic subsets are the maximal ones. A maximal parabolic subset corresponds to a set $J = J_r$, $1 \leq r \leq l$, which contains all the fundamental roots apart from α_r. The corresponding parabolic set P_{J_r} is maximal among the closed subsets and will be denoted P_r.

Standard parabolic subgroups of the Chevalley group $G = G(\Phi, K)$ are the subgroups which contain the standard Borel subgroup $B = B(\Phi, K)$ while *parabolic subgroups* are those conjugated to the standard parabolic ones. A classical theorem of Tits describes the lattice $\mathcal{L}(B, G)$.

Theorem. *The map $P_J \mapsto G(P_J)$ is a bijection between the sets of standard parabolic subsets and of standard parabolic subgroups.*

What is remarkable about this theorem is that there are no exceptions, i.e. it applies to *all* fields. This is quite unusual. Most of the analogous results for subgroups other than B break down for some small characteristics or cardinalities of the ground field.

Another theorem of Tits asserts that overgroups of B have remarkable properties with respect to conjugation.

Definition. *A subgroup B is called* **abnormal** *in G if for any $x \in G$ one has $x \in \langle B, xBx^{-1} \rangle$.*

It is easy to see that saying that B is abnormal in G is equivalent to saying that

i) Every subgroup containing B is self-normalizing,

ii) Any two distinct subgroups containing B arc not conjugate.

Now another main result of [T1] may be stated as follows.

Theorem. B *is abnormal in* G.

The first examples of abnormal subgroups of which one would think, are, of course,

– Sylow normalizers, i.e. normalizers of the Sylow subgroups in G.

– Carter subgroups of finite soluble groups.

Actually in the case when characteristic of K is positive, B is precisely the normalizer of U, which is a Sylow p-subgroup of G.

§ 5. SUBGROUPS OF THE BOREL SUBGROUP

The next example is that of overgroups of T in B. This description may be summarized in the following theorem. Throughout the section we assume that

$K \neq \mathbb{F}_2$ for $\Phi = A_l$, $l \geq 4$; $\Phi = E_l$;

$K \neq \mathbb{F}_2, \mathbb{F}_3$ for $\Phi = A_3$, B_l, C_l, $l \geq 2$, D_l, $l \geq 3$, F_4, G_2;

$K \neq \mathbb{F}_2, \mathbb{F}_4$ for $\Phi = A_2$;

K is perfect if char $K = 2$ and $\Phi = C_l$, $l \geq 1$.

Recall that $A_1 = C_1$, $B_2 = C_2$ and $D_2 = C_1 \times C_1$. In this paper we *always* treat the group $SL(2, K)$ as being *symplectic*, not linear. The reason is that like for other symplectic groups (but unlike $SL_{n \geq 3}$) conjugating a long root element $x_\alpha(\xi)$ by an element from T we can only multiply ξ by ε^2 for some $\varepsilon \in K^*$, whereas for all other cases we can multiply ξ by any element from K^*. This is why squares play such a role in the theory of symplectic groups and non-perfect fields of characteristic 2 always pop up as exceptions. Actually to state the first theorem as it is, we have to assume moreover that

char $K \neq 2$ if $\Phi = B_l$, C_l, $l \geq 2$, F_4,

char $K \neq 2, 3$ if $\Phi = G_2$.

However if we replace closed sets of roots by quasi-closed ones [BT], the theorem below remains valid without these last assumptions. The same applies to the two other theorems. The example which follows describes the lattice $\mathcal{L}(T, B)$.

Theorem. *The map* $S \mapsto G(S)$ *is a bijection between the set of closed subsets of* Φ^+ *and the set of subgroups of* B *containing* T.

This result was stated under different disguises in [S2], [CPS], [B1], [B2], [Sz1], [V1], [Az], [V21]. Analyzing the proofs of this result, we see that they give also the following result.

Theorem. *T is abnormal in B.*

The results show that the behaviour of T as a subgroup of B is very similar to the behaviour of B as a subgroup of G. The only difference is that while the results of Tits hold for *arbitrary* fields, here exceptions appear: it is easy to check that all excluded cases are actually exceptional.

Any overgroup of an abnormal subgroup is itself abnormal. Since every abnormal subgroup of G contained in B must contain T, the results above clearly imply

Theorem. *B is a minimal abnormal subgroup in G.*

Actually this result holds also for most of the excluded cases. For example, this is obviously true for $K = \mathbb{F}_2$, since in this case $B = U$ is nilpotent. The cases $K = \mathbb{F}_3, \mathbb{F}_4$ require somewhat more careful analysis.

The importance of Borel subgroups suggests that it might be interesting to study minimal abnormal subgroups. Every maximal subgroup is either normal or abnormal. Classification of minimal abnormal subgroups might give a new view of the maximal subgroups.

Problem. *Classify minimal abnormal subgroups of finite groups of Lie type up to conjugacy.*

§ 6. OVERGROUPS OF SPLIT MAXIMAL TORI

In § 4 and § 5 we described the lattices $\mathcal{L}(B, G)$ and $\mathcal{L}(T, B)$. Now we combine these results and describe the lattice $\mathcal{L}(T, G)$. This result is already *very* much more difficult, than the previous ones. In this and following sections we give just a vague idea of how it works. One can find many further details and an almost exhaustive bibliography in the surveys of A.E.Zalesskii [Z2], A.S.Kondratiev [Kn] and the author [V16], [V17] (see also [V10], [V14], [V18], [HV]).

Throughout the section we assume that

$|K| \geq 7$ if $\Phi = A_l$, $l \geq 2$, D_l, $l \geq 3$ or if G is adjoint of type $\neq E_8, F_4, G_2$;

$|K| \geq 13$ otherwise.

The numbers 7 and 13 in these conditions are explained as follows: in the proof we must pick up a nonzero element ε distinct from 5 prohibited values. But when we work with SL_2 the *square* of this element should also avoid these 5 values, this then what prohibits altogether eleven values. Thus the proof should work starting from the field of 12 elements. For safety we make it 13.

Moreover we make the same assumption on the characteristic of the ground field as in § 5. Again these assumptions are not necessary if we are ready to accept quasi-closed sets of roots (see [BT]) instead of the closed ones in the statement (of course, we must keep the condition from § 5 for the symplectic case).

Theorem. *For any intermediate subgroup F, $T \leq F \leq G$, there exists a unique closed set of roots $S \subseteq \Phi$ such that*

$$G(S) \leq F \leq N(S).$$

It is easy to check that the fields K, $|K| \leq 5$, are exceptions for all the cases and that the fields K, $|K| \leq 11$, are exceptions for the symplectic case (see [Ko1], [VD]). It is quite plausible though, that the theorem holds for all non-symplectic cases whenever $|K| \geq 7$ (this conjecture has been formulated in [V6], [V17]; in [S4] G.M.Seitz posed the problem of the description of overgroups of a split maximal torus for all fields K, $|K| \geq 4$).

In a classical paper [BT] by A.Borel and J.Tits this theorem was proven for the case of an algebraically closed field K. Of course in [BT] only *closed connected* subgroups of G were considered (without any restrictions on the characteristics whatsoever) and it was proven that they are exhausted by $G(S)$ for quasi-closed sets of roots $S \subseteq \Phi$. But in fact a classical theorem of Chevalley (see, for example [H1] or [Sp]) implies that for an algebraically closed field any subgroup containing T is closed in the Zariski topology.

For the group $\mathrm{GL}(n, K)$ (i.e. an extended Chevalley group of type A_{n-1}) this theorem was proven in 1976 in a paper of Z.I.Borewicz [B3], see also [BV1], [V2]. In [V4], [V8] other proofs of this (and somewhat stronger) results were proposed, based on the calculations with Bruhat and Bruhat-like decompositions.

The next important step was a paper by G.M.Seitz [S3] published in 1979, where he proved that the standard description holds for a finite field K such that char $K \neq 2$ and $|K| \geq 13$. The arguments in [S3] used the finiteness of the ground field in a very essential way and they could not be generalized to an infinite field. In fact such deep results as, say, theorems on 2-fusion were used (this explains the restriction imposed on the characteristic).

In 1979–80 analogous results were obtained by the author and E.V.Dybkova for extended Chevalley groups of types C_l and D_l, i.e. the general symplectic group $\mathrm{GSp}(2l, K)$ and the general orthogonal groups $\mathrm{GO}(2l, K)$ (see [VD], [V3]).

In the same paper [VD] the author and E.V.Dybkova proved the following reduction result for the symplectic group. Let char $K \neq 2$ and $|K| \geq 7$. Then the description of overgroups of a split maximal torus in $\mathrm{Sp}(2l, K)$ is standard if and only if the description of overgroups of a split maximal torus in $\mathrm{SL}(2, K)$ is standard.

Analogous results for the extended Chevalley group of type B_l i.e. for the odd orthogonal groups $GO(2l + 1, R)$ were obtained at the same time but were not published until [V14]. Actually it was noted in [V14] that the results of [V3] remain valid for the ordinary orthogonal groups $SO(n, R)$. Of course this is due to the fact that the groups $SO(n, R)$ are already extended by some (though not all in general) diagonal automorphisms.

The theorem for the adjoint groups of types E_6 and E_7 has been proven by the author. A sketch of its proof is presented in [V11], [V12], [V20] (see also [Va17] and references there).

A method which allowed us to prove the theorem for simply connected groups of non-symplectic types over an infinite field was developed by the author in the period 1983–1989. A crucial point was that for all non-symplectic groups one could reduce the problem to the *extended* Chevalley groups of smaller rank and for the extended groups the analogous problems had already been solved.

For the special linear group $SL(n, K)$, $n \geq 3$, over an infinite field K this proof has been published in the first part of [V6], while the subsequent parts contained a proof of standardness under assumption $|K| \geq 7$. For $\Phi = D_l$ standardness under this assumption is proven in [V18].

The case SL_2 was the last one to be solved and in § 9 we discuss the proof for this case. This proof shows that analysis of *abstract* subgroups over an *infinite* field may be quite a challenge even for such small groups as $SL(2, K)$.

§ 7. THE GROUP GL_n

Here we prove the theorem from § 6 for *adjoint* groups of type A_{n-1}. Instead of working in $PGL(n, K)$ itself we prefer to handle the corresponding *extended* simply connected group $GL(n, K)$. The *natural* proof for the general linear group, obtained in [Bo3], [BV1], [V2], works for all fields K, $|K| \geq 7$, (see also [V4], [V8], where this proof is transcribed in terms of Bruhat decomposition). Here we reproduce this proof in a slightly modified form.

Let $G = GL(n, K)$ be the general linear group and $D = D(n, K)$ be its subgroup of diagonal matrices. It is easy to prove that if there is a counter-example to the standard description of subgroups in G containing D, then there is a *primitive irreducible* counter-example. Thus it remains only to prove that if H is a primitive irreducible subgroup in G which contains D but is not contained in N, then it coincides with G. By McLaughlin's theorem it suffices to prove that H contains a non-trivial elementary transvection (then H contains a root subgroup and the normal subgroup in H, generated by it, either coincides with $SL(n, K)$, or is conjugate to $Sp(n, K)$, but no conjugate of $GSp(n, K)$ is normalized by D for $n \geq 4$).

Thus it remains to prove that if x is a non-monomial matrix, then the subgroup $X = \langle D, x \rangle$, generated by D and x contains an elementary transvection. The proof is based on the analysis of pseudo-reflections contained in X. Recall that e is the identity matrix of degree n and e_{ij}, $1 \leq i, j \leq n$, is a standard matrix unit, i.e. the matrix which has 1 in the position (i, j) and zeros elsewhere. Let now $d_r(\varepsilon) = e + (\varepsilon - 1)e_{rr}$, $1 \leq r \leq n$, $\varepsilon \in K^*$, be an 'elementary' pseudo-reflection. Fix some r and consider the conjugate $y = y(\varepsilon) = x d_r(\varepsilon) x^{-1}$ of $d_r(\varepsilon)$ by x:

$$
y(\varepsilon) = \begin{pmatrix}
1 + \alpha_1(\varepsilon - 1)\beta_1 & \alpha_1(\varepsilon - 1)\beta_2 & \cdots & \alpha_1(\varepsilon - 1)\beta_n \\
\alpha_2(\varepsilon - 1)\beta_1 & 1 + \alpha_2(\varepsilon - 1)\beta_2 & \cdots & \alpha_2(\varepsilon - 1)\beta_n \\
\vdots & \vdots & & \vdots \\
\alpha_n(\varepsilon - 1)\beta_1 & \alpha_n(\varepsilon - 1)\beta_2 & \cdots & 1 + \alpha_n(\varepsilon - 1)\beta_n
\end{pmatrix}
$$

where $\alpha_i = x_{ir}$ are the entries of the r-th column of x, while $\beta_j = x'_{rj}$ are the entries of the r-th row of the inverse matrix $x^{-1} = (x'_{ij})$.

Now fix some s, $1 \leq s \leq n$. The s-th rows of matrices $x d_r(\varepsilon) x^{-1}$ and $x d_r(\eta) x^{-1}$ are proportional, apart from the diagonal entries. But the diagonal entries may be corrected by multiplication by some $d_s(\theta)$. Indeed, choose $\varepsilon, \eta \in K^*$ in such a way, that $\varepsilon - 1, \eta - 1, 1 + \alpha_s(\varepsilon - 1)\beta_s, 1 + \alpha_s(\eta - 1)\beta_s$, $\varepsilon - \eta, \varepsilon - \eta^{-1}$ are all distinct from zero (we can do this, since our field contains at least 7 elements). Now set

$$
\theta = (\varepsilon - 1)(\eta - 1)^{-1}(1 + \alpha_s(\varepsilon - 1)\beta_s)^{-1}(1 + \alpha_s(\eta - 1)\beta_s).
$$

A direct calculation shows that the product

$$
z = \left(x d_r(\varepsilon) x^{-1} \right) d_s(\theta) \left(x d_r(\eta) x^{-1} \right)^{-1}
$$

has the same s-th row as the identity matrix and thus lies in a proper parabolic subgroup. Clearly, $t_{is}(z_{is}) \in X$ for all $i \neq s$. On the other side z_{is}, $i \neq s$, differs from $\alpha_i \beta_s$ only by an invertible factor. Thus either we get a nontrivial transvection in X, or (since we may vary both r and s) we have $x_{ir} x'_{rj} = 0$ for all $i \neq j$, in which case x is monomial, a contradiction.

§ 8. The group $SL_{n \geq 3}$

In this section we prove the theorem from § 6 for the case of *simply connected* Chevalley group $G = SL(n, K)$ of type A_{n-1}, $n \geq 3$, over an *infinite* field K, see [V6], Part I. Actually the proof works also for *large* finite fields of cardinality, say $> 6n + 1$. However considerable additional efforts are needed to prove the result in *natural* generality, i.e. for all fields K, $|K| \geq 7$ (see [V6], Parts II – IV).

Let $D = SD(n, K)$. As in the case of the general linear group it is easy[2] to reduce the proof to the analysis of a primitive irreducible counterexample. In other words, as before we have only to prove that if x is a non-monomial matrix, then the group $X = \langle D, x \rangle$ contains an elementary transvection.

Let $d_{ij}(\varepsilon)$, $1 \leq i \neq j \leq n$, $\varepsilon \in K^*$ be a 'two-dimensional semisimple element'

$$d_{ij}(\varepsilon) = e + (\varepsilon - 1)e_{ii} + (\varepsilon^{-1} - 1)e_{jj} = d_i(\varepsilon)d_j(\varepsilon^{-1}).$$

From the point of view of Chevalley groups $d_{ij}(\varepsilon)$ is a semisimple root element in $SL(n, K)$. The 'natural' proof is based on the analysis of the conjugates of such element in a counterexample.

However the analysis of an infinite (or a large finite) field is easier since in this case G contains non-trivial homologies, i.e. matrices with $n - 1$ equal eigenvalues. Namely, take any $\varepsilon \in K^*$ such that $\varepsilon^n \neq 1$. Then the matrix

$$d^r(\varepsilon) = \mathrm{diag}(\varepsilon, \dots, \varepsilon, \varepsilon^{1-n}, \varepsilon, \dots, \varepsilon) = \varepsilon d_r(\varepsilon^{-n}) \in G$$

is noncentral and has $n - 1$ equal eigenvalues.

The conjugates of such matrices look – up to a scalar factor – exactly like the conjugates of $d_r(\varepsilon) \in GL(n, K)$. We may again fix an r, $1 \leq r \leq n$, and an $\varepsilon \in K^*$, $\varepsilon^n \neq 1$ and consider the conjugate of $d^r(\varepsilon)$ by x:

$$y(\varepsilon) = xd^r(\varepsilon)x^{-1} = \varepsilon\big(\delta_{ij} + \alpha_i(\varepsilon^{-n} - 1)\beta_j\big), \qquad 1 \leq i, j, \leq n,$$

where $\alpha_i = x_{ir}$ and $\beta_j = x'_{rj}$, as before.

Fix some $1 \leq i \neq j \leq n$ and take an index $h \neq i, j$. We may assume that $\alpha_h, \beta_h, \alpha_i, \beta_i \neq 0$ (otherwise the problem is reduced to a proper parabolic subgroup). Set

$$\eta = \alpha_i(\varepsilon^{-n} - 1)\beta_i\big(1 + \alpha_h(\varepsilon^{-n} - 1)\beta_h\big)^{-1},$$

where invertibility of $1 + \alpha_h(\varepsilon^{-n} - 1)\beta_h$ prohibits at most n further values of ε. Prohibiting not more than $2n$ further values of ε we may even assume, that $\eta \neq \pm 1$. Now we may consider the matrix

$$z = y(\varepsilon)d_{hi}(\eta)y(\varepsilon^{-1}) \in X.$$

---

[2]This 'easy' refers to large fields. For $|K| \geq 7$ the reduction uses description of overgroups of $D(n, K)$ in $GL(n, K)$, see [V21]. This is a general pattern: in the proofs for simply connected groups quite often we have to use the corresponding results for adjoint groups.

A straightforward calculation shows that $z_{hi} = 0$ and

$$z_{ij} = (\eta^{-1} - \eta)\alpha_i(\varepsilon^{-n} - 1)\beta_j.$$

It is easy to check, that $z_{hi} = 0$ implies that $t_{ij}(z_{ij}z'_{jj}) \in X$. If ε is not a root of another polynomial of degree $2n$, we can assume that $z'_{jj} \neq 0$, and thus $t_{ij}(\alpha_i\beta_j) \in X$. Now the proof is completed exactly as in the case of GL_n.

The proof for the case $|K| \geq 7$ follows roughly the same lines, but there are some further complications. For example, the group G does not in general contain non-trivial homologies. As a result, it is impossible to conclude directly from the fact that $x_{ij} = 0$ for some $x \in X$ that x is monomial. This is why one is forced to consider more complicated dependencies among the entries of matrices from a counterexample (see [V6], III, IV).

The proofs of the theorem from § 6 for groups of other types are based on roughly the same ideas as the ones reproduced in this section and the preceding one. Of course, they are technically much more complicated. For classical groups one can use their minimal modules. For the exceptional ones we have to play with the Bruhat decomposition of semi-simple root elements and the like, see [V17] and references there. A sketch of the proof for the adjoint case may be found in [V11], [V12], [V20].

§ 9. THE GROUP SL_2

For the group $\mathrm{SL}(2, K)$ the theorem from § 6 asserts that there are only three proper intermediate subgroups between $\mathrm{SD}(2, K)$ and $\mathrm{SL}(2, K)$:

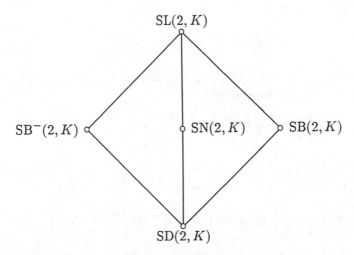

where $SB(2, K)$ and $SB^-(2, K)$ are the upper and the lower standard Borel subgroups in $SL(2, K)$, while $SN(2, K)$ is the monomial subgroup.

Of course for a finite field K we know *all* subgroups of $SL(2, K)$, but for an infinite field this result turned out to be incredibly tough. In [VD] the author and E.V.Dybkova proved the theorem for $SL(2, K)$ (and thus also for $Sp(2l, K)$) under the additional assumption that $|K^*| > |K^*/K^{*2}|$ and observed that $SN(2, K)$ was maximal in $SL(2, K)$ whenever $-1 \in K^{*2}$. In [K1] O.King proved the maximality of $SN(2, K)$ in full generality (compare also [L5] where the maximality of $SN(2, K)$ was proved under essentially the same additional assumptions $|K^*/K^{*2}| < \infty$ or $-1 \in K^{*2}$ as in [VD]). In 1986 the author noticed that a slight modification of this proof allows to prove the theorem for $SL(2, K)$ (and thus also for $Sp(2l, K)$) under the additional assumption that $-1 \in K^{*2}$ (see [V10], [V16]). Finally in [K2] O.King was able to prove the theorem for $SL(2, K)$ in full generality. Actually in this paper O.King describes overgroups of $SD(2, K)$ also for fields of characteristic 2 in terms of the K^2-submodules in K.

Here we reproduce a proof of O.King [K2]. Let F be a subgroup of $SL(2, K)$ which contains $SD(2, K)$ but is not contained in the three subgroups above. Passing from F to a subgroup which is conjugate to F by a diagonal matrix from $GL(2, K)$, we may assume that F contains a matrix of the form

$$f = \begin{pmatrix} 1 & 1 \\ \alpha & 1 + \alpha \end{pmatrix}, \quad \alpha \in K^*.$$

Passing, if necessary, from F to $\langle T, fTf^{-1} \rangle$, we may assume further that α does not belong to a fixed *finite* subset of K^*. We denote the root element $h_\alpha(\varepsilon)$ simply by $h(\varepsilon)$:

$$h(\varepsilon) = \begin{pmatrix} \varepsilon & 0 \\ 0 & \varepsilon^{-1} \end{pmatrix}, \quad \varepsilon \in K^*.$$

Now if char $K \neq 2, 3, 5$, O.King chooses an $\alpha \neq 0, 1, \pm2, \pm3, -4, 5, -6$ and forms the following product:

$$g = f^{-1}h\left(\frac{2}{\alpha}\right) f^{-1}h\left(\frac{\alpha+4}{\alpha+2}\right) fh\left(\frac{3}{4}\right) f^{-1}h\left(\frac{3}{\alpha-3}\right) f^{-1}$$

$$h\left(\frac{\alpha+6}{\alpha+3}\right) fh\left(\frac{5}{6}\right) f^{-1}h\left(\frac{\alpha-2}{\alpha-5}\right) fh(3)f^{-1}h\left(\frac{\alpha+3}{\alpha+2}\right) fh\left(\frac{2}{3}\right)$$

$$f^{-1}h\left(\frac{\alpha-1}{\alpha-2}\right) fh(\alpha-3) fh\left(\frac{4}{3}\right) f^{-1}h\left(\frac{\alpha+2}{\alpha+4}\right) fh\left(\frac{\alpha}{2}\right) f.$$

Then a direct calculation shows that $g_{21} = 0$ while g_{12} is a nonzero polynomial in α and thus equals zero only for finitely many values of α. This shows that some subgroup conjugate to F by a diagonal matrix contains a nontrivial elementary transvection. Thus $F = \mathrm{SL}(2, K)$. For char $K = 3, 5$ O.King constructs analogous (but shorter) formulae expressing transvections in terms of f and $h(\varepsilon)$. For char $K = 2$ both the proof and the statement must be modified considerably, since for a nonperfect field of characteristic 2 there may be infinitely many intermediate subgroups.

§ 10. PRONORMAL SUBGROUPS

Now we know what the analogue of the classification theorem for overgroups of T is. But what is the analogue of the conjugacy theorem? Before T was abnormal in B and B was abnormal in G. But T does not coincide with its normalizer N in G, so it cannot be abnormal.

Definition. *A subgroup T of a group G is called* **pronormal** *in G if for any $x \in G$ there exists $y \in \langle T, xTx^{-1} \rangle$ such that $yTy^{-1} = xTx^{-1}$.*

In other words, this means that any two subgroups conjugate to T in G are conjugate already in the subgroup they generate.

Let $N = N_G(T)$ be the normalizer of T in G. It is easy to see that saying that T is pronormal in G is equivalent to saying that

i) $N_G(H) = HN_N(H)$ for any subgroup H containing T;

ii) any two conjugate subgroups H and F containing T are conjugate by an element of N.

Actually both conditions may be conveniently fused in the following condition: for every two conjugate subgroups $T \le F, H \le G$ any $x \in G$ such that $xFx^{-1} = H$ may be expressed in the form $x = wy$ for appropriate $w \in N$ and $y \in F$.

Now we can state the analogue of conjugacy theorem. We keep the same assumptions on K as in § 6.

Theorem. *T is pronormal in G.*

The first examples of pronormal subgroups which one would think, are, of course,

    – Sylow subgroups of (normal subgroups in) finite groups,
    – Hall subgroups of soluble (normal subgroups in) finite groups,
    – U in G (see [S1]),
    – maximal tori of algebraic groups (see [H1], [Sp]).

More generally a *weakly closed* subgroup of a Sylow subgroup T of G is pronormal in G, etc.

Pronormality is a common generalization of normality and abnormality. The normalizer of a pronormal subgroup is abnormal. In particular this theorem implies that N is abnormal in G. However, it is not, generally speaking, a minimal abnormal subgroup.

It turns out that pronormality plays a very important role in the description of the lattice of subgroups of an abstract group [B6]. In fact for any pronormal subgroup T in G the lattice of intermediate subgroups looks as follows. It is a disjoint union

$$\mathcal{L}(T,G) = \bigcup^{\bullet} \mathcal{L}(F, N_G(F)),$$

taken over all subgroups F generated by some conjugates of T (either in G or in F itself, in this case this does not matter, see § 14 for the details). Intuitively the groups F are 'connected' with respect to T, in [B6] they are called T-complete. The normal closure T^H of T in any intermediate subgroup $H \in \mathcal{L}(T, G)$ coincides with one of the groups F above, while the normalizer $N_G(H)$ is contained in $N_G(T^H)$.

Thus to describe overgroups of a pronormal subgroup, one has to classify the T-complete subgroups F and to calculate the factor-groups $N_G(T)/T$.

However, as we shall see later, there are many further subgroups, for which $\mathcal{L}(T, G)$ may be decomposed in this way. Among these subgroups pronormal ones are characterized by the fact that they have large normalizers, see properties i) and ii) above. In fact, for a pronormal subgroup F the following two characteristic properties hold:

– every sandwich $\mathcal{L}(F, N_G(F))$ is isomorphic to an interval inside $\mathcal{L}(T, N)$ (in other words $N_G(F)/F$ is a section of N/T);

– two sandwiches $\mathcal{L}(F_1, N_G(F_1))$ and $\mathcal{L}(F_2, N_G(F_2))$ are conjugate if and only if they are conjugate by an element of N: $F_1 \sim_G F_2$ if and only if $F_1 \sim_N F_2$.

In particular, for a pronormal subgroup the calculation of all factor-groups $N_G(F)/F$ is reduced to the calculation of the 'Weyl group' N/T.

Thus, there are good chances for a nice description of overgroups of Sylow subgroup $S_p(G)$ in a finite group G. This is essentially done in [A5] for sporadic groups.

Problem. *Describe the overgroups of Sylow subgroups in finite groups of Lie type.*

As was pointed out by the referee, for $p = 2$ this problem has been partially solved. In fact [LSa1] and [Ka4] determine the *maximal* overgroups of the Sylow 2-subgroups. This gives a good start for a complete solution of the problem in the case of $p = 2$.

The following question is purely speculative, since one cannot really expect a reasonable answer. However even new examples of pronormal subgroups could be of great value for understanding the subgroup structure of finite groups.

Problem. *Classify pronormal subgroups of finite groups of Lie type.*

§ 11. GENERALIZATIONS TO RINGS

In this section we mention how the results of the previous sections may be generalized to rings. Of course here the answers should take into account the structure of the ideals of the ground ring.

Definition. *A* **net** *of ideals in R of type Φ is a family $\sigma = (\sigma_\alpha)$, $\alpha \in \Phi$, of ideals σ_α in R such that $\sigma_\alpha \sigma_\beta \subseteq \sigma_{\alpha+\beta}$ whenever $\alpha, \beta \in \Phi$ are such that $\alpha + \beta \in \Phi$.*

Now we can assign to a net σ of type Φ in R subgroups $E(\sigma)$, $G(\sigma)$ and $N(\sigma)$ in $G(\Phi, R)$ by the same formulae, as in § 3, for a closed set of roots S. Namely, we set

$$E(\sigma) = \langle x_\alpha(\xi), \ \alpha \in \Phi, \ \xi \in \sigma_\alpha \rangle.$$

Now we can define $G(\sigma)$ as $T \cdot E(\sigma)$ and $N(\sigma)$ as the normalizer of $G(\sigma)$ in G. Since K_1 of a ring is non-trivial, these are *not* the "correct" definitions of $G(\sigma)$ and $N(\sigma)$, but they suffice for the instances in which we use them. In fact $G(\sigma)$ should be defined by certain linear congruences on entries of matrices describing the action of elements from G on a minimal module, see § 13 for the case of the general linear group.

In the next two sections we can avoid this problem by *defining $N(\sigma)$* as $N_G(E(\sigma))$. In fact a considerable part of real work consists in proving that the correctly defined $N(\sigma)$ actually *does* coincide with the normalizer of $E(\sigma)$ for the cases we are interested in. This result is, for instance, a very wide generalization of the normality theorem of Suslin stating that the elementary subgroup $\mathrm{E}(n, R)$ is normal in $\mathrm{GL}(n, R)$, when R is commutative and $n \geq 3$. We do not discuss these aspects of the theory here, see [BV3], [V16] for the details and further references.

Nets of type A_l were introduced by N.S.Romanovskii [Ro] in a special case and by Z.I.Borewicz in general [Bo1] – [Bo5], [BV1] – [BV3]. Corresponding subgroups in GL_n, SL_n, Sp_{2n} were studied in detail by Z.I.Borewicz and his students (we cannot reproduce a complete bibliography here, consult [Ne] for the early history of these ideas and [Z2], [VD], [V14], [V15], [V18] for further references). For arbitrary Φ, nets were defined by K.Suzuki [Sz1], [Sz2] in a special case and by the author in general [V1], [V5]. These papers dealt with rings close to fields (like semilocal ones, see below), where the above definitions of $G(\sigma)$ and $N(\sigma)$ are satisfactory. For arbitrary rings the "correct" groups were defined in [VP].

With these definitions we can generalize the contents of the previous sections to some classes of rings. In fact $\mathcal{L}(T, B)$ can be described for an almost arbitrary ring R – the only requirement is that R should have enough units: R should be additively generated by R^* (by R^{*2} in the symplectic case) and the ideal generated by $\varepsilon-1$ (or ε^2-1, ε^3-1), where $\varepsilon \in R^*$, should coincide with R. Under these conditions every subgroup in B containing T coincides with $G(\sigma)$ for an appropriate net σ and T is pronormal in B (see [B1], [B2], [V1]).

It is somewhat more complicated with $\mathcal{L}(B,G)$. In fact there is no analogue of the Bruhat decomposition for rings other then direct sums of (skew-)fields[3]. For rings the lattice $\mathcal{L}(B,G)$ was studied in [Ro], [B1], [B2], [Sz1], [Sz2], [V1], and several further papers (see [V5] for a complete bibliography). In these papers $\mathcal{L}(D,G)$ was completely described when R is a "zero-dimensional" ring satisfying the restrictions above. Not to go into details, we note that a safe example of zero-dimensional rings are semilocal ones.

Recall that a ring R is called *semilocal* if its factor-ring modulo the Jacobson radical J is artinian. For a commutative ring this is equivalent to saying that R has finitely many maximal ideals, or that $R/J \cong K_1 \oplus \ldots \oplus K_t$ is a direct sum of finitely many fields (in particular R is local if $t=1$, i.e. if $R/J \cong K$ is a field). For semilocal rings the conditions imposed on their units amount simply to saying that $K_i \neq \mathbb{F}_2, \mathbb{F}_3, \mathbb{F}_4$, etc. For such rings again all intermediate subgroups H, $B \leq H \leq G$, are of the form $G(\sigma)$ and B is abnormal in G. (This is not the case in general however, even for such *very* nice rings as Hasse domains the answer is much more complicated).

These results applied to *all* types. As for the lattice $\mathcal{L}(T,G)$ so far only the *adjoint* (or extended) *classical* groups were considered. Let R again be semilocal. Then only the small direct summands of R/J may cause trouble. If all K_i are distinct from $\mathbb{F}_2$, $\mathbb{F}_3$, $\mathbb{F}_4$ and $\mathbb{F}_5$ (and char $K \neq 2$ if $\Phi = \mathrm{B}_l$ or C_l), then for any intermediate subgroup H, $T \leq H \leq G$, there exists a unique net σ of ideals in R of type Φ such that $G(\sigma) \leq H \leq N(\sigma)$, see [B3], [BV1], [V2], [V3], [VD], [V14]. Moreover if R is *local* then T is pronormal in G (for a general semilocal ring T need not be pronormal – it is only *paranormal*, compare § 14), see [V9], [VD], [V14]. For the general linear group over *non-commutative* semilocal rings one has to prohibit one more direct summand of R/J, namely the ring $\mathrm{M}(2,\mathbb{F}_2)$ of matrices of degree 2 over the field of two elements.

There are presently no approaches to the description of overgroups of T for local rings, not only for the exceptional groups, but even for such innocent looking creatures as SL_n. The only case for which it was possible to go through for $\mathrm{SL}_{n \geq 3}$ are the so called *uniserial* rings, for which ideals are linearly ordered by inclusion [Ha]. Here we state the part of the result which works for the adjoint classical groups and does not work in other

---

[3]This applies to groups of finite degree. For the stable elementary group $\mathrm{E}(\Lambda) = \varinjlim \mathrm{E}(n,\Lambda)$ there is an analogue of Bruhat decomposition which holds for *any* ring Λ – Sharpe's decomposition.

situations.

Let R be any commutative ring, J be its Jacobson radical. Consider the group B_J generated by B and all $x_\alpha(\xi)$, where $\alpha \in \Phi^-$, $\xi \in J$. In other words B_J is precisely the group $G(\tau)$ for the net τ defined by $\tau_\alpha = R$ if $\alpha \in \Phi^+$ and $\tau_\alpha = J$ otherwise. Then for the adjoint classical groups one can describe the lattice $\mathcal{L}(T, B_J)$ for almost arbitrary commutative rings R. Namely R should have enough units in the same sense as in the description of $\mathcal{L}(T, B)$ above. Assume that R is additively generated by R^* and that there exist such $\varepsilon, \eta \in R^*$ that $\varepsilon - 1$, $\eta - 1$, $\varepsilon - \eta$, $\varepsilon\eta - 1 \in R^*$. Then every intermediate subgroup H, $T \le H \le B_J$ is again of the form $G(\sigma)$ for an appropriate σ and T is abnormal in B_J. This is precisely what is still lacking for other groups, including $\mathrm{SL}_{n \ge 3}$.

The reason why the description of overgroups of T in G works only for *zero-dimensional* rings is that T consists entirely of semisimple elements. Of course, B has lots of unipotent elements, but their geometry is very poor. Semisimple elements may be very upsetting even for fields. Unipotents are much nicer and behave in a more predictable way. In the next sections we shall see in particular that when D has enough unipotents it is almost always possible to describe $\mathcal{L}(D, G)$ for *arbitrary* commutative rings.

Problem. *Describe $\mathcal{L}(T, G)$ for all Chevalley groups when R is a semilocal ring.*

§ 12. OVERGROUPS OF SUBSYSTEM SUBGROUPS

In this section we introduce another pattern in the distribution of intermediate subgroups. Namely, we study the overgroups of *regularly embedded* semisimple subgroups in the sense of E.B.Dynkin. Such subgroups were christened *subsystem subgroups* by M.W.Liebeck and G.M.Seitz.

Namely let $\Delta \subseteq \Phi$ be an embedding of root systems. This embedding defines a regular embedding of Chevalley groups $G(\Delta, R) \le G(\Phi, R)$ over any commutative ring R with 1. We are interested in the description of overgroups of $E(\Delta, R)$ in $G(\Phi, R)$. Of course to get such a description one has to assume something about Δ (and possibly also about R): the case $\Delta = \emptyset$ is upsetting.

The assumption which seems reasonable in all cases is that $\Delta^\perp = \emptyset$, i.e. that there are no roots in Φ orthogonal to all roots in Δ. However for commutative rings one usually needs stronger assumptions.

Now we state known results for the *classical* groups. First of all recall that the root subsystems of classical systems are as follows:

$$\Phi = A_l: \qquad \Delta = A_{k_1} + \ldots + A_{k_r},$$
$$\Phi = B_l: \qquad \Delta = A_{k_1} + \ldots + A_{k_r} + D_{l_1} + \ldots + D_{l_{s-1}} + B_{l_s},$$

$$\Phi = C_l: \qquad \Delta = A_{k_1} + \ldots + A_{k_r} + C_{l_1} + \ldots + C_{l_s},$$

$$\Phi = D_l: \qquad \Delta = A_{k_1} + \ldots + A_{k_r} + D_{l_1} + \ldots + D_{l_s},$$

Here $p+q = l$, $p = (k_1+1)+\ldots+(k_r+1)$, $q = l_1+\ldots+l_s$, and $k_1, \ldots, k_r \geq 0$, $l_1, \ldots, l_s \geq 0$ (see [Dn], table 9). It is assumed here that $A_0 = B_0 = D_1 = \emptyset$ is the empty root system. Set $n_i = 2l_i$ if the corresponding summand is C_{l_i} or D_{l_i} and $n_i = 2l_i + 1$ if it is B_{l_i}.

Then the corresponding groups $E(\Delta, R)$ are isomorphic to one of the following groups: to the group

$$\mathrm{E}(k_1 + 1, R) \oplus \ldots \oplus \mathrm{E}(k_r + 1, R)$$

in the case $\Phi = A_l$; to the group

$$\mathrm{E}(k_1 + 1, R) \oplus \ldots \oplus \mathrm{E}(k_r + 1, R) \oplus \mathrm{Epin}(n_1, R) \oplus \ldots \oplus \mathrm{Epin}(n_s, R)$$

in the cases $\Phi = B_l$ and $\Phi = D_l$; and, finally, to the group

$$\mathrm{E}(k_1 + 1, R) \oplus \ldots \oplus \mathrm{E}(k_r + 1, R) \oplus \mathrm{Ep}(2l_1, R) \oplus \ldots \oplus \mathrm{Ep}(2l_s, R)$$

in the case $\Phi = C_l$. Here $\mathrm{E}(n, R) = \mathrm{E}(A_{n-1}, R)$ is the elementary subgroup of $\mathrm{SL}(n, R)$ generated by all elementary transvections (see [HO]), while $\mathrm{Epin}(2l + 1, R) = \mathrm{E}(B_l, R)$, $\mathrm{Epin}(2l, R) = \mathrm{E}(D_l, R)$, and $\mathrm{Ep}(2l, R) = \mathrm{E}(C_l, R)$ are analogous subgroups in $\mathrm{Spin}(n, R)$ and $\mathrm{Sp}(2l, R)$ respectively. For the orthogonal group $\mathrm{SO}(n, R)$ the elementary subgroups $\mathrm{Epin}(n, R)$ should be replaced by $\mathrm{EO}(n, R)$.

Now we are ready to state the results about overgroups of subsystem subgroups in the classical groups. Assume that $\Phi = A_l, B_l, C_l$ or D_l and that R is a commutative ring, such that $2 \in R^*$ if Φ has roots of different length. Assume further that Δ is a subsystem in Φ such that $k_1, \ldots, k_r, l_1, \ldots, l_s$ of its irreducible summands are at least 2. In the orthogonal case assume moreover that all $k_1, \ldots, k_r$ are at least 4.

Theorem. *For any subgroup H in $G(\Phi, R)$ containing $E(\Delta, R)$ there exists a unique net σ of ideals in R such that*

$$E(\sigma) \leq H \leq N(\sigma).$$

In particular, for fields this theorem says that there exists a unique closed set of roots $S \subseteq \Phi$ such that $E(S) \leq H \leq N(S)$. Of course $E(\Delta, K)$ contains root subgroups and thus for fields this theorem may be deduced from the results of J.Mclaughlin, B.S.Stark, W.Kantor, Li Shang Zhi and F.Timmesfeld describing irreducible subgroups of classical groups generated by root subgroups (see [Ka1], [Ka2], [Ti1] – [Ti4] and references there).

However a much easier proof than that works for all commutative rings. We sketch this proof for the case $\Phi = A_l$ in the next section.

Now we discuss the bound on k_i, l_j in the theorem above, limiting ourselves to the case $\Phi = A_l$ (analogous remarks apply to other series as well). Observe that the theorem is not true any more if there are at least two summands of rank ≤ 1. Clearly, the case when we have at least two summands of rank 0 is intractable, so we assume that $k_i \geq 1$ for all i. Consider, for example the subsystem $\Delta = 2A_1$ in $\Phi = A_3$. Then clearly $E(2, K) \oplus E(2, K)$ is contained in $\mathrm{Sp}(4, K)$ which does not have the form predicted by the theorem.

For fields $\mathrm{Sp}(2l, K)$ is the only primitive irreducible counterexample *up to conjugacy*, but it is quite an exercise to explicitly list all the possibilities for *reducible* subgroups containing $E(\Delta, K)$ (see [Ko5]). Even worse, one immediately discovers that description of intermediate subgroups in our sense and their description up to conjugacy is not quite the same thing. For example, there exist infinite parametric families of intermediate subgroups conjugate to $\mathrm{Ep}(2l, K)$ (unlike the case of $k_i \geq 2$, when there are only finitely many $E(\Delta, K)$-complete intermediate subgroups).

On the other hand for *rings* the case when the irreducible summands of Δ have rank 1 is essentially intractable. In fact the theorem above would imply a description of subgroups in $G(\Delta, R)$ normalized by $E(\Delta, R)$ (compare [Go], [Sv]). As is well known, there is no hope to describe the normal structure of the groups $\mathrm{GL}(2, R)$ or $\mathrm{SL}(2, R)$ over general commutative rings (compare [W]). Even for such *very* nice rings as semilocal ones, or Hasse domains with infinite multiplicative groups, a complete description of normal subgroups was obtained only quite recently and involves much more complicated structures, than ideals [CK].

The 7-dimensional representation of the Chevalley group of type G_2 and the 8-dimensional representation of the Chevalley group of type B_3 are responsible for the stronger restrictions on the ranks of the irreducible summands of Δ in the orthogonal case. In fact, the embeddings $A_2 \subset G_2$ and $A_3 \subset B_3$ furnish counter-examples to the conclusion of the theorem above with the ranks of irreducible summands equal to two or to three.

There is no doubt that results analogous to the theorem above hold also for large subsystem subgroups of exceptional groups and that all the necessary tools are contained in [V19].

§ 13. THE GROUP GL_n OVER COMMUTATIVE RINGS.

Here, in the example of the general linear group $\mathrm{GL}(n, R)$, we explain the correct definition of $G(\sigma)$ and $N(\sigma)$ (see [B1] – [B5], [BV1] – [BV3]) and show how the proof of the preceding theorem works. The proof we reproduce below is a simplified version of the proof from [BV3].

Let Λ be an arbitrary associative ring with 1. A square array $\sigma = (\sigma_{ij})$, $1 \leq i,j \leq n$, of two-sided ideals σ_{ij} in Λ is called a net of ideals in Λ of degree n if $\sigma_{ir}\sigma_{rj} \subseteq \sigma_{ij}$ for all values of the indices i,j,r. A net σ is called a D-net if $\sigma_{ii} = \Lambda$ for all i. Clearly D-nets of degree n are precisely the nets of type A_{n-1} defined in § 11.

As in § 11 we can associate with σ the elementary net subgroup $E(\sigma)$ generated by all elementary transvections $t_{ij}(\xi) = e + \xi e_{ij}$, where $1 \leq i \neq j \leq n$, $\xi \in \sigma_{ij}$.

For a given net σ we denote by $M(\sigma)$ the corresponding subring in the full matrix ring $M(n,\Lambda)$, consisting of all matrices $x = (x_{ij})$ congruent to 0 modulo σ:

$$M(\sigma) = \left\{ x = (x_{ij}) \in M(n,\Lambda) \mid x_{ij} \in \sigma_{ij},\ 1 \leq i,j \leq n \right\}.$$

Then $e + M(\sigma)$ is a multiplicative system. The largest subgroup $G(\sigma)$ of the general linear group $G = \mathrm{GL}(n,\Lambda)$, contained in $e + M(\sigma)$ is called the net subgroup, corresponding to σ and is denoted by $G(\sigma)$. For a commutative ring R one has $G(\sigma) = G \cap (e + M(\sigma))$.

As before $N(\sigma)$ is the normalizer of $G(\sigma)$ in G. In many important situations, in particular for the nets appearing in the theorem above, $N(\sigma)$ consists of the matrices $x \in G$ such that $x_{ir}\sigma_{rs}x'_{sj} \subseteq \sigma_{ij}$ for all i,j,r,s.

Let ν be an equivalence relation on the index set $\{1,\dots,n\}$. We can associate with ν a D-net $[\nu]$ by setting $[\nu]_{ij} = \Lambda$ if $i \underset{\sim}{\nu} j$ and $[\nu]_{ij} = 0$ otherwise. Then the corresponding group $E(\nu) = E([\nu])$ is exactly one of the elementary subsystem subgroups $E(\Delta)$ considered in the preceding section. The condition on ranks of irreducible summands of Δ amounts to saying that all equivalence classes of ν contain at least *three* elements.

Now assume that R is commutative and $H \geq E(\nu)$. For $i \neq j$ denote by σ_{ij} the set of all $\xi \in R$ such that $t_{ij}(\xi) \in H$. Set, moreover, $\sigma_{ii} = R$. Clearly σ is the largest D-net of ideals such that $E(\sigma) \leq H$. We have to prove that $H \leq N(\sigma)$. This amounts to verifying the conditions $x_{ir}\sigma_{rs}x'_{sj} \subseteq \sigma_{ij}$ for all $x \in H$ and all $i,j,r,s,\ r \neq s,\ i \nsim j$.

Take any pair of indices $r \neq s$ and any $\xi \in \sigma_{rs}$ and consider the following transvection $y = xt_{rs}(\xi)x^{-1} \in H$. We have to prove that all these transvections belong to $E(\sigma)$. Fix an index j and take two indices p,q equivalent to j and such that j,p,q are pairwise distinct. Consider the matrix

$$z = yt_{hp}(a_{qr})t_{hq}(-a_{pr})y^{-1} \in H.$$

Explicit calculations using the commutativity of R show that $z_{ih} =' \delta_{ij}$ for all $h \neq p,q$, while $z_{ip} = \delta_{ip} + b_{ij}a_{qr}$ and $z_{iq} = \delta_{iq} + b_{ij}a_{pr}$. Another commutation with matrices from $E(\nu)$ shows that $y_{ij}x_{pr} \in \sigma_{ij}$ for all $p \sim j$, $p \neq j$.

These inclusions hold for *all* transvections of the form $y = xt_{rs}(\xi)x^{-1}$, $x \in H$. But this then readily implies that all these transvections belong to $E(\sigma)$. Indeed consider another transvection $u = yt_{jp}(1)y^{-1} \in H$. The inclusions above applied to u instead of y show that $u_{ip}y_{jj} \in \sigma_{ip} = \sigma_{ij}$. But $u_{ip} = y_{ip}y'_{pp}$, where $y'_{pp} = 1 - x_{pr}\xi x'_{sp}$, so that $u_{ip} \equiv y_{ij}(\mod \sigma_{ij})$. Thus $y_{ij}y_{jj} \in \sigma_{ij}$. Applying this last inclusion to u we get $u_{ip}u_{pp} \in \sigma_{ip} = \sigma_{ij}$. But $u_{ip}y_{pp} \equiv y_{ij}(\mod \sigma_{ij})$. Thus, finally, $y_{ij} \in \sigma_{ij}$ for all i, j.

Now to prove polynormality of $E(\nu)$ one has to prove that $E(\sigma)$ is normal in $N(\sigma)$ and that all conjugates of $E(\nu)$ contained in $N(\sigma)$ in fact lie already in $E(\sigma)$. These are relatively deep facts. For example, they contain Suslin's normality theorem as a very special case, see [BV3] for details.

Essentially the same proofs work for other classical cases as well (see [V10], [V15], [V16] for details and further references). In fact this result is a very wide generalization of the description of normal subgroups in classical groups over commutative rings (see references in [V19]).

§ 14. PARANORMAL AND POLYNORMAL SUBGROUPS

We have seen in § 12 that for the subsystem subgroups the lattice of intermediate subgroups admits the same nice description as for the split maximal tori. However the subsystem subgroups are not pronormal in general, simply because there normalizer might be too small.

Take for example the subgroup

$$D = E(m, K) \oplus E(m, K) \oplus E(2m, K) \le GL(4m, K).$$

Then clearly one cannot present all the matrices from the normalizer of $H = E(2m, K) \oplus E(2m, K) \ge D$ in the form hw, where $h \in H$ and $w \in N_G(D)$.

It turns out though, that the subsystem subgroups satisfy a weaker property, which guarantees the standard description of the intermediate subgroups. We introduce some notation first. Let $D \le G$ and denote by

$$\Omega_G(D) = \{xDx^{-1}, \ x \in G\}$$

the set of all conjugates of D in G.

Definition. *A subgroup F, $D \le F \le G$, is called D-complete if it is generated by subgroups which are conjugate to D in F*

$$F = D^F = \langle X \in \Omega_F(D) \rangle.$$

The following very important notion was introduced by Z.I.Borewicz and O.N.Macedońska in [BM].

Definition. *A subgroup D of G is called* **paranormal** *if for any $x \in G$ one has*

$$D^{\langle D, xDx^{-1} \rangle} = \langle D, xDx^{-1} \rangle.$$

In other words, a subgroup D is paranormal if for any $x \in G$ the subgroup $H = \langle D, xDx^{-1} \rangle$ is D-complete. Clearly it follows that *all* subgroups generated by some conjugates of D in G are D-complete. It is obvious that a pronormal subgroup is paranormal, but as we have seen the converse is not true.

Theorem. *Under the assumptions of the above theorems the elementary subsystem subgroup $E(\Delta, R)$ is paranormal in $G(\Phi, K)$.*

Other remarkable examples of paranormal subgroups, which are not, generally speaking, pronormal, are

   – maximal tori in *finite* Chevalley groups, [S6];
   – diagonal subgroups (*split* maximal tori) in *classical* groups over semilocal rings, [BV1], [V3], [VD], [V14].

Paranormal subgroups exhibit the same behaviour of intermediate subgroups as pronormal ones, apart from normalizers and conjugacy. Indeed, for any paranormal subgroup D in G the lattice of intermediate subgroups is a disjoint union

$$\mathcal{L}(D, G) = \bigcup{}^{\emptyset} \mathcal{L}(F, N_G(F)), \qquad F = \langle X \in \Omega_G(D), \ X \leq F \rangle.$$

taken over all subgroups F generated by some conjugates of D in G. As in the case of pronormal subgroups the normal closure D^H of D in any intermediate subgroup $H \in \mathcal{L}(D, G)$ coincides with one of the groups F above, while the normalizer $N_G(H)$ is contained in $N_G(D^H)$.

In the paper [BM] another, a still wider, generalization of pronormal subgroups is introduced.

Definition. *A subgroup D of G is called* **polynormal** *if for any subgroup $H \leq G$ one has*

$$D^{D^H} = D^H.$$

Immediately from the definition it follows that for any polynormal subgroup D in G the lattice of intermediate subgroups is a disjoint union

$$\mathcal{L}(D, G) = \bigcup{}^{\emptyset} \mathcal{L}(F, N_G(F)), \qquad F = D^F,$$

taken over all D-complete subgroups F. The sandwiches for the polynormal subgroups are a priori *larger*, than for the paranormal ones. Moreover it is possible that one subgroup, generated by some conjugates of D lies between another such subgroup and the normalizer of the latter in G. This is counter-intuitive and cannot arise for nice subgroups, like, say, groups of points of *connected* algebraic groups over *large* fields. This is why the author feels that this further generalization is less useful than the notion of a paranormal subgroup.

To the contrary paranormal subgroups arise quite naturally from the pronormal ones. Namely, if T is a pronormal subgroup in G, then a T-complete subgroup D may have small normalizer and is not necessarily pronormal in G, while on the other hand it is *always* paranormal.

Once more to describe overgroups of a paranormal or a polynormal subgroup D, one has to classify the D-complete subgroups F and to calculate the factor-groups $N_G(F)/F$. However, unlike for the case of pronormal subgroups, the calculation of these factors is in no way reduced to the calculation of N/D. This is precisely why G.Seitz has to change his torus when calculating the normalizer of an intermediate subgroup [S6].

§ 15. FIELD EXTENSION SUBGROUPS

In this section we start to discuss the groups related to the Aschbacher class $\mathcal{C}_3$. Let L be an extension of the ground field K of degree r. Then a vector space V of dimension m over L may be considered as a vector space of dimension $n = mr$ over K. Every L-linear endomorphism of the space is K-linear. Picking up a K-base in V we get an embedding $\mathrm{GL}(m, L) < \mathrm{GL}(n, K)$ (compare the next section for a more detailed description of the case $m = 1$).

One can define analogous embeddings for other groups of Lie type; sometimes there are several different patterns. For example, $G = \mathrm{Sp}(2n, K)$ has not only subgroups $\mathrm{Sp}(2m, L)$, where as above L/K is an extension of degree $r = n/m$, but also subgroups $\mathrm{U}(n, L)$, where L is a quadratic extension of K. For the *finite* classical groups one can find complete lists of all such embeddings in [A1], [KL2]. However the main goal of these works is the search for maximal subgroups. This is why they concentrate on the case when L/K is an extension of *prime* degree.

In this section we will be interested in the case when $m \geq 2$, i.e. when $\mathrm{GL}(m, L)$ contains unipotent elements. The image of a transvection from $\mathrm{GL}(m, L)$ is a *quadratic* unipotent element. The case $m = 1$ is *very* different and will be discussed in the next three sections.

We are interested in subgroups of $\mathrm{GL}(n, K)$ containing $\mathrm{SL}(m, L)$. For $m \geq 3$ the answer is quite uniform and may be stated as follows.

Theorem. *Let L/K be any field extension of degree r, $m \geq 3$, $n = mr$. Then for any subgroup $\mathrm{SL}(m, L) \leq H \leq G = \mathrm{GL}(n, K)$ there exists a unique* *intermediate subfield* $K \leq E \leq L$, $[L : E] = d$ *such that*

$$SL(md, E) \leq H \leq N_G(SL(md, E)).$$

This is a corollary from a result of Li Shang Zhi [L4]. In the case $m = 2$, $|K| \neq \mathbb{F}_2$, the description is similar, but there is another series of complete intermediate subgroups, namely the groups $Sp(2d, E)$ for all intermediate subfields $K \leq E \leq L$, $[L : E] = d$. In the exceptional case of the $Sp(4, \mathbb{F}_2)$ two further examples arise.

Li Shang Zhi has proven analogous results for overgroups of $Sp(m, L)$ in $GL(n, K)$. Nothing changes as compared with the case of overgroups of $SL(m, L)$, except, of course, that now the groups $Sp(md, E)$ always appear as complete intermediate subgroups, and not just for $m = 2$. He also announced analogous results for overgroups of $\Omega(m, L)$.

For the *finite* case analogous results were obtained independently by R.W.Dye, see [D1] – [D3] and references there. Basically he was interested only in the extensions of prime degree. On the other hand R.W.Dye considered also the case of tori, which Li Shang Zhi had to exclude for infinite fields (for a very good reason, as we shall see in the following sections). In the language of these works the field extension subgroups are stabilizers of spreads. The method of proof is also geometric.

It seems plausible that using ideas analogous to those described in § 11 and § 13 one can obtain analogous results for commutative rings. For the fields of char $\neq 2$ the answer follows also from the theory of quadratic pairs and its infinite analogue developed by F.Timmesfeld.

§ 16. Maximal tori

In this section we describe what maximal tori look like for the case of the general linear group and briefly outline the current situation concerning description of their overgroups. In principle what we call a maximal torus is the group of K-points of a maximal torus in the corresponding algebraic group.

Maximal tori are extremely important subgroups, especially in the finite case. They play a crucial role both in the structure theory (since they control conjugacy classes of semisimple elements) and in the representation theory. One may find detailed information concerning tori in the finite case in [SS], [C2], [C3], [S5], [S6], [Kn].

The situation over infinite fields is more complicated. Our description below is utterly naive, for example we do not worry about separability. One can find a complete cohomological classification of maximal tori in the classical groups over *infinite* fields in the paper [Km] by K.Kariyama.

First consider the case $n = 1$ in the construction of a field extension subgroup from the preceding section. In other words we consider inclusion

$$L^* \le \mathrm{GL}(n, K), \qquad [L : K] = n.$$

In fact, let $\omega_1, \ldots, \omega_n$ be a base of L over K. Then for any $\alpha \in L$. One has

$$\alpha \omega_j = \sum_{i=1}^n \alpha_{ij} \omega_i, \qquad \alpha_{ij} \in K, \quad 1 \le j \le n.$$

Now the above embedding is described by the correspondence $\alpha \mapsto (\alpha_{ij})$.

Problem. *Describe overgroups of L^* in $\mathrm{GL}(n, K)$.*

As we shall soon see this problem is *extremely* difficult in general, and provides examples for description of subgroups which is much more complicated than anything we encountered so far. However to make it still more difficult we may combine it with the Aschbacher class $\mathcal{C}_1$.

Namely, consider a partition $n = n_1 + \ldots + n_t$. Let $L_1, \ldots, L_t$ be extensions of K of degrees $n_1, \ldots, n_t$ respectively. Then one can consider

$$T = L_1^* \oplus \ldots \oplus L_t^* \le \mathrm{GL}(n, K).$$

Such a subgroup T is a maximal torus in $\mathrm{GL}(n, K)$.

Problem. *Describe overgroups of $L_1^* \oplus \ldots \oplus L_t^*$ in $\mathrm{GL}(n, K)$.*

When $t = n$ all the fields L_i coincide with K and we get precisely the problem considered in § 6.

Roughly speaking the situation with classification of overgroups of maximal tori is as follows. Here G is a simple K-defined algebraic group over a field K and T is an arbitrary K-defined maximal torus. We consider the lattice $\mathcal{L}(T_K, G_K)$ of subgroups lying between the groups of K points of T and G. For algebraically closed fields all tori are split and the results described in § 6 apply. The situation for the finite case will be outlined in the next section.

For the field $\mathbb{R}$ of real numbers all subgroups containing a maximal torus are closed in the real topology [Dj], [P]. In particular, there are only finitely many intermediate subgroups and they are easily classified using the corresponding results for Lie algebras.

For some special tori over local fields the problem was considered by S.L.Krupetskii, see [Kr1] – [Kr3] and references in [V17]. The qualitative answer for the general case was obtained by V.P.Platonov [P]. Let K be a local field with valuation v. Then any intermediate subgroup H is an open (in v-adic topology) subgroup in the group $(\overline{H})_K$ of K-points of its closure (in Zariski topology). This reduces the explicit classification to Bruhat-Tits theory. In particular there are not more than countably many intermediate subgroups.

For a global field K only the case of a *cyclic* maximal torus in one of the groups GL_n or SL_n has been considered so far by V.A.Koibaev and V.P.Platonov. In particular, Chebotarev's density theorem implies that there are *uncountably* many intermediate subgroups.

For arbitrary fields only the case of non-split tori in GL_2 has been considered by V.A.Koibaev, Z.I.Borewicz and others; see § 18 below for a description of this result.

§ 17. OVERGROUPS OF MAXIMAL TORI: FINITE CASE

For a finite ground field there is an almost exhaustive description of subgroups in Chevalley groups which contain a maximal torus. This remarkable result is due to G.M.Seitz [S6]. The standing assumption in this section is that $K = \mathbb{F}_q$ be a finite field, char $K \neq 2, 3$, $q \geq 13$. Seitz' results apply to all finite groups of Lie type, but for simplicity we state them only for the groups of normal types.

Let $\overline{G} = G(\Phi, \overline{K})$ be a Chevalley group of type Φ over the algebraic closure $\overline{K}$ of a finite field $K = \mathbb{F}_q$. Recall that the standard Frobenius endomorphism $\sigma = \sigma_q$ of $GL(n, \overline{K})$ is obtained by raising the matrix entries to the q-th power, $\sigma : (x_{ij}) \mapsto (x_{ij}^q)$. A standard Frobenius endomorphism of $\overline{G}$ is the mapping $\sigma : \overline{G} \to \overline{G}$ induced by σ_q under some embedding $i : \overline{G} \to GL(n, \overline{K})$, see [SS], [C3].

A maximal torus T in a finite Chevalley group $G = G(\Phi, K)$ has the form $T = \overline{T} \cap G$, where $\overline{T}$ is a σ-*invariant* maximal torus. Conjugacy classes of maximal tori correspond to the conjugacy classes of the Weyl group $W(\Phi)$, see [C2], [C3], [SS]. Let further $\overline{X}_\alpha$, $\alpha \in \Phi$, be the corresponding root subgroups. Since $\overline{T}$ is σ-invariant, these subgroups are permuted by σ.

Let $\Delta = \{\overline{X}_1, \ldots, \overline{X}_t\}$ be a σ-orbit of root subgroups. We consider the span $\overline{X} = \overline{X}(\Delta) = \langle \overline{X}_1, \ldots, \overline{X}_t \rangle$. The groups $X = O^{p'}(\overline{X}^\sigma)$ are called T-root subgroups by G.M.Seitz. They play the same role in the description of overgroups of T as the usual root subgroup do in the description of overgroups of a split torus.

Every T-root subgroup is either unipotent, or a group of Lie type over a certain extension of the ground field. If T is split, then all the T-root subgroups are unipotent; when T is minisotropic, all of them are semisimple. In general there are T-root subgroups of both types.

Now let S be a set of T-root subgroups which is closed in the following sense: if X, Y, Z are three T-root subgroups such that $X, Y \in S$ and $Z \leq \langle X, Y \rangle$, then $Z \in S$. Let $G(T, S)$ be the group generated by T and all $X \in S$. Let further $N(T, S)$ be the normalizer of $G(T, S)$ in G. Then the result of G.M.Seitz may be stated as follows.

Theorem. *Let T be an arbitrary maximal torus in $G = G(\Phi, K)$. Then for any subgroup H of G that contains T there exists a unique closed set S of T-root subgroups such that*

$$G(T, S) \leq H \leq N(T, S).$$

Actually Seitz has shown that the group $G(T, S)$ appearing in the theorem is is exactly the subgroup T^H generated by conjugates of T contained in H. This, plus standard arguments about the normalizers $N(T, S)$, amounts to the following result.

Theorem. *T is a paranormal subgroup of G.*

Another important result of [S6] asserts that the notion of a T-complete subgroup does not depend on the choice of a maximal torus T.

Theorem. *If H contains two maximal tori T_1 and T_2, then $T_1^H = T_2^H$.*

The proof in [S6] uses the Classification. In [S5], [Kn] one can find an explicit description of the maximal tori and T-root subgroups for the finite cases. Independently of the work of G.M.Seitz only the overgroups of Singer cycles in the classical groups (and their large subgroups) have been studied, see [Ka2], [He], [D1] – [D3], [De], [Wa].

§ 18. OVERGROUPS OF MAXIMAL TORI: GL_2

In this section we describe the work of V.A. Koibaev, Z.I.Borewicz and others on the subgroups of the group $G = GL(2, K)$ over an arbitrary field K which contain a maximal torus, see [Ko7], [BKH], [BK2], [Ko8]. The contents of this section show that for infinite fields the structure of the lattice of intermediate subgroups may be very different from the finite patterns.

Assume that K is an *infinite* field such that char $K \neq 2$ and L/K is a quadratic extension of K. We may assume that $L = K(\sqrt{d})$, $d \in K \setminus K^2$. Then the corresponding maximal torus in G has the form

$$T = T(d) = \left\{ \begin{pmatrix} x & yd \\ y & x \end{pmatrix}, \quad x, y \in K, \ (x, y) \neq (0, 0) \right\}.$$

We are interested in the lattice $\mathcal{L}(T, G)$. For a split maximal torus we were practically done once we could prove that our intermediate subgroup H contains an elementary transvection. Here the situation is strictly opposite: at this point the real work just starts.

Theorem. *Let* $T \leq H \leq G$. *Then either* $H \leq N_G(T)$, *or* H *contains an elementary transvection.*

We can associate with any intermediate subgroup H an additive subgroup $A \subseteq K^+$ setting:

$$A = A(H) = \left\{ \alpha \in K \mid \begin{pmatrix} 1 & 0 \\ \alpha & 1 \end{pmatrix} \in H \right\}.$$

This subgroup is called the *module of transvections* of the subgroup H. Further we consider the corresponding *ring of multipliers*

$$R = R(H) = \{ \lambda \in K, \ \lambda A \subseteq A \}.$$

It is clear that R is a ring and A is an R-module.

The first natural question is which R and A do arise? It turns out that there is some absolute ring of multipliers R_0 and a subring $R \leq K$ has the form $R(H)$ if and only if it contains R_0. This R_0 may be defined as follows: it is the subring in K generated by all elements of the form

$$\frac{d}{x^2 - d}, \qquad x \in K.$$

For example, for $K = \mathbb{Q}$ this R_0 has the form $R_0 = P_0{}^{-1}\mathbb{Z}$, where $P_0 = P$,

$$P = \left\{ p \text{ odd prime} , \ \left(\frac{d}{p}\right) = 1 \right\},$$

if $d \not\equiv 1(\mod 8)$ and $P_0 = P \cup \{2\}$ if $d \equiv 1(\mod 8)$.

At this point the following technical condition is imposed on the torus T: we assume that $2 \in R_0{}^*$. Under this condition the lattice of intermediate subgroups is completely described in [Ko8] (see also [BK2] for some preliminary results and [Ko7], [BKH] for a complete analysis of the case $K = \mathbb{Q}$). We reproduce this description below. Without this condition the answer is much more complicated.

Theorem. *For a subring* $R \subseteq K$ *and an* R-*submodule* $A \subseteq K$ *to have the form* $R = R(H)$, $A = A(H)$ *for some intermediate subgroup* H, $T \leq H \leq G$, *it is necessary and sufficient that* $R \supseteq R_0$ *and* $dA^2 \subseteq R$.

For the field $K = \mathbb{Q}$ this result holds also without assumption $2 \in R_0$. Thus, in particular, for *any* d there are $\aleph_1$ intermediate subgroups (as compared to only *five* of them for the case of a split maximal torus). Clearly, only *three* of these subgroups (the torus T itself, its normalizer and the whole group G) are algebraic, all others are Zariski dense in G.

The next question is to describe all intermediate subgroups with given R and A. Denote this sublattice of $\mathcal{L}(T, G)$ by $\mathcal{L}_{R,A} = \mathcal{L}_{R,A}(T, G)$ It turns out that there exists a unique smallest subgroup and a unique largest subgroup in $\mathcal{L}_{R,A}$.

Theorem. *The smallest subgroup in $\mathcal{L}_{R,A}$ is the group*

$$F = \left\langle T, \begin{pmatrix} 1 & 0 \\ \alpha & 1 \end{pmatrix}, \ \alpha \in A \right\rangle = T \begin{pmatrix} 1 & 0 \\ A & \Omega_0(A) \end{pmatrix},$$

where the subgroup $\Omega_0(A) \leq R^*$ is defined as follows:

$$\Omega_0(A) = \left\langle \frac{(d\alpha + x)^2 - d}{x^2 - d}, \ x \in K, \ \alpha \in A \right\rangle.$$

Now consider the ideal Q in R generated by the elements

$$\frac{dx}{x^2 - d}, \qquad x \in K,$$

and the set

$$S = \{ s \in R, \ Qs \subseteq dA \}.$$

Clearly S is an ideal in R such that $QA \subseteq S \subseteq A$.

Theorem. *The largest subgroup in $\mathcal{L}_{R,A}$ is the group*

$$F^0 = T \begin{pmatrix} 1 & 0 \\ A & \Omega^0(A) \end{pmatrix},$$

where the subgroup $\Omega^0(A) \leq R^*$ is defined as follows:

$$\Omega^0(A) = \{ \theta \in R^*, \ \theta^2 - 1 \in S \}.$$

Now we can explicitly describe the lattice $\mathcal{L}_{R,A}$.

Theorem. *Every subgroup H from $\mathcal{L}_{R,A}$ has the form*

$$H = T \begin{pmatrix} 1 & 0 \\ A & \Delta \end{pmatrix},$$

where Δ is a subgroup of R^* such that $\Omega_0(A) \leq \Delta \leq \Omega^0(A)$ and for all $x \in K$ and $\delta \in \Delta$ one has

$$1 + \frac{1}{\theta^2} \frac{d}{x^2 - d} (\theta^2 - 1) \in \Delta.$$

The final theorem of this section shows that the usual operations of passing to the normalizer and to the normal closure of T inside an intermediate subgroup leave $\mathcal{L}_{R,A}$ invariant.

Theorem. *For any $H \in \mathcal{L}_{R,A}$ one has $N_G(H), T^H \in \mathcal{L}_{R,A}$.*

Of course this pattern in the description of intermediate subgroups is more complicated, than anything we have encountered so far, since passage from H to T^H may generate *infinite* descending chains. The paper [BKH] and unpublished works by V.A.Koibaev show that still more complicated patterns arise when $2 \notin R_0^*$ and for cyclic tori in the general linear groups $GL(n, K)$.

§ 19. OVERGROUPS OF TENSORED SUBGROUPS

Let K be a field and $U_1, \dots, U_t$ be vector-spaces over K of dimensions $n_1, \dots, n_t$ respectively. Then their tensor product $V = U_1 \otimes \dots \otimes U_t$ has dimension $n = n_1 \dots n_t$. Thus one gets a natural embedding

$$D = \mathrm{GL}(n_1, K) \otimes \dots \otimes \mathrm{GL}(n_t, K) \le G = \mathrm{GL}(n, K).$$

These groups are obtained by fusing the Aschbacher classes $\mathcal{C}_4$ and $\mathcal{C}_7$. We are interested in the lattice $\mathcal{L}(D, G)$.

Of course once more we are not concerned about their maximality, we just want them to be large enough. If one wants to get maximal subgroups one must either assume that $t = 2$ and $n_1 \ne n_2$ (this is the class $\mathcal{C}_4$), or else that $n_1 = \dots = n_t = m$ and take the normalizer of D (this is the class $\mathcal{C}_7$). This normalizer is denoted by $\mathrm{GL}(m, K) \uparrow S_t$. As an abstract group it is isomorphic to the wreath product $\mathrm{GL}(n, K) \wr S_t$, but in a primitive representation.

One can construct analogous subgroups also for other groups of Lie type. Sometimes they fall into different families. For example, both $\mathrm{Sp}(l, K) \otimes \mathrm{Sp}(m, K)$ and $\mathrm{SO}(l, K) \otimes \mathrm{SO}(m, K)$ lie in the orthogonal group $\mathrm{O}(n, K)$, $n = lm$. For the finite case all the details about such embeddings may be found in [KL2].

Again the group D contains quadratic unipotent elements, so that for a finite field of characteristic $\ne 2$ description of *irreducible* subgroups containing D *up to conjugacy* may be deduced from the theory of quadratic pairs.

We restrict ourselves to a special case of a result by Li Shang Zhi [L9].

Theorem. Let $D = \mathrm{SL}(l, K) \otimes \mathrm{SL}(m, K)$, where $l, m \ge 3$. Let further, $G = \mathrm{GL}(n, K)$, $n = lm$. Then any subgroup H, $D \le H \le G$ either *normalizes* D *or* $\mathrm{SL}(n, K)$.

When $l = 2$ or $m = 2$ there are additional examples of complete intermediate subgroups. Li Shang Zhi has obtained also analogous results for overgroups of $\mathrm{Sp}(l, K) \otimes \mathrm{Sp}(m, K)$, $l, m \ge 4$, in $\mathrm{GL}(n, K)$, $n = lm$, and announced similar results for other classical groups.

There is no doubt that using ideas similar to those described in § 11 and § 13 one can obtain analogous results for an arbitrary commutative ring R and arbitrary number of factors t under assumption $n_1, \dots, n_t \ge 3$. The answer involves ideals of R and is similar to, but easier than, that for the case of subsystem subgroups. As already mentioned, for the fields of char $\ne 2$ this answer should follow also from the theory of quadratic pairs and its infinite analogue developed by F. Timmesfeld.

§ 20. OVERGROUPS OF CLASSICAL SUBFIELD SUBGROUPS

In this section we state results about the overgroups of the classical subgroups over subfields. These results touch Aschbacher classes C_5 and C_8. In the form we reproduce them below they are due to E.L.Bashkirov [Ba1] – [Ba7], but of course many special cases were known before.

Theorem. *Let K be an algebraic extension of a subfield L. Assume that either $n \geq 3$ or $n = 2$ and char $K \neq 2$. Then for any subgroup H, such that $\mathrm{SL}(n, L) \leq H \leq G = \mathrm{GL}(n, K)$, there exists a unique intermediate subfield E, $L \leq E \leq K$, such that*

$$\mathrm{SL}(n, E) \leq H \leq N_G(\mathrm{SL}(n, E)).$$

The proof of this result in based on the following remarkable fact which is a generalization of Dickson's lemma. Recall that any subgroup in $\mathrm{GL}(n, K)$ conjugate to $\{t_{12}(\alpha), \ \alpha \in L\}$ is called an L-root subgroup. The following theorem essentially answers the question what can be generated by two L-root subgroups.

Theorem. *Let L be an infinite field and λ be an element algebraic over L. If char $L = 2$ suppose that λ is separable over L. Then*

$$\left\langle \begin{pmatrix} 1 & L \\ 0 & 1 \end{pmatrix}, \begin{pmatrix} 1 & 0 \\ \lambda L & 1 \end{pmatrix} \right\rangle = \mathrm{SL}(2, L(\lambda)).$$

When λ is *transcendental* over L a theorem of J.Tits asserts that the group on the left hand side is isomorphic to the free product of two copies of L^+ (see [T3]). This lemma was first proven by E.L.Bashkirov by very elaborate explicit calculations [Ba1], [Ba4]. Later V.I.Chernousov found a much shorter algebro-geometric proof. Unfortunately this proof is not published, see [Z5] for an exposition.

Analogous results hold also for overgroups of other classical groups over subfields. As a pattern we state the following results from [Ba3] and [Ba7].

Theorem. *Let K be an algebraic extension of a subfield L. Assume that either $n = 2l \geq 4$ and char $K \neq 2$. Then for any subgroup H, $\mathrm{Sp}(n, L) \leq H \leq G = \mathrm{GL}(n, K)$, there exists a unique intermediate subfield E, $L \leq E \leq K$, such that one of the following holds*

(1) $\mathrm{SL}(n, E) \leq H \leq N_G(\mathrm{SL}(n, E))$;

(2) $\mathrm{Sp}(n, E) \leq H \leq N_G(\mathrm{Sp}(n, E))$;

(3) $\mathrm{SU}(n, E) \leq H \leq N_G(\mathrm{SU}(n, E))$.

This theorem essentially contains description of irreducible subgroups in $GL(n, K)$ generated by L-root subgroups. For $L \not\cong \mathbb{F}_3, \mathbb{F}_9$ any such subgroup is conjugate to one of the following groups: $SL(n, E)$, $SU(n, E)$, $Sp(n, E)$. Analogous result holds also for the case when char $K = 2$, but it is much more delicate, since in the non-separable case it involves subgroups defined by *pairs* of intermediate subfields, see [Ba6].

Theorem. *Let K be an algebraic extension of a subfield L. Assume that either $n = 2l \geq 4$ and char $K \neq 2$. Then for any subgroup H, $\Omega(n, L) \leq H \leq G = GL(n, K)$, one of the following holds:*

(1) H contains an L-root subgroup;

(2) there exists a unique intermediate subfield E, $L \leq E \leq K$, such that

$$\Omega(n, E) \leq H \leq N_G(\Omega(n, E)).$$

These results generalize a number of preceding results, including in particular results of R.W.Dye, O.H.King and Li Shang Zhi (see references in [KL], [L8], [L10], [L11]). Compare also [B5], [Nu].

In fact one can consider a more general problem. Let Φ be an arbitrary root system and $R \leq S$ be an arbitrary embedding of rings with 1. Describe subgroups intermediate between $G(\Phi, R)$ and $G(\Phi, S)$. This problem is solved in several special cases. One of the typical situations considered by many authors is when S is a good ring, like, say a Dedekind domain, and R is its field of fractions. This situation was studied by N.S.Romanovskii, R.A.Schmidt, J.E.Cremona, A.E.Zalesskii and others, see [Ro], [Sch], [Cr], [Z4]. A.V.Stepanov established much more general results stated in terms of stability conditions (see references in [Sv]).

§ 21. Concluding remarks

In the preceding sections we treated only several typical situations. In fact for specific fields (like finite, algebraically closed, local, global, etc.) and for various classes of rings (commutative, semilocal, subject to stability conditions, Dedekind domains, Hasse domains, etc.) there are *many* further results in the same spirit.

In this survey we presented almost exclusively results of *geometric* nature, where the answer does not essentially depend on the ground field. (Apart possibly from very few small exceptions). For example, the description of overgroups of a *split* maximal torus is a geometric result, saying that everything looks exactly as in the absolute case. The theorem of Seitz from § 17 is another example of a geometric result: the answer is different from that for the absolute case, but again it does not depend on the *finite* ground field and may be modeled at the absolute level. On the other hand § 18 gives an example of an *arithmetic* result: the absolute case provides no insight whatsoever.

One explanation of the geometric answers is that in reality everything happens in a linear object, like an associative algebra, a Lie algebra or a Jordan algebra. Let us illustrate this in the example of $GL(n, R)$. In the examples considered in §§ 4 – 17 the basic subgroups were always multiplicative groups of certain R-subalgebras in $M(n, R)$ (in fact subalgebras defined by *linear* congruences). The situation in § 18 was quite different.

Observe though that for rings this distinction becomes somewhat shaky even in such utterly geometric results as the description of overgroups of a subsystem subgroup. Of course the *qualitative* answer does not essentially depend on the ground ring: only on its lattice of ideals. But as soon as one tries to actually *calculate* the factor-groups like $N(\sigma)/E(\sigma)$ one inevitably encounters K-functors. But explicit calculations of K-functors depend in a critical way on the *arithmetic* nature of a ring.

So far we have not touched the class $\mathcal{S}$. In this context one may observe that *simple* irreducible groups defined over the ground *field* rarely have many overgroups. For an algebraically closed field this was proven by E.B.Dynkin in characteristic 0 and by G.M.Seitz [S11] in the general case. Exceptional groups tend to have few simple subgroups which act irreducibly on their minimal modules, see [Te], [S15]. For the finite case see [LSS], [STe], [S14]. Some results of this sort have been known before, especially for finite fields, see, for example [Co] or [Su1], [Su2]. It is probable that results of such type remain valid for arbitrary fields provided the subgroup is isotropic enough.

On the other hand description of overgroups of a reducible subgroup H may be quite complicated even when H itself has a very transparent structure. One of the arising difficulties is that it is not always immediately clear how to connect the *irreducible* overgroups of H with the structure of the irreducible components of H. For example, I.D.Suprunenko classified subgroups of $GL(n, p^m)$ containing $SL(2, p)$ in a representation with two irreducible components. As a recent example we could mention [DP1], [DP2].

ACKNOWLEDGEMENTS

I could repeat here everything I said in [V16], [V17], [V19]. Even not doing so I must acknowledge the influence of constant discussions and correspondence with Z.I.Borewicz, V.A.Koibaev, G.M.Seitz and A.E.Zalesskii on the development of my views on the subject.

I would like to express sincere gratitude to my *Gastgeber* A.Bak for friendly attention and support during my stay in Germany. Many German colleagues, of whom I must mention H.Abels, B.Fischer, H.Helling,

J.Mennicke, U.Rehmann, K.Ringel and R.Scharlau in Bielefeld, O.Kegel, G.Malle and F.Timmesfeld outside of it, helped me in a number of ways. It was very useful to discuss the subject with V.P.Platonov, who visited Bielefeld in 1993.

I very much appreciate the help of C.Parker at the final stage of the preparation of this work. He carefully read the manuscript and corrected it in many places (English and otherwise).

BIBLIOGRAPHY

[A1] Aschbacher M. On the maximal subgroups of the finite classical groups. – *Invent. Math.* – 1984. – V.76, N.3. – P.469–514.

[A2] Aschbacher M. Subgroup structure of finite groups. – *Proc. Rutgers Group Theory Year. 1983–84.* – Cambridge Univ.Press., 1984.

[A3] Aschbacher M. Finite group theory. – *Cambridge Studies in Advanced Math.* – Cambridge Univ.Press., 1986, 274P.

[A4] Aschbacher M. Finite simple groups and their subgroups. – *Lecture Notes Math.* – 1986. – V.1185. – P.1–57.

[A5] Aschbacher M. Overgroups of Sylow subgroups in sporadic groups. – *Mem. Amer. Math. Soc.* – 1986. – V.343. – 235P.

[Az] Azad H. Root groups. – *J. Algebra.* – 1982. – V.76, N.1. – P.211–213.

[BB] Ba M.S., Borewicz Z.I. On the arrangement of intermediate subgroups. – *Rings and Linear Groups.* – Krasnodar, 1988. – P.14–41 (In Russian).

[Ba1] Bashkirov E.L. On subgroups of $SL_2(K)$ over an infinite field K. – *Vesti Akad. Nauk BSSR, ser. fiz.–mat. nauk.* – 1985. – N.2. – P.112–113; *Manuscript deposited in VINITI.* – 1984 – N.7369–84. – 64P. (In Russian).

[Ba2] Bashkirov E.L. On linear group containing the special unitary group of non-zero index. – *Vesti Akad. Nauk BSSR, ser. fiz.–mat. nauk.* – 1986. – N.5. – P.122–123; *Manuscript deposited in VINITI.* – 1985 – N.5897–85. – 36P. (In Russian).

[Ba3] Bashkirov E.L. On linear group containing the symplectic group. – *Vesti Akad. Nauk BSSR, ser. fiz.–mat. nauk.* – 1987. – N.3. – P.116–117; *Manuscript deposited in VINITI.* – 1986 – N.2616–86. – 18P. (In Russian).

[Ba4] Bashkirov E.L. On subgroups of the group $SL_2(K)$ over a non-perfect field K of characteristic 2. – *Vesti Akad. Nauk BSSR, ser. fiz.–mat. nauk.* – 1987. – N.5. – P.115–116; *Manuscript deposited in VINITI.* – 1986 – N.3248–86. – 25P. (In Russian).

[Ba5] Bashkirov E.L. Some subgroups of the special linear group over a field. – *Vesti Akad. Nauk BSSR, ser. fiz.–mat. nauk.* – 1991. – N.1. – P.31–35. (In Russian).

[Ba6] Bashkirov E.L. Linear groups that contain the group $Sp_n(K)$ over a field of characteristic 2. – *Vesti Akad. Nauk BSSR, ser. fiz.–mat. nauk.* – 1991. – N.4. – P.21–26. (In Russian).

[Ba7] Bashkirov E.L. On linear groups containing the commutator subgroup of the orthogonal group of index at least 1. – *to appear in Siberian Math. J.*

[Ba8] Bashkirov E.L. English summary of works by E.L.Bashkirov. – *unpublished.*

[BMo] Berman S., Moody R. Extensions of Chevalley groups. – *Israel J. Math.* – 1975. – V.22, N.1. – P.42–51.

[Bo]   Borel A. Properties and linear representations of Chevalley groups. – *Lecture Notes Math.* – 1970. – V.131. – P.1–55.

[BT]   Borel A., Tits J. Groupes réductifs. – *Publ. Math. Inst. Hautes Et. Sci.*, 1965. – N.27. – P.55–150.

[B1]   Borewicz Z.I. On parabolic subgroups in linear groups over a semilocal ring.   *Vestnik Leningr. Univ. Math.* – 1981. – V.9. – P.187–196.

[B2]   Borewicz Z.I. On parabolic subgroups of the special linear group over a semilocal ring. – *Vestnik Leningr. Univ. Math.* – 1981. – V.9. – P.245–251.

[B3]   Borewicz Z.I. Description of subgroups of the general linear group that contain the group of diagonal matrices. – *J. Sov. Math.* – 1981. – V.17, N.2. – P.1718–1730.

[B4]   Borewicz Z.I. Some subgroups of the general linear group. – *J Sov. Math.* – 1982. – V.20, N.6. – P.2528–2532.

[B5]   Borewicz Z.I. Description of subgroups of general linear groups which are full of transvections. – *J. Sov. Math.* – 1987. – V.37. – P.928–934.

[B6]   Borewicz Z.I. On the arrangement of subgroups. – *J Sov. Math.* – 1981. – V.19, N.1. – P.977–981.

[BK1] Borewicz Z.I., Koibaev V.A. Subgroups of the general linear group over the field of five elements. – In: *Algebra and Number Theory*, Ordzhonikidze (Vladikavkaz). – 1978. – N.3. – P.9–32. (In Russian)

[BK2] Borewicz Z.I., Koibaev V.A. Subgroups of $GL(2, K)$ containing a non-split maximal torus. – *Vestnik. Sankt-Peterburg. Univ.* – 1993. – N.2.

[BKH] Borewicz Z.I., Koibaev V.A., Tran Ngoc Hoi Lattices of subgroups in $GL(2, \mathbb{Q})$ containing a non-split torus. – *J. Sov. Math.* – 1993. – V.63, N.6. – P.622–634.

[BKo] Borewicz Z.I., Kolotilina L.Yu. Normalizers of intermediate subgroups in the general linear group over a ring. – *Vestn. Leningr. Univ., Math.* – 1988. – V.21, N.1. – P.1–4.

[BM]   Borewicz Z.I., Macedońska O.N. On the lattice of subgroups. – *J. Sov. Math.* – 1984. – V.24, N.4. – P.395–399.

[BV1] Borewicz Z.I., Vavilov N.A. Subgroups of the general linear group over a semilocal ring containing the group of diagonal matrices. – *Proc. Steklov Inst. Math.* – 1980. – Issue 4. – P.41–54.

[BV2] Borewicz Z.I., Vavilov N.A. On the definition of a net subgroup. – *J. Sov. Math.* – 1985. – V.30. – P.1810–1816.

[BV3] Borewicz Z.I., Vavilov N.A. The distribution of subgroups in the general linear group over a commutative ring – *Proc. Steklov. Inst. Math.* – 1985. – N.3. – P.27–46.

[Bu] Bourbaki N. *Groupes et algèbres de Lie. Ch. 4–6.* – Hermann: Paris. – 1968. – 288P.

[BGL] Burgoyne N., Griess R.L., Lyons R. Maximal subgroups and automorphisms of Chevalley groups. – *Pacif. J. Math.* – 1977. – V.71, N.2. – P.365–403.

[C1] Carter R.W. *Simple groups of Lie type.* – London et al.: Wiley. – 1972. – 331P.

[C2] Carter R.W. Conjugacy classes in the Weyl group. – *Compositio Math.* – 1972. – V.25, N.1. – P.1–59.

[C3] Carter R.W. *Finite groups of Lie type: Conjugacy classes and complex characters.* – London et al.: Wiley. – 1985. – 544P.

[CPS] Cline E., Parshall B., Scott L. Minimal elements of $N(H,p)$ and conjugacy of Levi complements of finite Chevalley groups. – *J. Algebra.* – 1975. – V.34, N.3. – P.521–523.

[Co] Cooperstein B.N. Nearly maximal representations for the special linear group. – *Michigan Math. J.* – 1980. – V.27, N.1. – P.3–19.

[CK] Costa D.L., Keller G.E. The $E(2, A)$ sections of $SL(2, A)$. – *Ann. Math.* – 1991. – V.134. – P.159–188.

[Cr] Cremona J.E. On $GL(n)$ of Dedekind domains. – *Quart. J. Math., 2nd ser.* – 1988. – V.39, N.156. – P.423–426.

[De] Dempwolff U. Linear groups with large cyclic subgroups and translation planes. – *Rend. Sem. Math. Univ. Padova.* – 1987. – V.77, N.1. – P.69–113.

[Dj] Djoković D.Ž. Subgroups of compact Lie groups containing a maximal torus are closed. – *Proc. Amer. Math. Soc.* – 1981. – V.83., N.1. – P.431–432.

[DP1] Djoković D.Ž., Platonov V.P. Algebraic groups and linear preserver problems. – *Comptes Rendus Acad. Sci. Paris., Sér. 1.* – 1993. – T.317. – P.925–930.

[DP2] Djoković D.Ž., Platonov V.P. Linear preserver problems and algebraic groups. – *to appear.*

[D1] Dye R.W. Maximal subgroups of symplectic groups stabilizing spreads. I, II. – *J. Algebra.* – 1984. – V.87, N.2. – P.493–509; *J. London. Math. Soc.* – 1989. – V.40, N.2. – P.215–226.

[D2] Dye R.W. Maximal subgroups of finite orthogonal groups stabilizing spreads of lines. – *J. London. Math. Soc.* – 1986. – V.33, N.3. – P.279–293.

[D3] Dye R.W. Spreads and classes of maximal subgroups of $GL_n(q)$, $SL_n(q)$, $PGL_n(q)$ and $PSL_n(q)$. – *Ann. Math. Pura Appli..* – 1991. – V.158. – P.33–50.

[Dn] Dynkin E.B. Semi-simple subalgebras of semi-simple Lie algebras. – *Amer. Math. Soc. Transl. Ser.* – 1957. – V.6. – P.111–244.

[Fr] Friedland S. Maximality of the monomial subgroup. – *Linear Multilinear Algebra.* – 1985. – V.18. – P.1–7.

[Go] Golubchik I.Z. On the subgroups of the general linear group $GL_n(R)$ over an associative ring R. – *Russian Math. Surveys.* – 1984. – V.**39**, N.1. – P.157–158.

[HO] Hahn A.J., O'Meara O.T. *The classical groups and K-theory.* – Springer: Berlin et al. – 1989. – 576P.

[Ha] Hamdan I. *Subgroups of the general linear group over the field of fractions of a semilocal ring.* – Ph. D. Thesis, Leningrad State Univ. – 1987. – 79P. (In Russian).

[HV] Harebov A.L., Vavilov N.A. On the lattice of subgroups of Chevalley groups containing a split maximal torus. – *Preprint Univ. Warwick.* – 1993. – N.14. – 27P.

[He] Hering Ch. Transitive linear groups and linear groups which contain irreducible subgroup of prime order. I, II. – *Geom. dedic.* – 1974. – V.2, N.4. – P.425–460; *J. Algebra.* – 1985. – V.93, N.1. – P.151–164.

[Ho] Hołubowski W. Subgroups of isotropic orthogonal groups containing the centralizer of a maximal split torus. – *J. Sov. Math.* – 1993. – V.63, N.6. – P.653–656.

[H1] Humphreys J.E. *Linear algebraic groups.* – Springer: New York et al. – 1975. – 247P.

[H2] Humphreys J.E. *Introduction to Lie algebras and representation theory, 3rd Printing.* – Springer: New York et al. – 1980. – 171P.

[Km] Kariyama K. On conjugacy classes of maximal tori in classical groups. – *J. Algebra.* – 1989. – V.125, N.1. – P.133–149.

[Ka1] Kantor W.M. Subgroups of classical groups generated by long root elements. – *Trans. Amer. Math. Soc..* – 1979. – V.248, N.2. – P.347–379.

[Ka2] Kantor W.M. Linear groups containing a Singer cycle. – *J. Algebra.* – 1980. – V.62, N.1. – P.232–234.

[Ka3] Kantor W.M. Generation of linear groups. – In *The geometric Vein: Coxeter Festschrift*. – Springer Verlag. – Berlin et al., 1981. – P.497–509.

[Ka4] Kantor W.M. Primitive permutation groups of odd degree, and an application to finite projective planes. – *J. Algebra*. – 1987. – V.106, N.1. – P.15–45.

[Ke1] Key J.D. Some maximal subgroups of $PSL(n, q)$, $n \geq 3$, $q = 2^r$. – *Geom. Dedic.* – 1975. – V.4, N.2–4. – P.377–386; erratum – *ibid.* – 1977. – V.6, N.3. – P.389.

[Ke2] Key J.D. Some maximal subgroups of certain projective unimodular groups. – *J. London Math. Soc.* – 1979. – V.19, N.2. – P.219–230.

[K1] King O.H. On subgroups of the special linear group containing the special orthogonal group. – *J. Algebra*. – 1985. – V.96, N.1. – P.178–193.

[K2] King O.H. On subgroups of the special linear group containing the diagonal subgroup. – *J. Algebra*. – 1990. – V.132, N.1. – P.198–204.

[K3] King O.H. The subgroup structure of the classical groups. – *Contemp. Math.* – 1992. – V.131, Part.1. – P.209–215.

[KL1] Kleidman P., Liebeck M.W. A survey of the maximal subgroups of the finite simple groups. – *Geom. dedic.* – 1988. – V.25. – P.375–389.

[KL2] Kleidman P., Liebeck M.W. *The subgroup structure of the finite classical groups*. – Cambridge Univ. Press. – 1990. – 303P.

[Ko1] Koibaev V.A. Some examples of non-monomial linear groups without transvections. – *J. Sov. Math.* – 1982. – V.20, N.6. – P.2610–2611.

[Ko2] Koibaev V.A. Subgroups of the full linear group over the field of four elements. – In: *Algebra and Number Theory*, Nalchik. – 1979. – N.4. – P.21–31. (In Russian).

[Ko3] Koibaev V.A. A description of D-complete subgroups of the full linear group over the field of three elements. – *J. Sov. Math.* – 1984. – V.24, N.4. – P.434–436.

[Ko4] Koibaev V.A. Subgroups of the full linear group over the field of three elements. – In: *Structural Properties of Algebraic Systems*, Nalchik. – 1981. – P.56–68. (In Russian).

[Ko5] Koibaev V.A. On subgroups of the full linear group that contain the group of elementary block-diagonal matrices. – *Vestn. Leningr. Univ., Math.* – 1982. – V.15. – P.169–177.

[Ko6] Koibaev V.A. Subgroups of the special linear group over fields of four or five elements which contain the group of diagonal matrices. – *Arithmetic and Geometry of Varieties*. – Kujbyshev, 1989. – P.78–91. (In Russian).

[Ko7] Koibaev V.A. Subgroups of $GL(2, \mathbb{Q})$ containing a non-split maximal torus. – *Dokl. Akad. Nauk SSSR*. – 1990. – V.41, N.3. – P.414–416.

[Ko8] Koibaev V.A. Subgroups of GL(2, K) containing a non-split maximal torus. − *Zap. Nauchn. Seminarov. Sankt-Peterburg. Matemat. Inst.* − 1993. − V.211.

[Kn] Kondratiev A.S. Subgroups of finite Chevalley groups. − *Russian Math. Surv.* − 1986. − V.41. − P.65–118.

[Kr1] Krupetskii S.L. Subgroups of the unitary group over a local field. − *J. Sov. Math.* − 1982. − V.19. − P.1027–1041.

[Kr2] Krupetskii S.L. Subgroups of the unitary group over a dyadic local field. − *J. Sov. Math.* − 1984. − V.24. − P.436–442.

[Kr3] Krupetskii S.L. Intermediate subgroups in the unitary group over a the skew field of quaternions. − *J. Sov. Math.* − 1984. − V.26. − P.1894–1897.

[L1] Li Shang Zhi Maximal subgroups in classical groups over arbitrary fields. − *Proc. Symp. Pure Math.* − 1987. − V.47, Part 2. − P.487–493.

[L2] Li Shang Zhi Maximality of symplectic groups over fields in linear groups. − *Kexue Tongbao.* − 1988. − V.33, N.21. − P.1608–1610. (Chinese).

[L3] Li Shang Zhi Overgroups of certain subgroups in the classical groups over division rings. − *Contemp. Math.* − 1989. − V.82. − P.53–57.

[L4] Li Shang Zhi Overgroups in GL(nr, F) of certain subgroups of SL(n, K). I. − *J. Algebra.* − 1989. − V.125, N.1. − P.215–235.

[L5] Li Shang Zhi The maximality of monomial subgroups of linear groups over division rings. − *J. Algebra.* − 1989. − V.127, N.1. − P.22–39.

[L6] Li Shang Zhi Overgroups of certain subgroups in the classical groups. − *Contemp. Math.* − 1989. − V.82. − P.53–57.

[L7] Li Shang Zhi Overgroups of SL(n, K) in GL(n, F) ($K \subset F$). − *Acta. Math. Sinica.* − 1990. − V.33, N.6. − P.774–778. (Chinese).

[L8] Li Shang Zhi Overgroups of SU(n, K, f) or $\Omega(n, K, f)$ in GL(n, K). − *Geom. dedic.* − 1990. − V.33, N.3. − P.241–250.

[L9] Li Shang Zhi Overgroups in GL($U \otimes W$) of certain subgroups of GL(U)$\otimes$ GL(W). I. − *J. Algebra.* − 1991. − V.137, N.2. − P.338–368.

[L10] Li Shang Zhi A new type of classical groups over skew-fields of characteristic 2. − *J. Algebra.* − 1991. − V.138, N.1. − P.399–419.

[L11] Li Shang Zhi Overgroups of a unitary group in GL(2, K). − *J. Algebra.* − 1992. − V.149, N.2. − P.275–286.

[LSa1] Liebeck M.W., Saxl J. The primitive permutation groups of odd degree. − *J. London Math. Soc.* − 1985. − V.31. − P.250–264.

[LSa2] Liebeck M.W., Saxl J. Maximal subgroups of finite simple groups and their automorphism groups. − *Contemp. Math.* − 1992. − V.131, Part 1. − P.243–259.

[LSS] Liebeck M.W., Saxl J., Seitz G.M. On the overgroups of irreducible sub-groups of the finite classical groups. – *Proc. London Math. Soc.* – 1987. – V.55, N.3. – P.507-537.

[LSe1] Liebeck M.W., Seitz G.M. Maximal subgroups of exceptional groups of Lie type, finite and algebraic. – *Geom. dedic.* – 1990. – V.35. – P.353–387.

[LSe2] Liebeck M.W., Seitz G.M. Subgroups generated by root elements in groups of Lie type. – *to appear.*

[LSe3] Liebeck M.W., Seitz G.M. Reductive subgroups of exceptional algebraic groups. – *to appear.*

[Ne] Newman M. *Integral matrices.* – N.–Y. – London, 1972. – 224P.

[Nu] Nuzhin Y.N. Groups contained between groups of Lie over various fields. – *Algebra and Logic.* – 1983. – V.22. – P.378–389.

[P] Platonov V.P. Subgroups of algebraic groups over local and global fields containing a maximal torus. – *submitted to Compt. Rendus. Acad. Sci. Paris.*

[PR] Platonov V.P., Rapinchuk A.S. *Algebraic groups and number theory.* – Moscow, Nauka. – 1991. – 656P. (In Russian, English Transl. in Springer).

[Ro] Romanovskii N.S. On subgroups of the general and the special linear groups over a ring. – *Math. Notes.* – 1971. – V.9. – P699–708.

[Sch] Schmidt R.A. Subgroups of the general linear groups over the field of a Dedekind ring. – *J. Sov. Math.* – 1982. – V.19. – P.1052–1059.

[S1] Seitz G.M. Flag-transitive subgroups of Chevalley groups. – *Ann. Math.* – 1973. – V.57, N.1. – P.27–56.

[S2] Seitz G.M. Small rank permutation representations of finite Chevalley groups. – *J. Algebra.* – 1974. – V.28, N.3. – P.508–517.

[S3] Seitz G.M. Subgroups of finite groups of Lie type. – *J. Algebra.* – 1979. – V.61, N.1. – P.16–27.

[S4] Seitz G.M. Properties of the known simple groups. – *Proc. Symp. Pure. Math.* – 1980. – V.37. – P.231–237.

[S5] Seitz G.M. On the subgroup structure of classical groups. – *Comm. Algebra.* – 1982. – V.10, N.8. – P.875–885.

[S6] Seitz G.M. Root subgroups for maximal tori in finite groups of Lie type. – *Pacif. J. Math.* – 1983. – V.106, N.1. – P.153–244.

[S7] Seitz G.M. Parabolic subgroups containing the centralizer of a unipotent element. – *J. Algebra.* – 1983. – V.184, N.1. – P.240–252.

[S8] Seitz G.M. Unipotent subgroups of groups of Lie type. – *J. Algebra.* – 1983. – V.84, N.1. – P.253–278.

[S9] Seitz G.M. Overgroups of irreducible linear groups. – *Proc. Rutgers Group Theory Year, 1983/84.* – Cambridge Univ. Press, 1984. – P.95–106.

[S10] Seitz G.M. Representations and maximal subgroups. – *Proc. Symp. Pure Math.* – 1987. – V.47. – P.275–287.

[S11] Seitz G.M. Maximal subgroups of classical algebraic groups. – *Mem. Amer. Math. Soc.* – 1987. – V.67, N.365. – 286P.

[S12] Seitz G.M. Representations and maximal subgroups of finite groups of Lie type. – *Geom. dedic.* – 1988. – V.25. – P.391–406.

[S13] Seitz G.M. Maximal subgroups of exceptional groups. – *Contemp. Math.* – 1989. – V.82. – P.143–157.

[S14] Seitz G.M. Cross-characteristic embeddings of finite groups of Lie type. – *Proc. London. Math. Soc.* – 1990. – V.60, N.1. – P.166-200.

[S15] Seitz G.M. Maximal subgroups of exceptional algebraic groups. – *Mem. Amer. Math. Soc.* – 1991. – V.90, N.441. – 197P.

[S16] Seitz G.M. Subgroups of finite and algebraic groups. – *Groups, Combinatorics and Geometry, (Durham – 1990)*. – Cambridge Univ. Press, 1992. – P.316–326.

[SST] Seitz G.M., Solomon R., Turull A. Chains of subgroups in groups of Lie type. II. – *J. London Math. Soc.* – 1990. – V.42, N.1. – P.93–100.

[STe] Seitz G.M., Testerman D.M. Extending morphisms from finite to algebraic groups. – *J. Algebra.* – 1990. – V.131. – P.559–574.

[STu] Solomon R., Turull A. Chains of subgroups in groups of Lie type. I. – *J. Algebra.* – 1990. – V.132, N.1. – P.174–184.

[Sp] Springer T.A. *Linear algebraic groups.* 2nd ed. – Birkhäuser: Boston et al.,1981. – 320P.

[SS] Springer T.A., Steinberg R. Conjugacy classes. – *Lecture Notes Math.* – 1970. – V.131. – P.167–266.

[St] Steinberg R. *Lectures on Chevalley groups.* – Yale University. – 1968. – 277P.

[Sv] Stepanov A.V. On the distribution of subgroups normalized by a given subgroup. – *J. Sov. Math.* – 1993. – V.198. – P.769–776.

[SK] Subbotin I.Ya., Kuzennyj N.F. On groups with fan subgroups. – *Contemp. Math.* – 1992. – V.131, Part. 1. – P.383–388.

[Su1] Suprunenko I.D. Subgroups of $GL(n,p)$ containing $SL(2,p)$ in the irreducible representation of degree n. – *Math. USSR Sbornik.* – 1980. – V.37. – P.425–440.

[Su2] Suprunenko I.D. Subgroups of $GL(n,p^m)$ containing $SL(2,p)$ in the irreducible representation of degree n. I, II. – *Izv. Akad. Nauk. BSSR, Ser. Fiz.–Math.* – 1979. – N.1. – P.18–24; N.2. – P.11–16. (In Russian).

[Sz1] Suzuki K. On parabolic subgroups of Chevalley groups over local rings. – *Tôhoku Math. J.* – 1976. – V.28, N.1. – P.57–66.

[Sz2] Suzuki K. On parabolic subgroups of Chevalley groups over commutative rings. – *Sci. Repts. Tokyo Kyoiku Daigaku, Sect.A.* – 1977. – N.366–382. – P.225–232.

[Te] Testerman D. Irreducible subgroups of exceptional algebraic groups. – *Mem. Amer. Math. Soc.* – 1988. – V.75, N.340. – 190P.

[Ti1] Timmesfeld F.G. On the identification of natural modules for symplectic and linear groups defined over arbitrary fields. – *Geom. dedic.* – 1990. – V.35, N.1. – P.127–142.

[Ti2] Timmesfeld F.G. Groups generated by k-transvections. – *Invent. Math.* – 1990. – V.100. – P.167–206.

[Ti3] Timmesfeld F.G. Groups generated by k-root subgroups. – *Invent. Math.* – 1991. – V.106. – P.575–666.

[Ti4] Timmesfeld F.G. Groups generated by k-root subgroups – a survey. – *Groups, Combinatorics and Geometry, (Durham – 1990).* – Cambridge Univ. Press, 1992. – P.183–204.

[T1] Tits J. Théorème de Bruhat et sous-groupes paraboliques. – *C. R. Acad. Sci. Paris.* – 1962. – V.254. – P.2910–2912.

[T2] Tits J. Groupes semi-simples isotropes. – *Colloq. Théorie des groupes algèbriques (Bruxelles, 1962).* – Paris, 1962. – P.137–147.

[T3] Tits J. Ensembles ordonnés, immeubles et sommes amalgamées. – *Bull. Soc. Math. Belg., Sér. A.* – 1986. – V.38. – P.367–387.

[Tu] Tuma J. Intervals in subgroup lattices of infinite groups. – *J. Algebra.* – 1989. – V.125. – P.367–399.

[V1] Vavilov N.A. Parabolic subgroups of Chevalley groups over a semi-local ring. – *J. Sov. Math.* – 1987. – V.37. – P.942–952.

[V2] Vavilov N.A. On subgroups of the general linear group over a semi-local ring that contain the group of diagonal matrices. – *Vestnik Leningr. Univ., Math.* – 1981. – V.14. – P.9–15.

[V3] Vavilov N.A. On subgroups of split orthogonal groups in even dimensions. – *Bull. Acad. Polon. Sci., Ser. Sci. Math.* – 1981. – V.29, N.9–10. – P.425–429.

[V4] Vavilov N.A. Bruhat decomposition for subgroups containing the group of diagonal matrices. I, II. – *J. Sov. Math.* – 1984. – V.24. – P.399–406; 1984. – V.27. – P.2865–2874.

[V5] Vavilov N.A. Parabolic subgroups of Chevalley groups over a commutative ring. – *J. Sov. Math.* – 1984. – V.26, N.3. – P.1848–1860.

[V6] Vavilov N.A. On subgroups of the special linear group that contain the group of diagonal matrices. I, II, III, IV. – *Vestnik Leningr. Univ., Math.*

– 1985. – V.18, N.4. – P.3–7; 1986. – V.19. – P.9–15; 1987. – V.20. – P.1–8; 1988. – N.3. – P.7–15.

[V7] Vavilov N.A. Maximal subgroups of Chevalley groups containing a split maximal torus. – Transl. Amer. Math. Soc., 2nd Ser. – 1991. – V.149. – P.53–59.

[V8] Vavilov N.A. Bruhat decomposition of one-dimensional transformations. – Vestnik Leningr. Univ., Math. – 1986. – V.19. – P.17–24.

[V9] Vavilov N.A. A conjugacy theorem for subgroups of GL_n containing the group of diagonal matrices. – Colloq. Math. – 1987. – V.54, N.1. – P.9–14.

[V10] Vavilov N.A. Subgroups of split classical groups. – Dr. Sci. Thesis (Habilitationsschrift), Leningrad State Univ. – 1987. – 334P. (In Russian).

[V11] Vavilov N.A. Weight elements of Chevalley groups. – Soviet Math. Dokl. – 1988. – V.37, N.1. – P.92–95.

[V12] Vavilov N.A. Conjugacy theorems for subgroups of extended Chevalley groups containing a split maximal torus. – Soviet Math. Dokl. – 1988. – V.37, N.2. – P.360–363.

[V13] Vavilov N.A. Structure of split classical groups over commutative rings. – Soviet Math. Dokl. – 1988. – V.37. – P.550–553.

[V14] Vavilov N.A. On subgroups of split orthogonal groups. I, II. – Siberian Math. J. – 1988. – V.29, N.3. – P.341–352; ibid, submitted.

[V15] Vavilov N.A. On subgroups of split orthogonal groups over a ring. – Siberian Math. J. – 1988. – V.29, N.4. – P.537–547.

[V16] Vavilov N.A. On subgroups of the split classical groups. – Proc. Math. Inst. Steklov. – 1991. – N.4. – P.27–41.

[V17] Vavilov N.A. Subgroups of Chevalley groups containing a maximal torus. – Transl. Amer. Math. Soc. – 1993. – V.155. – P.59–100.

[V18] Vavilov N.A. On subgroups of the spinor group containing a split maximal torus. I, II. – J. Sov. Math. – 1993. – V.63, N.6. – P.638–653.

[V19] Vavilov N.A. Structure of Chevalley groups over commutative rings. – Proc. Conf. Non-associative algebras and related topics, Hiroshima – 1990. – World Sci. Publ.: Singapore et al. – 1991. – P. 219–335.

[V20] Vavilov N.A. Unipotent elements in subgroups of extended Chevalley groups containing a split maximal torus. – Russian Math. Doklady. – 1993. – V.328, N.5. – P.536–539.

[V21] Vavilov N.A. Unipotent elements in subgroups which contain a split maximal torus. – Preprint Univ. Warwick. – 1993. – N.13. – 10P.

[VD] Vavilov N.A., Dybkova E.V. Subgroups of the general symplectic group containing the group of diagonal matrices. I, II. – J. Sov. Math. – 1984. – V.24. – P.406–416; 1985. – V.30. – P.1823–1832.

[VP] Vavilov N.A., Plotkin E.B. Net subgroups of Chevalley groups. I, II. – J. Sov. Math. – 1982. – V.19. – P.1000–1006; 1984. – V.27. – P.2874–2885.

[Wa] Walter J.H. Rigid cyclic subgroups in Chevalley groups. I. – J. Algebra. – 1990. – v.131, N.2. – P.688–702.

[W] Wilson J.S. Economical generating sets for finite simple groups. – This Volume.

[Z1] Zalesskii A.E. Linear groups. – Russian Math. Surveys. – 1981. – V.36, N.5. – P.56–107.

[Z2] Zalesskii A.E. Linear groups. – J. Sov. Math. – 1985. – V.31, N.3. – P.2974–3004.

[Z3] Zalesskii A.E. Linear groups. – Itogi Nauki i Techniki, Fundamental Trends of Math.. – 1989. – V.37. – P.114–228.

[Z4] Zalesskii A.E. On maximal subgroups of Chevalley groups. – unpublished.

[Z5] Zalesskii A.E. Subgroups of $SL_2(P)$ containing $SL_2(F)$, where P/F is an algebraic field extension. – unpublished.

THOMAS WEIGEL

On a certain class of Frattini extensions of finite Chevalley groups.

INTRODUCTION

Let X denote some simply connected simple Chevalley group scheme, i.e., $X = A_l,\ B_l,\ C_l,\ D_l,\ E_{6...8},\ F_4,\ G_2$. By $\mathbb{F}_q$ we will denote a finite field with q elements and we define $\mathbb{F}$ to be its algebraic closure. The purpose of this note is to consider extensions

$$1 \longrightarrow N \longrightarrow H \longrightarrow X(\mathbb{F}_q) \longrightarrow 1, \qquad (*)$$

where N is isomorphic to $\mathfrak{L}_X(\mathbb{F}_q) := \mathfrak{L}_X \otimes \mathbb{F}_q$ as $\mathbb{F}_q X(\mathbb{F}_q)$-module and $\mathfrak{L}_X$ denotes the $\mathbb{Z}$-Lie ring associated to a Chevalley basis.

In the first section we study a particular extension $(**)$ which is defined via the algebraic integers of some p-adic numberfield of characteristic 0. It will turn out that apart from a finite number of examples $(**)$ is a Frattini extension.

The study of extensions mentioned above is equivalent to the analysis of the second degree cohomology group $H^2(X(\mathbb{F}_q), \mathfrak{L}_X(\mathbb{F}_q))$. In the second section we will follow the approach given by E.Cline, B.Parshall, L.Scott and W.van der Kallen [3] to calculate this cohomology group in certain particular cases. It will turn out that "generically" the extension $(**)$ is the unique non-split extension.

In section 3 we will give some conclusions concerning the profinite completion $\widehat{X(A)}$ of the arithmetic group $X(A)$, where A is the ring of algebraic integers of some numberfield $\mathbb{K}$.

This note contains only sketches of proofs. Full proofs of the statements will appear in [9].

1. EXAMPLES ARISING FROM ARITHMETIC GROUPS.

Apart from the split extension there exists another "canonical" extension: Let $K/\mathbb{Q}_p$ be the unramified extension of $\mathbb{Q}_p$ with $res(K) \simeq \mathbb{F}_q$, let $\mathcal{O}$ denote the ring of algebraic integers of K and $\wp := p.\mathcal{O}$ the unique maximal ideal in $\mathcal{O}$. Put $R := \mathcal{O}/\wp^2$. Then $X(R) \simeq X(\mathcal{O})/M_2$, where M_k denotes the kth

congruence subgroup [7, Cor. 4.4.]. Recall that

$$M_k := \{ g \in X(\mathcal{O}) \mid \phi(g) \equiv id_W \,(\mathrm{mod}\ \wp^k) \}$$

for some appropriate faithful rational representation ϕ: $X(K) \longrightarrow GL_K(V)$, $im(\phi|_{X(\mathcal{O})}) \leq GL_{\mathcal{O}}(W)$ and for some free $\mathcal{O}$-module $W \leq V$ of maximal rank. Then $X(\mathcal{O})/M_1 \simeq X(\mathbb{F}_q)$ as abstract groups and $M_1/M_2 \simeq \mathfrak{L}_X(\mathbb{F}_q)$ as $\mathbb{F}_q X(\mathbb{F}_q)$-modules (cf. [9, (3.5.)]). For example if $X = A_{n-1} = SL_n$ and $K = \mathbb{Q}_p$, one has the extension

$$1 \longrightarrow \mathfrak{sl}_n(\mathbb{F}_p) \longrightarrow SL_n(\mathbb{Z}/p^2.\mathbb{Z}) \longrightarrow SL_n(\mathbb{F}_p) \longrightarrow 1$$

For these extensions one can show the following:

Theorem 1. *Let X and R be as above.*
(a) Let $(X,q) \neq (A_1,3), (A_2,2), (G_2,2)$. Then

$$1 \longrightarrow \mathfrak{L}_X(\mathbb{F}_q) \longrightarrow X(R) \longrightarrow X(\mathbb{F}_q) \longrightarrow 1 \qquad\qquad (**)$$

is a non-split extension.
(b) Let $(X,q) \neq (A_1,2), (A_1,3), (A_1,4), (A_2,2), (A_3,2), (B_2,2), (B_3,2),$ $(B_4,2), (C_3,2), (D_4,2), (F_4,2), (G_2,2)$. Then $(**)$ is a Frattini extension. Furthermore, if $(X,q) = (A_1,2), (A_1,3), (A_1,4), (A_2,2), (A_3,2), (B_2,2),$ $(B_3,2), (D_4,2), (G_2,2)$, then $(**)$ is a non-Frattini extension.

Remarks: (1) For $X = A_l$ and q prime the above theorem follows from some work of B.Beisiegel [1].
(2) For $p \geq 5$ this was remarked by R.L.Griess (Lecture given at the Santa Cruz Conference (1979)).
(3) Three cases remain to be considered: $(X,q) = (B_4,2), (C_3,2)$ or $(F_4,2)$.

Idea of the Proof. (a) The proof will be done in 2 steps. First one proves the assertion for some easier particular cases, namely $X = A_l$ or $(X,q) = (C_2,3)$, second one reduces the general case to one of these particular cases.
Let $H := SL_{l+1}(R)$, $l \geq 1$, and W the rational free RH-module of R-rank $l+1$. Then the following holds:

(i) $p \neq 2, 3$. Then there exists no element of order p in H, which acts as transvection on $W/p.W$.
(ii) $p = 3, l \geq 2$. Then there exist no elements $g, h \in H$ of order p, $[g,h] = 1$, which act as transvection on $W/p.W$.
(iii) $p = 3$, $l = 1$, $q > p$. Let $g \in H$ be an element of order p acting as transvection on $W/p.W$. Let $P \in Syl_p(C_H(g))$. Then P contains only 2 elements of order 3.

(iv) $p = 2$, $l \geq 3$. There exist no elements $\alpha_{i,j}$, $i \neq j$, $i,j \in \{1,\ldots,l+1\}$ of order 2 in H which act as transvection on $W/p.W$, and satisfy $[\alpha_{i,j},\alpha_{k,m}] = 1$ for $k \neq j$ and $[\alpha_{i,j},\alpha_{j,m}] = \alpha_{i,k}$.

(v) $p = 2$, $l = 3$, $q > p$. Let $\alpha \in H$ be an element of order 2 acting as transvection on $W/p.W$ with center $\mathfrak{C}$ and axis $\mathfrak{A}$. Then for every element $\beta \in H$ of order 2 acting on $W/p.W$ as transvection with center $\mathfrak{C}$ and axis $\mathfrak{A}$, one has $\alpha N = \beta N$, where $N = \ker(X(R) \to X(\mathbb{F}_q))$.

(vi) $p = 2$, $l = 1$. Then H does not contain elements of order 2 acting as transvection on $W/p.W$.

Thus the facts (i),..,(vi) imply that $(\ast\ast)$ is a non-split extension or $(X,q) = (A_1, 3)$, $(A_2, 2)$. Similar arguments as previously mentioned show that for $(X,q) = (C_2, 3)$, $(\ast\ast)$ is a non-split extension.

For the reduction step the following elementary lemma will be useful.

Lemma 2. (cf. [9, Lemma 4.4.]) *Let $C \hookrightarrow A \twoheadrightarrow B$ be a split extension of a finite group B by some finite elementary abelian p-group C. Assume that U is a subgroup of A such that $U \cap C$ has an U-invariant complement in C. Then $C \cap U \hookrightarrow U \twoheadrightarrow UC/C$ splits aswell.*

In our case this yields the following:

Lemma 3. (cf. [9, Lemma 4.6.]) *Let $Y \longrightarrow X$ be a simple simply-connected scheme corresponding to a subrootsystem Ψ of the rootsystem Φ of X. Let Δ denote a base of Ψ and Π a base of Φ. Assume that*

(1) $$rk(A_{sr})_{s \in \Pi, r \in \Delta} = rk(A_{rr'})_{r,r' \in \Delta} \quad or \quad rk(\Phi) = rk(\Psi),$$

and

(2) $r \in \Psi, s \in \Phi \setminus \Psi, ir + s \in \Phi$ *implies* $ir + s \notin \Psi$ *or* $|M_{r,s,i}| \equiv 0 (mod\ p)$,

where A_{rs} is the cartan matrix and $|M_{r,s,i}| = \binom{P+i}{i}$, where P denotes the maximal non-negative number k such that $s - kr \in \Phi$. Then a splitting of the extension $\mathcal{L}_X(\mathbb{F}_q) \hookrightarrow X(R) \twoheadrightarrow X(\mathbb{F}_q)$ induces a splitting of the extension $\mathcal{L}_Y(\mathbb{F}_q) \hookrightarrow Y(R) \twoheadrightarrow Y(\mathbb{F}_q)$.

The lemma allows a reduction to one of the previously mentioned cases or $(X,q) = (G_2, 2)$. One may choose the subscheme Y as listed in Table 1. This completes the proof of part (a).

For (b) additional information on the Loewy series of the $\mathbb{F}_q X(\mathbb{F}_q)$-module $\mathcal{L}_X(\mathbb{F}_q)$ is necessary.

Table 1

Φ		Ψ		Conditions on p
A_l,	$l \geq 1$	A_k,	$k < l$	$p \nmid (k+1)$
B_l,	$l \geq 2$	B_k,	$k < l$	$p \neq 2$
		A_k,	$k < l$	$p \nmid (k+1)$
B_2		A_1,	(long)	$*$
B_3		A_3,	(long)	$*$
B_4		A_3		$*$
C_l,	$l \geq 3$	C_k,	$k < l$	$p \neq 2$
		A_k,	$k < l$	$p \nmid (k+1)$
D_l,	$l \geq 4$	D_k,	$k < l$	$p \neq 2$
		A_k,	$k < l$	$p \nmid (k+1)$
D_4		A_3		$*$
E_l,	$l = 6, 7, 8$	D_5		$p \neq 2$
		A_4		$p \neq 5$
F_4		B_4		$*$
G_2		A_1,	(long)	$p \neq 2$
		A_2,	(long)	$*$

Lemma 4. (G.Hiß [4], G.Hogeweij [5], J.Hurley [6]) *Let* $(X, p) \neq (B_l, 2)$, $(C_l, 2)$, $(F_4, 2)$, $(G_2, 3)$. *Then* $soc(\mathfrak{L}_X(\mathbb{F}_q)) = Z(\mathfrak{L}_X(\mathbb{F}_q))$ *and* $Z(\mathfrak{L}_X(\mathbb{F}_q))$ *is a trivial* $X(\mathbb{F}_q)$-*module of* $\mathbb{F}_q$-*dimension* $rk(\Phi) - rk((A_{rs})_{r,s \in \Phi})$. *Furthermore,* $\mathfrak{L}_X(\mathbb{F}_q)/Z(\mathfrak{L}_X(\mathbb{F}_q))$ *is an absolutely irreducible* $\mathbb{F}_q X(\mathbb{F}_q)$-*module.*

Using Lemma 4 and the knowledge of the p-part of Schur multipliers of finite Chevalley groups one obtains that for $(X, p) \neq (B_l, 2)$, $(C_l, 2)$, $(F_4, 2)$, $(G_2, 3)$ and $(X, q) \neq (A_1, 2)$, $(A_1, 3)$, $(A_1, 4)$, $(A_2, 2)$, $(A_3, 2)$, $(D_4, 2)$, $(G_2, 2)$, $(**)$ is a Frattini extension. For the remaining cases one has to use a similar reduction argument as the previously mentioned one (cf. [9, Lemma 4.10.]). Finally one can show the following embeddings:

$$A_1(\mathbb{F}_3) \hookrightarrow A_1(\mathbb{Z}_3),$$
$$A_2(\mathbb{F}_2) \hookrightarrow A_2(\mathbb{Z}_2),$$
$$G_2(\mathbb{F}_2) \hookrightarrow G_2(\mathbb{Z}_2),$$
$$2.A_1(\mathbb{F}_2) \hookrightarrow A_1(\mathbb{Z}_2),$$
$$2.A_1(\mathbb{F}_4) \hookrightarrow A_1(\mathcal{O}),$$
$$2.B_3(\mathbb{F}_2) \hookrightarrow B_3(\mathbb{Z}_2),$$
$$2^2.D_4(\mathbb{F}_2) \hookrightarrow D_4(\mathbb{Z}_2),$$

where $\mathbb{Z}_p$ denote the p-adic integers and $\mathcal{O}$ the algebraic integers of the field

$K = \mathbb{Q}_2(\xi)$, where $\xi^2 + \xi + 1 = 0$. Furthermore, $2.A_3(\mathbb{F}_2) \hookrightarrow A_3(\mathbb{Z}/4.\mathbb{Z})$ and $2.B_2(\mathbb{F}_2) \leq B_2(\mathbb{Z}/4.\mathbb{Z})$. This completes the proof of the theorem.   ∎

2. REMARKS ON COHOMOLOGY.

The major source for our considerations concerning the cohomology group $H^2(X(\mathbb{F}_q), \mathfrak{L}_X(\mathbb{F}_q))$ is the following theorem of E.Cline, B.Parshall, L.Scott and W. van der Kallen [3, (6.6.)].

Theorem 5. *Let M be a finite dimensional rational $\mathbb{F}X(\mathbb{F})$-module and $n \in \mathbb{N}_0$. For a non-negative number e let $M^{[p^e]}$ be the module obtained by "twisting" with the Frobenius endomorphism $x \rightarrow x^{p^e}$. Then there exist non-negative integers e_0 and f_0 such that for $q \geq p^{f_0}$ and $e \geq e_0$*

$$H^n_{rat}(X(\mathbb{F}), M^{[p^e]}) \simeq H^n(X(\mathbb{F}_q), M^{[p^e]}) \simeq H^n(X(\mathbb{F}_q), M).$$

Indeed, in their proof they also show a way how to determine the numbers e_0 and f_0 explicitly for a given module M and $n \in \mathbb{N}_0$. In our case one obtains the following

Proposition 6. *Let $q = p^f$, $f \geq 2$ and $q \neq 4, 8, 9, 16, 25$. Then*

$$H^2_{rat}(X(\mathbb{F}), \mathfrak{L}_X(\mathbb{F})^{[p]}) \simeq H^2(X(\mathbb{F}_q), \mathfrak{L}_X(\mathbb{F})).$$

In the case that $\mathfrak{L}_X(\mathbb{F})$ is an irreducible $\mathbb{F}X(\mathbb{F})$-module, the second degree cohomology with coefficient in $\mathfrak{L}_X(\mathbb{F})$ has been determined by J.B.Sullivan (cf. [8, Prop.3]).

Theorem 7. *Let $\mathfrak{L}_X(\mathbb{F})$ be an irreducible $\mathbb{F}X(\mathbb{F})$-module, i.e., $(X, p) \neq (A_l, p), p|(l+1), (B_l, 2), (C_l, 2), (D_l, 2), (F_4, 2), (E_6, 3), (E_7, 2), (G_2, 3)$. Then*

$$H^2_{rat}(X(\mathbb{F}), \mathfrak{L}_X(\mathbb{F})^{[p^e]}) \simeq \begin{cases} (0) & \text{for } e = 0 \\ \mathbb{F} & \text{for } e > 0 \end{cases}$$

Thus Proposition 6 and Theorem 7 imply the following

Proposition 8. *Let $(X, p) \neq (A_l, p), p|(l + 1), (B_l, 2), (C_l, 2), (D_l, 2), (F_4, 2), (E_6, 3), (E_7, 2), (G_2, 3)$ and $q = p^f$, $f \geq 2$, $q \neq 4, 8, 9, 16, 25$. Then $H^2(X(\mathbb{F}_q), \mathfrak{L}_X(\mathbb{F})) \simeq \mathbb{F}$.*

This treatment cannot give any information if $\mathbb{F}_q$ is a prime field. For "small" fields and small rank there are examples where $H^2(X(\mathbb{F}_q), \mathfrak{L}_X(\mathbb{F})) \not\simeq \mathbb{F}$, e.g., $H^2(A_1(\mathbb{F}_3), \mathfrak{L}_{A_1}(\mathbb{F})) = 0$, $H^2(A_2(\mathbb{F}_2), \mathfrak{L}_{A_2}(\mathbb{F})) = 0$. However, it seems most likely that the condition $f \geq 2$ can be removed (up to a finite number of exceptions) from the proposition above. This is also motivated by the fact that for all primes $p \neq 2, 3$, $H^2(A_1(\mathbb{F}_p), \mathfrak{L}_{A_1}(\mathbb{F})) \simeq \mathbb{F}$ (cf.[2]).

Let G be a finite group and M a finite $\mathbb{F}_q G$-module. Then $H^*(G, M) = H^*(\mathrm{Hom}_{\mathbb{F}_q G}(\mathbf{P}, M))$, where $\mathbf{P} = (P_i, \partial_i)$ denotes a free resolution of the trivial module $\mathbb{F}_q$. Thus

$$H^*(G, M) \otimes_{\mathbb{F}_q} \mathbb{F} = H^*(\mathrm{Hom}_{\mathbb{F}_q G}(\mathbf{P}, M) \otimes_{\mathbb{F}_q} \mathbb{F}).$$

Since each of the free modules P_i can be chosen to be finite dimensional over $\mathbb{F}_q$, one has

$$\mathrm{Hom}_{\mathbb{F}_q G}(P_i, M) \otimes_{\mathbb{F}_q} \simeq \mathrm{Hom}_{\mathbb{F} G}(P_i \otimes_{\mathbb{F}_q} \mathbb{F}, M \otimes_{\mathbb{F}_q} \mathbb{F}).$$

Since $P_i \otimes_{\mathbb{F}_q} \mathbb{F}$ is a free $\mathbb{F} G$-module, $\mathbf{P} \otimes_{\mathbb{F}_q} \mathbb{F} = (P_i \otimes_{\mathbb{F}_q} \mathbb{F}, \partial_i \otimes id)$ is a free resolution of the trivial $\mathbb{F} G$-module $\mathbb{F}$. Thus

$$H^*(G, M) \otimes_{\mathbb{F}_q} \mathbb{F} \simeq H^*(G, M \otimes_{\mathbb{F}_q} \mathbb{F}).$$

Under the hypothesis of Proposition 8, $\mathfrak{L}_X(\mathbb{F}_q)$ is an absolutely irreducible $\mathbb{F}_q X(\mathbb{F}_q)$-module and $\mathfrak{L}_X(\mathbb{F}_q) \otimes_{\mathbb{F}_q} \mathbb{F} \simeq \mathfrak{L}_X(\mathbb{F})$ as $\mathbb{F} X(\mathbb{F}_q)$-module. Thus it follows that

$$H^2(X(\mathbb{F}_q), \mathfrak{L}_X(\mathbb{F}_q)) \simeq \mathbb{F}_q.$$

In particular, in this case there exists a unique non-split extension

$$1 \longrightarrow \mathfrak{L}_X(\mathbb{F}_q) \longrightarrow H \longrightarrow X(\mathbb{F}_q) \longrightarrow 1$$

and Theorem 1 implies that $H \simeq X(R)$.

3. Remarks on Arithmetic groups.

Theorem 1 generalizes immediately to the arithmetic groups $X(K)$, where $K/\mathbb{Q}_p$ is a finite unramified extension. If K is any finite extension field of $\mathbb{Q}_p$ one obtains the following:

Corollary A. *Let $K/\mathbb{Q}_p$ be a finite extension of the p-adic numbers. Let $\mathcal{O}$ denote the ring of algebraic integers of K and $q := res(K)$. Consider the extension*

$$1 \longrightarrow M_1 \longrightarrow X(\mathcal{O}) \longrightarrow X(\mathbb{F}_q) \longrightarrow 1, \qquad (\dagger)$$

where M_1 denotes the first congruence subgroup of $X(\mathcal{O})$. If $K/\mathbb{Q}_p$ is unramified, then the extension (†) is a Frattini extension or

$$(X,F) = (A_1,\mathbb{F}_2),(A_1,\mathbb{F}_3),(A_1,\mathbb{F}_4),(A_2,\mathbb{F}_2),(A_3,\mathbb{F}_2),$$
$$(B_2,\mathbb{F}_2),(B_3,\mathbb{F}_2),(B_4,\mathbb{F}_2),(C_3,\mathbb{F}_2),(D_4,\mathbb{F}_2),(G_2,\mathbb{F}_2),(F_4,\mathbb{F}_2).$$

On the other hand if (†) is a Frattini extension, then $K/\mathbb{Q}_p$ has to be unramified and

$$(X,F) \neq (A_1,\mathbb{F}_2),(A_1,\mathbb{F}_3),(A_1,\mathbb{F}_4),(A_2,\mathbb{F}_2),(A_3,\mathbb{F}_2),$$
$$(B_2,\mathbb{F}_2),(B_3,\mathbb{F}_2),(D_4,\mathbb{F}_2),(G_2,\mathbb{F}_2).$$

There exists also a globalisation of Corollary A. Let $\mathbb{K}$ be a numberfield, i.e., $|\mathbb{K}/\mathbb{Q}| < \infty$, and let A denote the algebraic integers of $\mathbb{K}$. Then the strong-approximation-property implies that $\widetilde{X(A)} \simeq X(\bar{A})$, where $\widetilde{X(A)}$ denotes the completion of the discrete group $X(A)$ with respect to the topology defined by the congruence subgroups of $X(\Lambda)$, and $\bar{A}$ denotes the closure of A in the finite adele ring of $\mathbb{K}$. In particular, $\bar{A} \simeq \bigoplus_{v \in} A_v$, where $\mathfrak{V}$ is the set of all finite places of $\mathbb{K}$ and A_v is the completion of A with respect to v. One can use the solution of the congruence-subgroup-problem for Chevalley groups (cf. [7]) to translate the globalisation of Corollary A in the language of profinite groups. Then one obtains the following:

Corollary B. *Let $\mathbb{K}$ be a algebraic number field and let A denote the ring of algebraic integers in $\mathbb{K}$. Let $\widetilde{X(A)}$ denote the profinite completion of the discrete group $X(A)$. Assume further that $X \neq A_1, B_2$ and also that for $p = 2$, $X \neq A_2, A_3, B_3, B_4, C_3, D_4, G_2, F_4$. Then $\mathbb{K}/\mathbb{Q}$ is unramified in p if and only if $O_p(\widetilde{X(A)}) \leq Frat(\widetilde{X(A)})$.*

REFERENCES.

[1] B.BEISIEGEL: 'Die Automorphismengruppen homozyklischer p-Gruppen'; *Arch. Math.* **29**, (1977), no.4, 363-366.

[2] R.BRAUER, C.NESBITT: 'On the modular characters of groups', *Ann. of Math.* **42**, (1941), 556-590.

[3] E.CLINE, B.PARSHALL, L.SCOTT, W.VAN DER KALLEN: 'Rational and generic cohomology', *Invent. math.* **39**, (1977), 143-163.

[4] G.HISS : 'Über die Liftbarkeit modularer Darstellungen endlicher Gruppen'; *'Dissertation'*, Freiburg, (1983)

[5] G.M.HOGEWEIJ: 'Almost-classical Lie-Algebras I'; *Indag. Math.* **44**, (1982), 441-460.

[6] J.F.HURLEY: 'Centers of Chevalley Algebras'; *J. Math. Soc. Japan* **34**, (1982), 219-222.

[7]M.MATSUMOTO: 'Sur les sous-groupes arithmetiques des groupes semisimple
    déployés'; *Ann. E.N.S.* (4) **2**, (1969), 1-62.
[8]J.B.SULLIVAN: 'Frobenius operations on Hochschild cohomology', *Amer. J.
    Math.* **104**, No.4, (1979), 765-780.
[9]TH.WEIGEL: 'On the profinite completion of arithmetic groups of split type',
    submitted to *J. Pure App. Alg.*.

ECONOMICAL GENERATING SETS
FOR FINITE SIMPLE GROUPS

John S. Wilson

1. Generators for finite simple groups

The search for small generating sets for finite simple groups, and other closely related groups, dates back to the beginning of the twentieth century, and, while it is less fundamental than the study of subgroup structure, it has always provided a good test for the power of existing techniques and an impetus for the development of new ones. In his book 'Linear Groups with an Exposition of the Galois Field Theory', written in 1900, L. E. Dickson proved that for each odd prime power $q \neq 9$ the group $\mathrm{SL}_2(q)$ can be generated by the two matrices

$$\begin{pmatrix} 1 & 1 \\ 0 & 1 \end{pmatrix} \quad \text{and} \quad \begin{pmatrix} 1 & 0 \\ \lambda & 1 \end{pmatrix},$$

where λ is a field generator. In 1901, Miller [19] showed that each simple alternating group $\mathrm{Alt}(n)$ can be generated by two elements, and that the generators can be chosen to have orders 2 and 3 except when $n = 6, 7$ or 8. He also showed in a later paper [20] in 1928 that the generators can be chosen to have orders 2 and m for any integer $m \geq 4$ such that $\mathrm{Alt}(n)$ has elements of order m. In an important paper in 1962, Steinberg [21] showed that all finite simple groups of Lie type can be generated by two elements, and, as a consequence of results of Aschbacher and Guralnick [2] in 1984, each of the sporadic simple groups can be generated by two elements. Therefore one of the many remarkable facts to emerge from the classification of the finite simple groups is the statement that every finite simple group can be generated by two elements. Indeed, finite simple groups have many generating pairs: Dixon [9] has proved that the probability that a randomly chosen pair of elements of an alternating group G generates G tends to 1 as $|G| \to \infty$, and Kantor and Lubotsky [12] have proved the corresponding statement for finite simple classical groups. Thus one may hope to find generating pairs satisfying a variety of additional conditions. The additional condition that we shall consider here is that one generator has order 2 and the other has preassigned order. A discussion of generating pairs satisfying conditions of other types can be found in [5] and [11].

Steinberg wrote in his paper [21] on generation of Chevalley groups by pairs of elements "it is possible that one of the generators can be chosen of order 2 ... if true, [this result] would quite likely require methods much more detailed than those used here". Both parts of this prediction turned out to be true. It was proved in 1992 by Malle, Saxl and Weigel [16] that every finite simple group can be generated by an involution and one other element. This result was the culmination of work of many authors, starting with the work of Miller [19], [20] on the alternating groups, and a paper of Brahana [4] in 1930 in which appropriate generating pairs were given for all simple groups of order less than one million known at that time. Two important steps on the way were the treatment of the groups of type $PSL_n(q)$ by Albert and Thompson [1] in 1959 and the work of Aschbacher and Guralnick [2] which covers the sporadic groups and the groups of Lie type of rank 1. The results became progressively more difficult to prove, and the work in [16] relies on close knowledge of the subgroup structure of the finite simple groups and also makes use of the Deligne–Lusztig theory of characters of reductive groups. In view of the increasing complexity of the proofs of the results mentioned above, when considering to what extent the order of the second generator can be restricted we shall clearly have to be content with results which are less comprehensive.

A group G is said to be $(2, m)$-generated if it can be generated by an involution and an element of order m. Because the $(2, 2)$-generated groups are just the dihedral groups, the first interesting case concerns $(2, 3)$-generation. This case is of particular interest because the $(2, 3)$-generated groups are just the images of order at least 6 of the modular group $PSL_2(\mathbb{Z})$, since $PSL_2(\mathbb{Z})$ is isomorphic to the free product $C_2 * C_3$. For this reason the determination of the simple $(2, 3)$-generated groups has received a considerable amount of attention. We have already discussed the result of Miller [19] on the

$(2, 3)$-generation of alternating groups. The sporadic simple groups which are $(2, 3)$-generated were determined by Woldar [31]: $M_{11}, M_{22}, M_{23}, \mathrm{McL}$ are not $(2, 3)$-generated, and all of the other sporadic simple groups are. We shall therefore restrict our attention to the groups of Lie type. Clearly the Suzuki groups are not $(2, 3)$-generated since they have no elements of order 3. The groups $\mathrm{PSL}_2(9)$, $\mathrm{PSL}_4(2)$ are isomorphic to Alt (6), Alt (8) and so are not $(2, 3)$-generated. Two other classical groups which are not $(2, 3)$-generated are $\mathrm{PSL}_3(4)$ and $\mathrm{PSU}_3(3^2)$. The latter group is not $(2, 3)$-generated since it was shown by Wagner [29] that it cannot be generated by three involutions; on the other hand if $G = \langle a, b \rangle$ is perfect and $a^2 = b^3 = 1$ then clearly G is generated by the three conjugate involutions a, a^b, a^{b^2}. It has been shown by Malle [14], [15] that the Chevalley groups $G_2(q)$ and the twisted groups $^2G_2(q)$, $^3D_4(q)$ and $^2F_4(q)$ are $(2, 3)$-generated.

Conjecture. (Di Martino–Vavilov [6]) All finite simple groups of Lie type are $(2, 3)$-generated except for some groups of low rank in characteristics 2 and 3.

The principal evidence for this conjecture is provided by the following results:

Theorem 1. (a) (Tamburini, [22]) *For all q and all $n \geq 25$, the group* $\mathrm{PSL}_n(q)$ *is $(2, 3)$ generated.*

(b) (Tamburini–Wilson–Gavioli, [28]) *For $n \geq 37$, the following are $(2, 3)$-generated:* $\mathrm{PSp}_{2n}(q)$, *for q odd;* $\mathrm{P\Omega}^+_{2n}(q)$, *for all q;* $\mathrm{PSU}_{2n}(q^2)$, *for q odd.*

(c) (Tamburini–Wilson, [27]) *For $n \geq 55$, the following are $(2, 3)$-generated:* $\mathrm{P\Omega}_{2n+1}(q)$, *for all q;* $\mathrm{P\Omega}^-_{2n+2}(q)$, *for q odd;* $\mathrm{PSU}_{2n}(q^2)$, *for q even; and* $\mathrm{PSU}_{2n+1}(q^2)$, *for q odd.*

Therefore, of the simple classical groups of large rank, all groups of odd characteristic and most of even characteristic behave in accordance with the conjecture. It is likely that the remaining groups can be handled using similar methods. If this turns out to be the case, then in a sense the conjecture will be established. However, the task of finding accurate bounds on ranks in these results will remain, and this seems likely to be difficult.

2. Generators for $E_n(R)$

The proof of Theorem 1 depends on some methods for constructing generating sets developed by Chiara Tamburini and me in a series of papers [22], [25], [26], [27], [28], [30]. These methods are rather flexible and I want to give an illustration of their use. We need a concise and easily manipulated notation which is well adapted to the problems under consideration, and this is obtained by keeping as close as possible to permutation matrices, for which the notation for permutations is available, and by keeping track of information in diagrams like those in Figure 1 and Figure 2 below.

For simplicity I will concentrate on groups of type SL_n, but it is not necessary to consider only groups defined over finite fields. The extra generality introduced below will be justified by an application to finite classical groups at the end of Section 3.

Let R be any commutative ring with a 1. As usual we denote by e_{ij} a matrix with (i, j)-entry 1 and all other entries 0, and we denote by $E_n(R)$ the subgroup of $GL_n(R)$ generated by $\{1 + \lambda e_{ij} \mid i \neq j, \lambda \in R\}$. If R is Euclidean or semilocal (in particular, if R is a finite ring), then $E_n(R)$ coincides with the group $SL_n(R)$ of matrices of determinant 1.

We note that if $E_n(R)$ is generated by finitely many matrices then R is generated as a ring by the finitely many entries of these matrices. It was proved in [28] that if R is any finitely generated ring then $E_n(R)$ is $(2, 3)$-generated for all sufficiently large n:

Theorem 2. (Tamburini–Wilson–Gavioli, [28]) *Let the ring R be generated by $t_1, \ldots, t_d$, where t_1 is a unit of R of finite multiplicative order. Then $E_n(R)$ is $(2, 3)$-generated for all $n \geq 12d + 16$. Similar results hold for some other classical groups.*

Theorem 2 takes a particularly striking form in the case when $R = \mathbb{Z}$, because it shows that $PSL_n(\mathbb{Z})$ is an epimorphic image of $PSL_2(\mathbb{Z})$ for all $n \geq 28$. By contrast, if $r \geq 3$ then $SL_r(\mathbb{Z})$ has the congruence subgroup property (from

Mennicke [18] and Bass, Lazard and Serre [3]), so that all proper images of $\mathrm{PSL}_r(\mathbb{Z})$ are finite.

Instead of indicating a proof of Theorem 2, I will sketch the proof of an easier result.

Theorem 3. *If R is as in Theorem 1 and $m \geq 7$ then $E_n(R)$ is $(2, m)$-generated for all $n \geq (2m + 2d + 2)m + 1$.*

Assume that the hypotheses of Theorem 3 hold. We may write $n = hm + r$ where h is even and $1 \leq r \leq 2m$; thus $h \geq 2m + 2d + 2$. We regard $\mathrm{GL}_n(R)$ as acting in the usual way on a free R-module M having a basis Ω with n elements. To simplify the description of our matrix generators we label the elements of Ω as v_j^i for $1 \leq i \leq h, 1 \leq j \leq m$ and $y_1, \ldots, y_r$, and regard these elements as arranged in a diagram as follows:

$$v_1^1 \quad v_2^1 \quad \ldots \quad v_m^1$$
$$y_1 \qquad v_1^2 \quad v_2^2 \quad \cdots \quad v_m^2$$
$$[y_2] \qquad\quad v_1^3 \quad v_2^3 \quad \ldots \quad v_m^3$$
$$[y_3] \qquad\quad v_1^4 \ \cdots$$

$$\cdot$$

$$v_m^{h-1}$$
$$v_1^h \quad v_2^h \quad \ldots \quad v_m^h$$

Figure 1

Thus the elements v_j^i are arranged in h *rows* $R_i = \{ v_j^i \mid 1 \leq j \leq m \}$ of m elements, with the first element of R_{i+1} below the last element of R_i, and y_k is placed below below v_2^k for each appropriate k, so that there are $h - 1 + r$ *columns* of length 2. We let b, a_1 be permutations of Ω whose orbits of length greater than 1 are respectively the rows and the columns; thus b has order m and a_1 has order 2. This description determines a_1 completely, but we shall need to specify the action of b more precisely: let

$$b = \prod (v_1^i, v_2^i, \ldots, v_{m-2}^i, v_m^i, v_{m-1}^i).$$

We shall regard permutations of Ω as module automorphisms of M in the obvious way. Let a_2 be the module automorphism which acts with matrix

$$\begin{pmatrix} 1 & t_k \\ 0 & -1 \end{pmatrix}$$

on the submodule with basis $v_4^{2m+2k-1}, v_4^{2m+2k}$ for $1 \le k \le d$, maps v_4^h to $\pm v_4^h$ (where the choice of sign will be explained below) and fixes all other elements of Ω. It is clear that a_2 has order 2.

We shall make repeated use of the following observation. Suppose that $\Omega = \Omega_1 \cup \Omega_2$ and $\Omega_1 \cap \Omega_2 = \emptyset$ and let $\varphi_1, \varphi_2 \in GL(M)$. If φ_i acts as the identity on Ω_i and maps $\langle \Omega \backslash \Omega_i \rangle$ to itself for $i = 1, 2$, then φ_1, φ_2 commute.

Set $a = a_1 a_2$. From the above observation a_1, a_2 commute, so that $a^2 = b^3 = 1$. Since h is even, b is an even permutation and so lies in $SL_n(R)$. The signature of the permutation a_1 is $(-1)^{h-1+r}$, and so by making an appropriate choice of sign in the definition of a_2 we can ensure that $a \in SL_n(R)$. It is not hard to show that, subject to this choice, a, b are in fact in $E_n(R)$.

Now the points of Ω moved by a_2^b are just the images under b of the points moved by a_2. Thus a_2^b fixes all points except 5th elements in rows, and so from the observation above $[a, a_2^b] = 1$. Similarly a_1^b fixes all 4th elements in rows, and so $[a_2, a_1^b] = 1$. Therefore we have

$$[a, a^b] = [a, a_1^b] = [a_1, a_1^b], = c, \text{ say,}$$

and c is a permutation matrix. The only non-trivial orbits of c arise where the supports of a_1, a_1^b overlap. It is easy to check that these orbits are

$$\{y_1, v_2^1, v_3^1\}$$

and

$$\{v_2^j, y_j\} \quad \text{and} \quad \{v_3^j, v_{m-1}^{j-1}\} \qquad \text{for } 2 \le j \le r.$$

Therefore $c^2 = (y_1, v_2^1, v_3^1)^{\pm 1}$.

For $\Delta \subseteq \Omega$, write Alt (Δ) for the subgroup of $\mathrm{SL}_n(R)$ acting as even permutations on Δ and as the identity on $\Omega \setminus \Delta$. Choose $\Delta \subseteq \Omega$ maximal with respect to containing $\{y_1, v_2^1, v_3^1\}$ and satisfying

$$\mathrm{Alt}\,(\Delta) \leq \langle a, b \rangle.$$

Claim 1. If $R_i \cap \Delta \neq \emptyset$ then $R_i \subseteq \Delta$.

Since $(\mathrm{Alt}\,(\Delta))^b = \mathrm{Alt}\,(\Delta b)$ and $\Delta \cap \Delta b \neq \emptyset$, the group $\langle \mathrm{Alt}\,(\Delta), (\mathrm{Alt}\,(\Delta))^b \rangle$ is primitive on $\Delta \cup \Delta b$, and since it also contains a 3-cycle it must be Alt $(\Delta \cup \Delta b)$. The maximality of Δ now implies that $\Delta b = \Delta$, and since $\langle b \rangle$ is transitive on R_i the claim follows.

Claim 2. If $i \leq h - 1$ and $R_i \subseteq \Delta$ then $R_{i+1} \cap \Delta \neq \emptyset$.

Write $\Delta_1 = \Delta \setminus \{ v_4^j \mid j \leq h \}$. Then a_2 commutes with Alt (Δ_1), so that

$$\mathrm{Alt}\,(\Delta \cup \Delta_1 a_1) = \langle \mathrm{Alt}\,(\Delta), (\mathrm{Alt}\,(\Delta_1))^{a_1} \rangle = \langle \mathrm{Alt}\,(\Delta), (\mathrm{Alt}\,(\Delta_1))^a \rangle \leq \langle a, b \rangle.$$

The maximality of Δ implies that $\Delta_1 a_1 \subseteq \Delta$, and we have $v_1^{i+1} = v_m^i a_1 \in \Delta$.

Claims 1 and 2 show that Δ contains all rows, and a similar argument shows that $\Delta = \Omega$. In other words, $\langle a, b \rangle$ contains all even permutations. It follows that $\langle a, b \rangle$ contains either a_2 if a_1 is an even permutation, or the product of a_2 and an arbitrary transposition if a_1 is odd. It is now a routine matter to show, by taking commutators and conjugates repeatedly, starting with $[a_2, a_2^u]$ for suitable even permutations u if a_1 is even, and using variants of the commutator identity

$$[1 + \lambda e_{ij}, 1 + \mu e_{jk}] = 1 + \lambda \mu e_{ik}$$

for distinct i, j, k, to show that $\mathrm{E}_n(R) \leq \langle a, b \rangle$.

The proof of Theorem 2 is based on the use of a diagram like Figure 1, but it is substantially harder, because it is impossible to arrange that the permutations which arise have nearly disjoint supports.

We now state some more results which can be proved using similar methods.

Definition. Let $\mathcal{C}$ be a class of groups. A group G is *residually* $\mathcal{C}$ if for all $g \in G \setminus 1$ there exist a group $H \in \mathcal{C}$ and an epimorphism $\theta : G \to H$ with $g\theta \neq 1$.

Theorem 4. *Let R be a finitely generated Euclidean ring and let $\mathcal{C} = \{ \mathrm{PSL}_n(R) \mid n \geq 3 \}$ and $m \geq 7$. Then the free product $C_2 * C_m$ is residually $\mathcal{C}$.*

In the group $\langle a \rangle * \langle b \rangle$, where $a^2 = b^m = 1$, each non-trivial word is conjugate to a, b^i or to a product

$$b^{i_1} a b^{i_2} a \ldots b^{i_s} a$$

with $1 \leq i_j \leq m - 1$ for each j. To obtain an epimorphism to a group $\mathrm{PSL}_n(R)$ in which the image of the above word is non-trivial we modify the action of b on the rows in Figure 1 so that b^{i_j} takes the first point of R_j to the last point of R_j. Thus the image of this word in $\mathrm{E}_n(R)$ takes v_1^1 to v_1^{s+1} and so is non-trivial. Unfortunately we cannot now ensure easily that the image of the map to $\mathrm{E}_n(R)$ contains a 3-cycle, and so it is somewhat harder than in Theorem 3 to prove that this image is $\mathrm{E}_n(R)$. The proof of the corresponding result for $m = 3$ is harder still:

Theorem 5. (Tamburini–Wilson–Gavioli, [28]) *Let R be a finitely generated Euclidean ring and let $\mathcal{C} = \{ \mathrm{PSL}_n(R) \mid n \geq 3 \}$. Then the free product $C_2 * C_3$ is residually $\mathcal{C}$.*

Like Theorem 2, Theorem 5 takes a striking form when $R = \mathbb{Z}$ since it shows that $\mathrm{PSL}_2(\mathbb{Z})$ is residually $\{ \mathrm{PSL}_n(\mathbb{Z}) \mid n \geq 3 \}$. By contrast, if $r \geq 3$ then $\mathrm{PSL}_r(\mathbb{Z})$ certainly cannot be residually $\{ \mathrm{PSL}_n(\mathbb{Z}) \mid n \geq r + 1 \}$ because it has the congruence subgroup property.

We end this section by stating the most general result on generation of groups which has been proved using these methods.

Theorem 6. (Tamburini–Wilson, [26]) *Let A, B be finite non-trivial groups such that $|A||B| \geq 12$. Then for all $n \geq 4|A||B| + 12$ and all q, $\mathrm{SL}_n(q)$ can be generated by a copy of A and a copy of B.*

The strategy is essentially the same, except that now we need to group the rows which are the regular B-orbits into rectangles. Writing $|A| = l+1$ and $|B| = m$, we consider a diagram as shown below. The non-trivial orbits of A are the bottom $l+1$ points of each column having length at least $l+1$. There are many unpleasant technical difficulties of a rather elementary nature to be overcome when the orders of A, B are small and all vestiges of elegance disappear in the treatment of these cases.

y_1

$$
\begin{array}{cccccc}
v_{11}^1 & \cdots & v_{1r}^1 & \cdots & v_{1m}^1 \\
\vdots & & \vdots & & \vdots \\
v_{l1}^1 & \cdots & v_{lr}^1 & \cdots & v_{lm}^1 \\
& & v_{11}^2 & \cdots & v_{1r}^2 & \cdots & v_{1m}^2 \\
& & \vdots & & \vdots & & \vdots \\
& & v_{l1}^2 & \cdots & v_{lr}^2 & \cdots & v_{lm}^2 \\
& & [y_2] & & v_{11}^3 & \cdots & v_{1r}^3 & \cdots & v_{1m}^3 \\
& & & & \vdots & & \vdots & & \vdots \\
& & & & v_{l1}^3 & \cdots & v_{lr}^3 & \cdots & v_{lm}^3 \\
& & & & [y_3] & & v_{11}^4 & \cdots \\
& & & & & & \vdots \\
& & & & & & & \ddots
\end{array}
$$

Figure 2

3. $PSL_n(q)$ revisited

The techniques described in Section 2 give no information about groups of low rank. However, for the lowest dimensions, direct matrix calculations are possible. It was shown by Macbeath [13] that $PSL_2(q)$ is $(2,3)$-generated for all $q \neq 9$, by Garbe [10] that $SL_3(q)$ is $(2,3)$-generated for all $q \neq 4$ and by Tamburini and Vassallo [23], [24] that $SL_4(q)$ is $(2,3)$-generated for all $q \neq 2$. (Clearly $SL_2(q)$ cannot be $(2,3)$-generated for q odd, since the only involution is in the centre.) The matrix computations become progressively harder as the ranks of these groups increase, and the techniques of Section 2 only become available when the rank is quite large. However, the gap between the results for groups of low rank and groups of high rank has been narrowed in [24], where it is proved that $SL_n(q)$ is $(2,3)$-generated if $n \geq 13$ and either q is even or $n \neq 13, 14, 17$; and the gap has been closed almost entirely in odd characteristic:

Theorem 6. (Di Martino–Vavilov, [6], [7]) *Let q be odd and $q \neq 9$. Then $SL_n(q)$ is $(2,3)$-generated for $n \geq 5$.*

Theorem 6 also gives another approach to the proof that the groups $SL_n(q)$ with n large are $(2,3)$-generated. Like Theorem 2, it is constructive and allows one to write down explicit matrices generating $SL_n(q)$; in fact the generators given by the two results are very similar. The major difference is in the proof, which is linear rather than permutational in character. It relies on McLauchlin's classification [17] of finite irreducible linear groups generated by root subgroups and the theorem of Dickson mentioned in Section 1. It seems very probable that the ideas developed in the proof of Theorem 6 will also yield a proof that the exceptional Chevalley groups of types E_6, E_7, E_8 in odd characteristic are $(2,3)$-generated. The treatment of the groups $SL_n(q)$ with $n \leq 11$ in Theorem 6 involves a considerable amount of case by case analysis and the general arguments fail for the groups $SL_n(q)$ with $n = 8, 10$ and $q = 3, 5, 7$; that the proposed generators for these groups are indeed generators was checked with computer calculations using GAP 3.2.

We now return to classical groups of large rank. The information given by Theorem 2 about finite classical groups is much stronger than mere $(2,3)$-generation of these groups. We shall show that Theorem 2 implies that large direct powers of the groups $\mathrm{SL}_n(q)$ are $(2,3)$-generated.

We recall that if $\theta : R \rightarrow S$ is a homomorphism, then there is an induced homomorphism $\mathrm{GL}_n(R) \rightarrow \mathrm{GL}_n(S)$ defined by $(a_{ij}) \mapsto (a_{ij}\theta)$. If θ is surjective, so is the induced map $\mathrm{E}_n(R) \rightarrow \mathrm{E}_n(S)$. We note also that if $R \cong R_1 \oplus R_2$ then $\mathrm{E}_n(R) \cong \mathrm{E}_n(R_1) \times \mathrm{E}_n(R_2)$.

Let K be a finite field and consider the polynomial ring $R = K[x_1, x_2, \ldots, x_e]$. So R can be generated by $d = e + 1$ elements, of which one is a unit of finite order. Let I be the intersection of the kernels of all homomorphisms from R to K extending the identity map on K and let $\overline{R} = R/I$. So $\overline{R} \cong K^f$ where $f = |K|^e$, by the Chinese remainder theorem. Thus we have an epimorphism

$$\mathrm{E}_n(R) \rightarrow \mathrm{E}_n(\overline{R}) \cong (\mathrm{E}_n(K))^f = (\mathrm{SL}_n(K))^f.$$

We conclude from Theorem 2 that $(\mathrm{SL}_n(K))^f$ is $(2,3)$-generated if $n \geq 12d + 16 = 12e + 28$. This gives the following

Corollary. $(\mathrm{SL}_n(q))^r$ is $(2,3)$-generated, where $r = q^{[(n-28)/12]}$. Similarly for some other classical groups; e.g. for q odd, $(\mathrm{Sp}_{2n}(q))^s$ is $(2,3)$-generated, where $s = q^{[(n-37)/12]}$.

It is worth emphasising that the proof of the Corollary does not depend on the class-ification of the finite simple groups.

Write $g(n)$ for the largest integer m such that $(\mathrm{SL}_n(q))^{q^m}$ is $(2,3)$-generated for all q. From the Corollary and elementary considerations we have

$$\left[\frac{(n-28)}{12}\right] \leq g(n) \leq n^2 \qquad \text{for all } n,$$

and it is not hard to improve each of these bounds for $g(n)$ slightly. It would be interesting to know more about the asymptotic behaviour of $g(n)$ as $n \rightarrow \infty$.

REFERENCES

[1] A. A. ALBERT AND J. G. THOMPSON, Two element generation of the projective unimodular group, *Illinois J. Math.* **3** (1959), 421-439.

[2] M. ASCHBACHER AND R. GURALNICK, Some applications of the first cohomology group, *J. Algebra* **90** (1984), 446-460.

[3] H. BASS, M. LAZARD AND J-P. SERRE, Sous-groupes d'indice fini dans $SL(n, \mathbf{Z})$, *Bull. Amer. Math. Soc.* **70** (1964), 385-392.

[4] H. R. BRAHANA, Pairs of generators of the known simple groups whose orders are less than one million, *Ann. of Math.* (2) **31** (1930), 529-549.

[5] L. DI MARTINO AND M. C. TAMBURINI, 2-generation of finite simple groups and some related topics. In "Generators and Relations in Groups and Geometries", (editors A. Barlotti et al.), Kluwer Academic Publishers, 1991.

[6] L. DI MARTINO AND N. VAVILOV, (2, 3)-generation of $SL(n,q)$, I, cases $n = 5,6,7$, *Comm. in Algebra* **22** (1994), 1321-1347.

[7] L. DI MARTINO AND N. VAVILOV, (2, 3)-generation of $SL(n,q)$, II, *Comm. in Algebra*, to appear.

[8] L. E. DICKSON, "Linear Groups with an Exposition of The Galois Field Theory," Teubner, 1901; reprinted by Dover, 1958.

[9] J. D. DIXON, The probability of generating the symmetric group, *Math. Z.* **110** (1969), 199-205.

[10] D. GARBE, Über eine Klasse von arithmetisch definierbaren Normalteilern der Modulgruppe, *Math. Ann.* **235** (1978), 195-215.

[11] W. KANTOR, Some topics in asymptotic group theory. In "Groups, Combinatorics and Geometry", (editors M. W. Liebeck and J. Saxl), Cambridge University Press, 1992.

[12] W. KANTOR AND A. LUBOTSKY, The probability of generating a finite classical group, Geom. Dedicata 36 (1990), 67–87.

[13] A. M. MACBEATH, Generators of the linear fractional groups, Proc. Sympos. Pure Math. 12 (1967), 14–32.

[14] G. MALLE, Hurwitz groups and G_2, Canad. Math. Bull. 33, (1990), 349–357.

[15] G. MALLE, Small exceptional Hurwitz groups. In these proceedings.

[16] G. MALLE, J. SAXL AND T. WEIGEL, Generation of classical groups, Geom. Dedicata 49 (1994), 85–116.

[17] J. E. MCLAUGHLIN, Some groups generated by transvections, Arch. Math. (Basel) 18 (1969), 108–115.

[18] J. MENNICKE, Finite factor groups of the unimodular group, Ann. of Math. 81 (1965), 31–37.

[19] G. A. MILLER, On the groups generated by two operators, Bull. Amer. Math. Soc. 7 (1901), 424–426.

[20] G. A. MILLER, Possible orders of two generators of the alternating and of the symmetric group, Trans. Amer. Math. Soc. 30 (1928), 24–32.

[21] R. STEINBERG, Generators for simple groups, Canad. J. Math. 14 (1962), 277–283.

[22] M. C. TAMBURINI, Generation of certain simple groups by elements of small order, Rend. Istit. Lombardo Sci. (A) 121 (1987), 21–27.

[23] M. C. TAMBURINI AND F. VASSALLO, (2, 3)-generazione di SL(4, q) in caratteristica dispari e problemi collegati, *Boll. Un. Mat. Ital. A*, to appear.

[24] M. C. TAMBURINI AND F. VASSALLO, (2, 3) generazione di gruppi lineari. To appear.

[25] M. C. TAMBURINI AND J. S. WILSON, A residual property of free products, *Math. Z.* **186** (1984), 525–530.

[26] M. C. TAMBURINI AND J. S. WILSON, On the generation of finite simple groups by pairs of subgroups, *J. Algebra* **116** (1988), 316–333.

[27] M. C. TAMBURINI AND J. S. WILSON, On the (2, 3)-generation of some classical groups, II. In preparation.

[28] M. C. TAMBURINI, J. S. WILSON AND N. GAVIOLI, On the (2, 3)-generation of some classical groups, I, *J. Algebra*, to appear.

[29] A. WAGNER, The minimal number of involutions generating some three-dimensional groups, *Boll. Un. Mat. Ital. A* (5) **15** (1978), 431–439.

[30] J. S. WILSON, A residual property of free groups, *J. Algebra* **138** (1991), 36–47.

[31] A. J. WOLDAR, On Hurwitz generation and genus actions of sporadic groups, *Illinois J. Math.* **33** (1989), 416–437.